Collins Advanced Modular Scie

Biology AS

Series Editor: Mike Bailey

Mike Bailey

Keith Hirst

This book has been designed to support AQA Biology specification B.
It contains some material which has been added in order to clarify the
specification. The examination will be limited to material set out in the
specification document.

Published by HarperCollins *Publishers* Limited
77–85 Fulham Palace Road
Hammersmith
London
W6 8JB

www.CollinsEducation.com
Online support for schools and colleges

© Mike Bailey and Keith Hirst 2000
First published 2000
Reprinted 2000 (twice), 2001

ISBN 0 00 327751 8

Mike Bailey and Keith Hirst assert the moral right to be identified as the
authors of this work

All rights reserved. No part of this publication may be reproduced, stored
in a retrieval system, or transmitted in any form or by any means,
electronic, mechanical, photocopying, recording or otherwise, without
either the prior permission of the Publisher or a licence permitting
restricted copying in the United Kingdom issued by the Copyright
Licensing Agency Ltd., 90 Tottenham Court Road, London W1P 0LP.

British Library Cataloguing in Publication Data
A catalogue record for this publication is available from the British Library

Cover design by Chi Leung
Design by Glynis Edwards
Edited by Kathryn Senior
Illustrations by Barking Dog Art, Russell Birkett, Tom Cross, Jerry Fowler,
Hardlines, Illustrated Arts and Mark Jordan
Picture research by Caroline Thompson
Index by Julie Rimington
Production by Kathryn Botterill

Printed and bound in Great Britain by Scotprint

The publisher wishes to thank the Assessment and Qualifications Alliance
for permission to reproduce examination questions.

You might also like to visit
www.**fire**and**water**.com
The book lover's website

CORE PRINCIPLES

Chapter 1	**Cells**	6
Chapter 2	**Cells and their environment**	26
Chapter 3	**Exchanges**	40
Chapter 4	**Proteins and enzymes**	54
Chapter 5	**Food and digestion**	68

GENES AND GENETIC ENGINEERING

Chapter 6	**The genetic code**	88
Chapter 7	**The cell cycle**	106
Chapter 8	**Sex and reproduction**	118
Chapter 9	**Genetic engineering and microbes**	130
Chapter 10	**Genes and medicine**	144

PHYSIOLOGY AND TRANSPORT

Chapter 11	**Transport systems**	154
Chapter 12	**Energy and exercise**	168
Chapter 13	**Transport systems in plants**	184

| Answers to questions | 200 |
| Glossary | 210 |

Acknowledgements

Text and diagrams reproduced by kind permission of:
Ballantine Publishing Group; Blackwell Publishers; Harvard University Press; New Scientist; Scientific American; The Independent; John Wiley and Sons; Wadsworth Publishers; Wolfe Publishers.

Every effort has been made to contact the holders of copyright material, but if any have been inadvertently overlooked the publishers will be pleased to make the necessary arrangements at the first opportunity.

The publishers would like to thank the following for permission to reproduce photographs
(T = Top, B = Bottom, C = Centre, L = Left, R = Right):

Allsport, 26CR, S Forster, 44T, M Hewitt, 44B, G Mortimore, 118, M Powell, 183; Heather Angel, 36;
Aquarius Picture Library, 105;
Biophoto Associates, 7, 9, 10, 13, 15, 16, 20C, 21, 50, 75BR, TL&TR, 77TR, BR & CR, 84CL, 101L, 107CT, C, CB & B, 120BR, 125B, 189TC, 190, 192, 196;
Chris Bonington Photo Library, 53, 176;
Bruce Coleman Collection, 85, 100, 126, 184, J Burton, 41, F Bruemmer, 42L;
Bruce Coleman Inc. 42C, H Reinhard, 83BR, K Taylor, 83B, A G Potts, 189TL, E Pott, 188R; J Allan Cash Ltd, 62(inset);
Environmental Picture Library/V Miles, 131, T Adamson, 143;
Sally & Richard Greenhill Photo Library, 174;
Professor Don Grierson, 137;
Holt Studios International/N Cattlin, 111, 112;
Andrew Lambert, 28, 55;
NHPA/J Shaw, 188L;
Oxford Scientific Films Ltd/Babs & Bert Wells, 40(inset), M Colebeck, 40T, D Guravich, 40C;
PPL Therapeutics, 144;

Photos Horticultural, 110;
Planet Earth Pictures/S Bloom, 83C, T Brakefield, 83L, R Matthews, 125R;
Rex Features Ltd, 68, 168;
Roslin Institute, Edinburgh 106, 114, 145C;
Royal Holloway College/EM Unit, 17;
Science Photo Library, 6, 14, 20L, 27, 34, 39, 42B, 43, 59, 74, 75BL, 77TL & BL, 79, 84TR, 88, 90, 91, 92, 102, 107T, 117, 120BL, 122, 124, 127, 130, 132, 145T, 146, 150, 153, 159, 177, 178, 186;
Shout, 154;
South Eastern Technology Center, Augusta, Georgia, 54;
The Stock Market, 62L;
Wander Ltd, 26T;
The Wellcome Trust, 101R;

Front cover:
Paul Thiessen/John Hopkins University (top left)
Tony Stone Images (centre)
Science Photo Library (top right)

This book contains references to fictitious characters in fictitious case studies. For educational purposes only, photographs have been used to accompany these case studies. The juxtaposition of photographs and case studies is not intended to identify the individual in the photograph with the character in the case study. The publishers cannot accept any responsibility for any consequences resulting from this use of photographs and case studies except as expressly provided by law.

To the student

This book aims to make your study of advanced science successful and interesting. Science is constantly evolving and, wherever possible, modern issues and problems have been used to make your study stimulating and to encourage you to continue studying science after you complete your current course.

Using the book

Don't try to achieve too much in one reading session. Science is complex and some demanding ideas need to be supported with a lot of facts. Trying to take in too much at one time can make you lose sight of the most important ideas – all you see is a mass of information.

Each chapter starts by showing how the science you will learn is applied somewhere in the world. At other points in the chapter you may find more examples of the way the science you are covering is used. These detailed contexts are not needed for your examination but should help to strengthen your understanding of the subject.

The numbered questions in the main text allow you to check that you have understood what is being explained. These are all short and straightforward in style – there are no trick questions. Don't be tempted to pass over these questions, they will give you new insights into the work. Answers are given in the back of the book.

This book covers the content needed for the AQA Specification B in Biology at AS-level. The Key Facts for each section summarise the information you will need in your examination. However, the examination will test your ability to apply these facts rather than simply to remember them. The main text in the book explains these facts. The application boxes encourage you to apply them in new situations. Extension boxes provide extra detail not required for your examination. These are interesting to read and will support your studies beyond AS-level.

Words written in bold type appear in the glossary at the end of the book. If you don't know the meaning of one of these words check it out immediately – don't persevere, hoping all will become clear.

Past paper questions are included at the end of each chapter. These will help you to test yourself against the sorts of questions that will come up in your examination.

The Key Skill Assignments allow you to practise for any Key Skill assessments you may have. The assignments are often starting points for work and you will need to access other books and sources of information to complete the activity.

1 Cells

Screening for cervical cancer started in 1964 and had only a modest effect on death rates during the 14 years that followed. However, in 1988, when a national call system was introduced, this doubled the proportion of eligible women tested to 85 per cent and trebled the rate of fall in death rates. By 1997 only half as many women were dying of cervical cancer compared with 1950. A recent report in the British Medical Journal said that in 1997 alone, screening probably prevented 800 deaths from cervical cancer in women aged 25-54.

Most women have a smear test every 3 to 5 years. A spatula is used to scrape a few cells away from the surface of the cervix. These are transferred to a microscope slide, stained and examined with a light microscope. A trained technician can spot cells showing early cancerous changes by their enlarged appearance and bigger nuclei. Prompt treatment of a woman whose smear contains cancerous cells can prevent a life-threatening cancer.

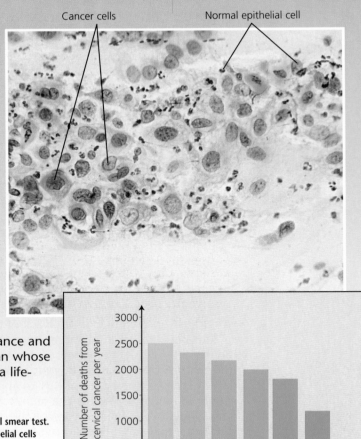

The photograph (above right) shows cells obtained by a cervical smear test. The orange-stained cancerous cells are larger than normal epithelial cells and their nuclei are more noticeable. Cancer occurs when the normal process of cell division and differentiation goes wrong and cells divide out of control.

The bar chart shows how the number of deaths from cervical cancer has fallen in Britian since 1950. Notice the large drop between 1987 and 1997, due to more widespread use of the cervical smear test.

1.1 Cells

At birth, a human baby has about 20 billion (20 000 000 000) cells. All have developed from a single cell, the fertilised egg. Normal growth over the 9 months of pregnancy depends on **cell division** to increase the numbers of cells and on a separate process called **differentiation**. This enables cells to specialise to carry out a particular function. When control of either of these crucial processes breaks down, the result can be cancer. Cancer cells lose the special features of the tissue they arise from and they divide much faster than the cells around them.

Looking inside cells

You have probably used a light microscope to look at cells from the lining of your cheek. Cheek cells are **epithelial cells**. They look very like the normal cervical cells you can see in the photograph at the top of this page. Fig. 1 on page 7 shows what cells look like when seen through a light microscope (Fig. 2). Staining techniques usually show the cell **nucleus** as a blob, and the **cytoplasm** as a paler substance that looks empty. In fact, the cytoplasm is packed with **organelles**, such as mitochondria and ribosomes.

Units of measurement

The two most common **units** used to describe the size of organelles are:

- the **micrometre** (μm), one thousandth of a millimetre (10^{-6} metre)
- the **nanometre (nm)**, one thousandth of a micrometre (10^{-9} metre).

Resolving power and cell size

You can only see the larger organelles in a cell with a light microscope because even the larger organelles are near the limit of a light microscope's **resolving power**.

Resolving power should not be confused with magnification. Resolving power is the ability to distinguish between two objects. Think of a car at night. When it is far away, the two headlights appear as a single light. As the car gets nearer you can see two headlights; at some point your eyes are able to *resolve* two lights, not one. With binoculars you could resolve the light into two headlights when the car was much further away.

Fig. 1 What can you see with a light microscope?

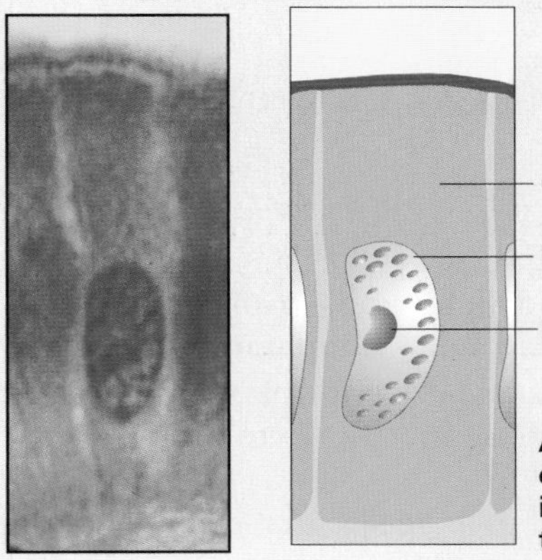

A light micrograph of epithelial cells from the human small intestine with diagram showing the main features visible.
- Cytoplasm
- Nucleus
- Nucleolus

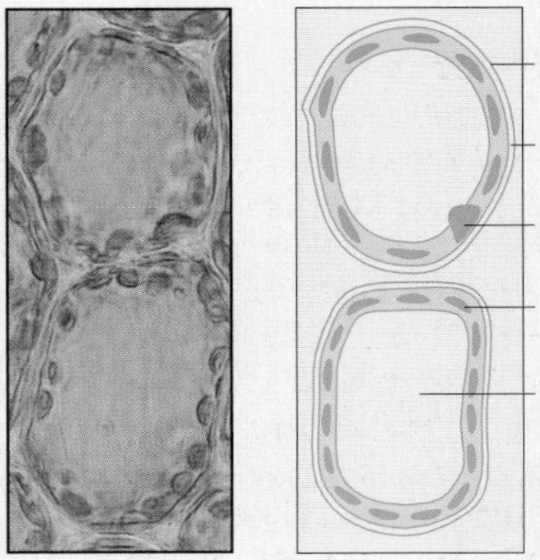

A light micrograph of mesophyll cells from a leaf with diagram showing the main features visible.
- Cell wall
- Cell membrane
- Nucleus
- Chloroplast
- Vacuole

Fig. 2 The light microscope

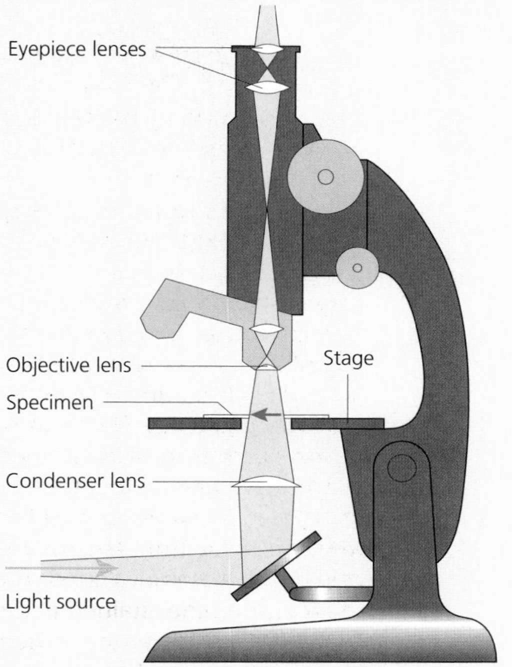

- Eyepiece lenses
- Objective lens
- Stage
- Specimen
- Condenser lens
- Light source

The basic structure of a standard light microscope is shown above. This is the sort of microscope that is commonly used for teaching purposes and that you have probably used in science lessons. It is also known as a compound microscope because it has two lenses. The eyepiece lens and the objective lens combine to produce a greater magnification than would be possible with only a single lens.

The total magnification is calculated by multiplying the magnification of the two lenses together. For example, if the eyepiece lens has a magnification of × 10 and the objective lens has a magnification of × 50, the total magnification of the microscope is × 500.

The limitations of the light microscope are not due to the construction of the equipment, but to the nature of light itself. Even the most powerful lens cannot resolve two dots that are separated by less than 250 nm. This is because no lens system can ever resolve two dots that are closer together than half the wavelength of the light used to view them. The wavelength of visible light is between 500 and 650 nm.

Since the lenses in a light microscope will not allow you to distinguish between two objects that are smaller than 250 nm (0.25 μm), the maximum magnification of a light microscope is × 1500. That's good enough to see animal cells (diameter 30-50 μm) and individual bacteria (length 5-8 μm). It is also sufficient to detect the changes in a cell that might be the beginning of cancer and to view larger organelles such as the nucleus (diameter 10 μm). But it's not powerful enough to see small organelles such as ribosomes (diameter 20 nm) or cell membranes (thickness 7-10 nm).

1 Why are there no light microscopes with a magnification of × 2000?

The electron microscope

The detailed **ultrastructure** of plant and animal cells was revealed in the 1950s when the electron microscope was first used. A specimen for the electron microscope has to be specially prepared; a simple smear of cells on a microscope slide would not work. Very thin slices of the specimen are cut, preserved and stained. The specimen is then placed in a chamber inside the electron microscope, which is sealed and air is sucked out to produce a vacuum. Electromagnets focus a beam of electrons that passes through the specimen and onto a viewing screen (Fig. 3). A modern electron microscope can magnify objects up to 500 000 times.

2 The wavelength of the beam of electrons that is used in an electron microscope is approximately 0.005nm. What is the theoretical limit of the electron microscope's resolving power?

The development of the electron microscope has had a huge impact on biology. It makes it possible to see the details of cell organelles and has actually allowed new organelles to be discovered. Look at the differences between the photographs of animal and plant cells taken with the light microscope and the electron microscope (Fig. 4).

The electron microscope reveals that cells contain many **membranes**. The cytoplasm is surrounded by the **cell surface membrane** or **plasma membrane**. Other organelles, such as

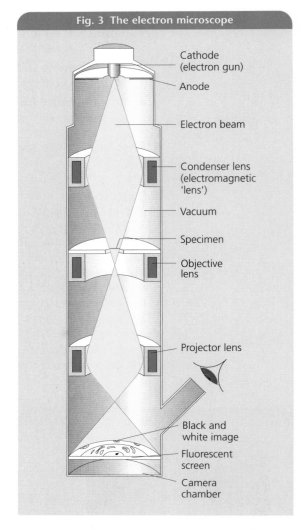

Fig. 3 The electron microscope

mitochondria and **chloroplasts** have an outer and an inner membrane. And a network of membranes, the **endoplasmic reticulum**, spreads throughout the cytoplasm.

3 List the organelles that can only be seen with the electron microscope:
a in both animal and plant cells.
b in plant cells only.

KEY FACTS

- A **cell** is the smallest independent unit of life; all living things are made up of one or more cells. The human body contains millions of cells.

- Cells contain **organelles,** compartments in the cytoplasm that are surrounded by their own membrane. They carry out specialised tasks in the cell. Most organelles are visible only when a cell is viewed with an electron microscope.

- A **tissue** is a group of cells with similar structure. Skeletal muscle, for example, is a tissue made up from skeletal muscle cells. The inner surface of the gut is lined with epithelial tissue, which is made up of a layer of epithelial cells.

- An **organ**, such as the pancreas contains several different tissues, all of which contribute to its overall function.

1 CELLS

Fig. 4 Animal and plant cells

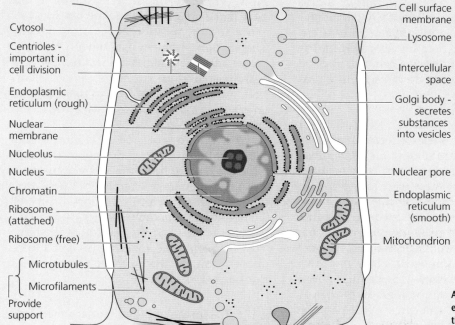

A generalised animal cell

Labels: Cytosol; Centrioles – important in cell division; Endoplasmic reticulum (rough); Nuclear membrane; Nucleolus; Nucleus; Chromatin; Ribosome (attached); Ribosome (free); Microtubules; Microfilaments; Provide support; Cell surface membrane; Lysosome; Intercellular space; Golgi body – secretes substances into vesicles; Nuclear pore; Endoplasmic reticulum (smooth); Mitochondrion

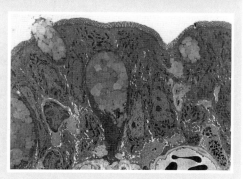

An electron micrograph of epithelial cells from the small intestine. Magnification × 2000.

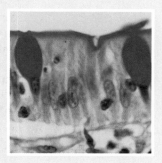

A light micrograph of epithelial cells from the small intestine. Magnification × 1000

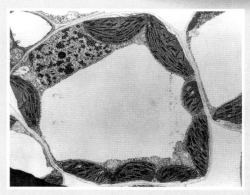

An electron micrograph of a single plant cell from a leaf. Magnification × 5000.

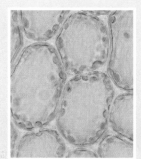

A light micrograph of parenchyma cells from a leaf, showing chloroplasts. Magnification × 1000.

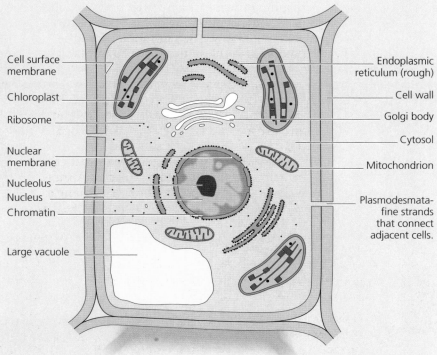

A generalised plant cell

Labels: Cell surface membrane; Chloroplast; Ribosome; Nuclear membrane; Nucleolus; Nucleus; Chromatin; Large vacuole; Endoplasmic reticulum (rough); Cell wall; Golgi body; Cytosol; Mitochondrion; Plasmodesmata – fine strands that connect adjacent cells.

1.2 Major cell organelles

Both animal and plant cells have a cell surface membrane, a nucleus, mitochondria, endoplasmic reticulum, ribosomes, a Golgi body and lysosomes. We look at all of these organelles in more detail in this section. In addition, plant cells have a cell wall and chloroplasts. These are described in the next section (page 18).

Membranes

The word membrane means 'very thin layer.' It is important that you understand the difference between a cell membrane and a body membrane such as the alveolar membrane in the lungs or the corneal membrane over the front of the eye. These membranes are made from a thin layer of whole cells. Cell membranes are parts of cells that consist of layers of **lipid** molecules with some **protein** molecules between them.

The photograph in Fig. 5 is a high power transmission electron micrograph of the cell surface membrane. You can see the membrane as two dark bands separated by a clear central area. It looks like this because the membrane is not a single layer; it consists of two layers of **phospholipid** molecules. Proteins are embedded in each thin layer of phospholipids, some spanning the membrane from one side to the other, others appearing on one face of the membrane only. The phospholipid and protein molecules fit together to form a continuous pattern like the tiles in a mosaic. However, unlike mosaic tiles, the molecules that make up a membrane are not fixed in place and the position of the proteins can change from moment to moment. For this reason, cell membranes are said to have a **fluid mosaic structure**. The fluid mosaic theory of cell membrane structure was first put forward in 1972, as a result of work with the electron microscope.

The protein molecules that form part of the membrane may occur only in the upper or lower layer of lipid molecules. Some proteins span the membrane completely. These proteins are often important in transport of substances across the membrane (see page 34). Proteins that protrude from the outside of the cell surface membrane are often involved in cell recognition, or they may act as receptor sites for hormones.

4 The image of the cell surface membrane is quite fuzzy at the magnification shown in the photograph in Fig. 5.

a Estimate the thickness of the cell surface membrane in nanometres.

b This thickness is approximately twice the length of a phospholipid molecule. Estimate the length of a phospholipid molecule in nanometres.

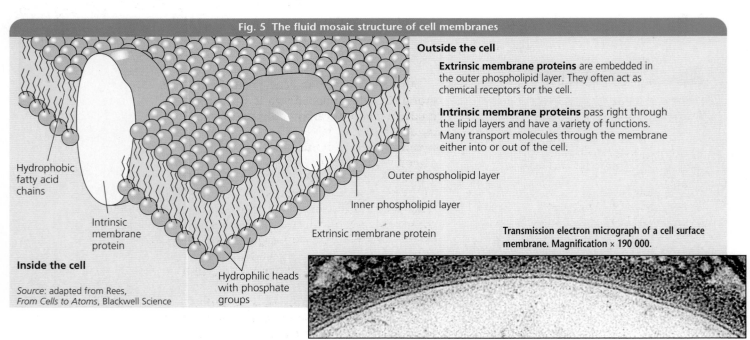

Fig. 5 The fluid mosaic structure of cell membranes

Outside the cell

Extrinsic membrane proteins are embedded in the outer phospholipid layer. They often act as chemical receptors for the cell.

Intrinsic membrane proteins pass right through the lipid layers and have a variety of functions. Many transport molecules through the membrane either into or out of the cell.

Outer phospholipid layer
Inner phospholipid layer
Extrinsic membrane protein

Hydrophobic fatty acid chains
Intrinsic membrane protein
Inside the cell
Hydrophilic heads with phosphate groups

Source: adapted from Rees, From Cells to Atoms, Blackwell Science

Transmission electron micrograph of a cell surface membrane. Magnification × 190 000.

Lipids

Phospholipids are a type of lipid and learning more about this important group of compounds will help you to understand membranes better.

In general, lipids contain the elements carbon, hydrogen and oxygen. Fats and oils, the other major types of lipid are important as energy stores in animals and plants. Lipids are large molecules that consist of triglycerides. These form when a condensation reaction joins three fatty acid molecules and one molecule of glycerol (Fig. 6).

A glycerol molecule contains a chain of three carbon atoms with attached hydrogen atoms and hydroxyl groups (OH). The glycerol molecule is common to all triglycerides so the properties of different fats depend on the nature of the fatty acids that are linked to it.

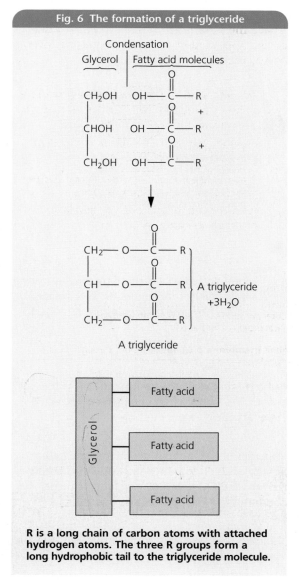

Fig. 6 The formation of a triglyceride

R is a long chain of carbon atoms with attached hydrogen atoms. The three R groups form a long hydrophobic tail to the triglyceride molecule.

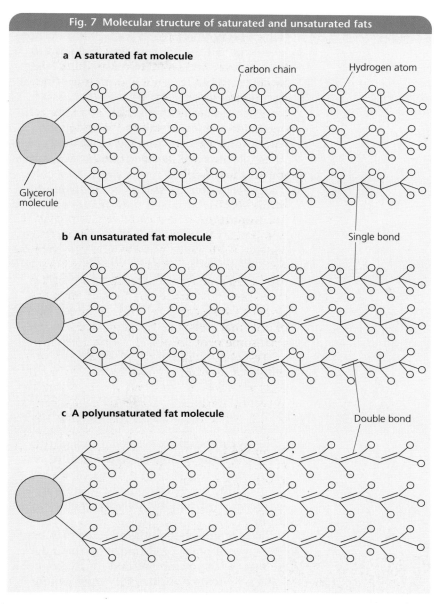

Fig. 7 Molecular structure of saturated and unsaturated fats

A single fatty acid molecule contains an acid group (COOH) attached to a **hydrocarbon chain**, a chain of carbon atoms with attached hydrogen atoms. We usually denote the hydrocarbon chain with the letter 'R', giving a general formula for a fatty acid as R–COOH. The hydrocarbon chain itself can contain up to twenty-four carbon atoms and forms a long 'tail' to the molecule. If every carbon atom in the chain is joined by single C–C bond, we say the fatty acid is **saturated** (Fig. 7a). If there is at least one double C=C somewhere in the hydrocarbon chain, we say the fatty acid is **unsaturated** (Fig. 7b). A chain with many double C=C bonds is **polyunsaturated** (Fig. 7c). Most animal fats are saturated while most plant fats are unsaturated.

Lipids and water

Lipids do not dissolve in water. Most do not even mix with water very well and tend to form a layer on top of it. If stirred into the water, the lipid molecules group together to form droplets.

The acid group that is part of a fatty acid molecule ionises in water to form H^+ and COO^- ions. This end of a fatty acid molecule is therefore attracted to water, but the long carbon tail at the other end repels water. Molecules that repel water are said to be **hydrophobic**, those that are attracted to water are said to be **hydrophilic**.

Phospholipids and membranes

Phospholipids are similar in structure to triglycerides, but one of the three fatty acids is replaced by a negatively charged phosphate group (Fig. 8a).

Like fatty acids, phospholipids do not dissolve in water. The tail of a phospholipid molecule is hydrophobic but the 'head' of the molecule is hydrophilic, it 'loves' water. So, when phospholipids are placed in water, they prefer to sit on the surface with their heads down and tails up, or they form spheres. These spheres can be either single-layered or they have the same double-layered structure as cell membranes (Fig. 8b). In both cases, the hydrophilic heads get to be near the water while the hydrophobic tails stay well away from it.

When phospholipids form a cell membrane, the layer of polar heads make up the outer surfaces of the membrane and the hydrophobic tails are enclosed inside. This arrangement forms a 'skin' around the cell, and around individual internal cell organelles.

Fig. 8 Phospholipids

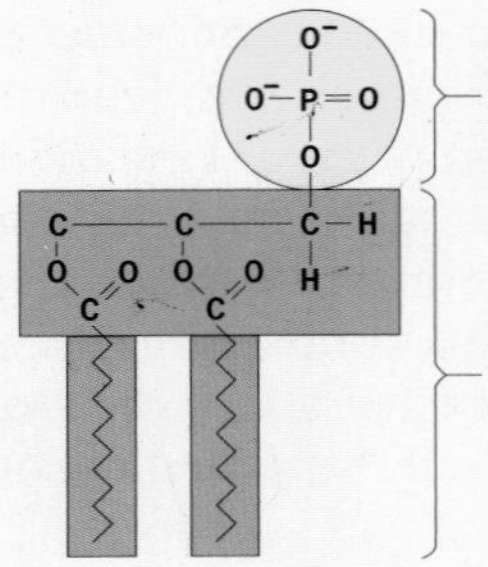

a The structure of a phospholipid. The phosphate gives the molecule a polar head and a non-polar tail.

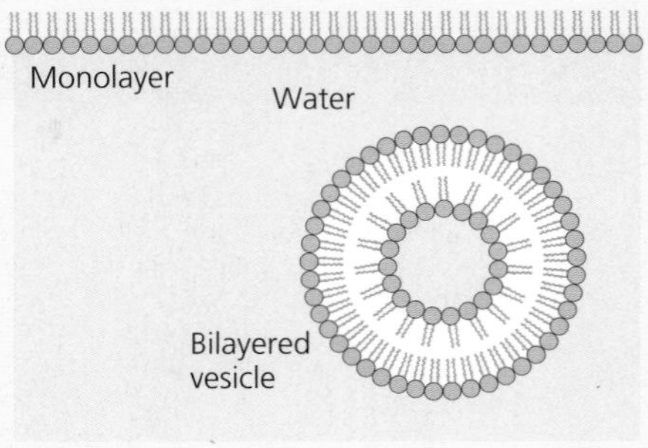

- Polar head contains a negatively charged phosphate group that is hydrophilic (water-loving)
- Non-polar tail is hydrophobic (water-hating)

b In water, the hydrophilic heads of the polar phospholipids face outwards while the hydrophobic tails point inwards. Double-layered vesicles form in the water and a phospholipid monolayer covers the surface.

5 When added to water, oil molecules form a single layer of molecules on the surface. Explain how the molecules are arranged.

KEY FACTS

- All cells are surrounded by a cell membrane and many organelles also have their own membranes;
- Cell membranes consist of a double layer of phospholipid molecules with embedded proteins. Some proteins are present in only one of the phospholipid layers, others span the membrane;
- The proteins are free to move within the phospholipid layer and membranes are said to have a fluid mosaic structure;
- Phospholipids are a type of lipid; lipids contain carbon, hydrogen and oxygen;
- Fats and oils are also lipids; their basic unit is the triglyceride. This forms when one glycerol molecule and three fatty acids are joined by a condensation reaction;
- The properties of fats and oils depend on the nature of their fatty acids. In saturated fats, fatty acid chains contain only single C–C bonds; the fatty acid chains of unsaturated fats contain at least one double C=C bond;
- Phospholipids, are similar to triglycerides but one of the fatty acids is replaced by a phosphate group.

The nucleus

The nucleus, shown in the diagram and photograph in Fig. 9, is the largest organelle in the cell. It is usually spherical and has a diameter of about 10 μm. Most of the cell's **deoxyribonucleic acid (DNA)** exists in the nucleus. This **nucleic acid** contains all the information required to make a new copy of the cell and to control the cell's activities. In a dividing cell, the DNA molecules are condensed into **chromosomes** that become clearly visible (see Chapter 7). At other times the nucleus has a grainy appearance because the DNA molecules extend throughout the nucleus as **chromatin**. Nuclei have one or more nucleoli that are visible in electron micrographs as darkly stained spheres. Nucleoli produce the RNA needed to make **ribosomes**, the organelles that produce proteins using an RNA template. The nucleus is bounded by a **nuclear membrane**, which is a phospholipid bilayer. The membrane contains pores large enough to allow big molecules such as messenger RNA to pass out of the nucleus and into the cytoplasm where they go to the ribosomes to be 'read' to produce proteins.

You can learn more about the two types of nucleic acids – DNA and RNA – in Chapter 6.

6 Suggest why the nuclear membrane has large pores and is so intimately associated with the rough endoplasmic reticulum.

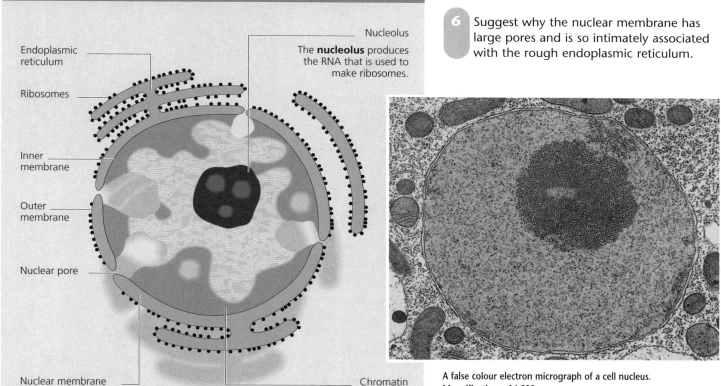

Fig. 9 A cell nucleus

The **nucleolus** produces the RNA that is used to make ribosomes.

The **nuclear membrane** is a double phospholipid membrane. Its large pores, not found in any other phospholipid membrane, allow large RNA molecules to pass through it.

When a cell is not dividing, the DNA is in the form of **chromatin**, which stains darkly.

A false colour electron micrograph of a cell nucleus. Magnification × 16 500

KEY FACTS

- The nucleus contains the cell's DNA, the information the cell needs to divide and to control its activities;
- In a non-dividing cell, DNA molecules are extended as chromatin; when the cell divides, the DNA molecules condense into visible chromosomes;
- The nucleus has its own membrane that has pores to allow outward traffic of RNA.
- Nucleoli in the nucleus make ribosomal RNA.

1 CELLS

APPLICATION

Magnification

Most biological illustrations have an attached scale, for example, × 60. This means that the object in the picture has been magnified 60 times. In exams you are often asked either to find the actual size of a specimen, or to calculate how many times it has been magnified.

You need to remember two formulae:

$$\text{Actual size} = \frac{\text{Image size}}{\text{Magnification}}$$

$$\text{Magnification} = \frac{\text{Image size}}{\text{Actual size}}$$

1. The photograph below shows a human egg covered with sperm. The egg is actually 0.1 mm across. Calculate:
   a. the magnification;
   b. the volume of the egg and a single sperm head in cubic μm;
   c. how many sperm heads would fit inside the egg.

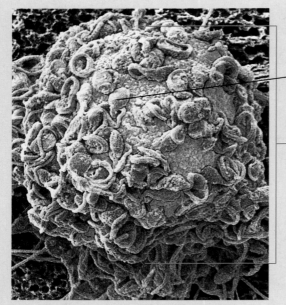

Sperm head

Egg

4. The ant in this photograph above is magnified 25 times. Calculate the actual length of its head in milimetres, from the top of its head to its biting mouthparts.

2. The photograph on the left shows bacteria magnified 9240 times (× 9240). Calculate the actual mean length of the bacteria in μm.

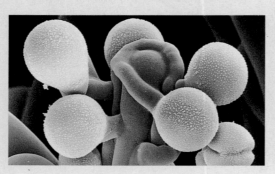

3. The photograph on the left shows some pollen grains on the pistil of a flower. It has been magnified 420 times. Choose three of the pollen grains and calculate their actual mean width in millimetres.

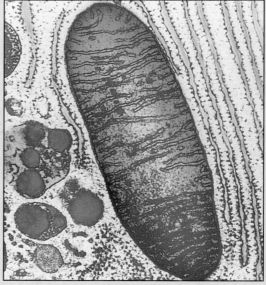

5. The photograph above shows a section through part of an animal cell magnified 80 000 times (× 80 000). Calculate the length and width of the mitochondrion in nanometres.

1 CELLS

Mitochondria

Mitochondria are found in all living plant and animal cells. They are the site of **aerobic respiration**, the biochemical process that oxidises glucose to release energy.

Each mitochondrion (Fig. 10) has two phospholipid membranes; an **outer membrane** that surrounds the entire organelle and a highly folded **inner membrane**. The **matrix**, the central fluid-filled space contains the free enzymes that catalyse reactions in the early stages of respiration. The **cristae** that result from the intricate folds of the inner phospholipid membrane have a large internal surface area and hold many of the enzymes concerned with the final stages in respiration in place. At the cristae, the transfer of energy to a molecule called **adenosine triphosphate (ATP)** occurs. ATP is the 'energy currency' of the cell – its energy is used to contract muscles, to build up large molecules from smaller units and to power **active transport** (see page 36).

7 Look at Fig. 10. Use information from the diagram to suggest the advantage that having a highly folded inner membrane gives to the mitochondrion.

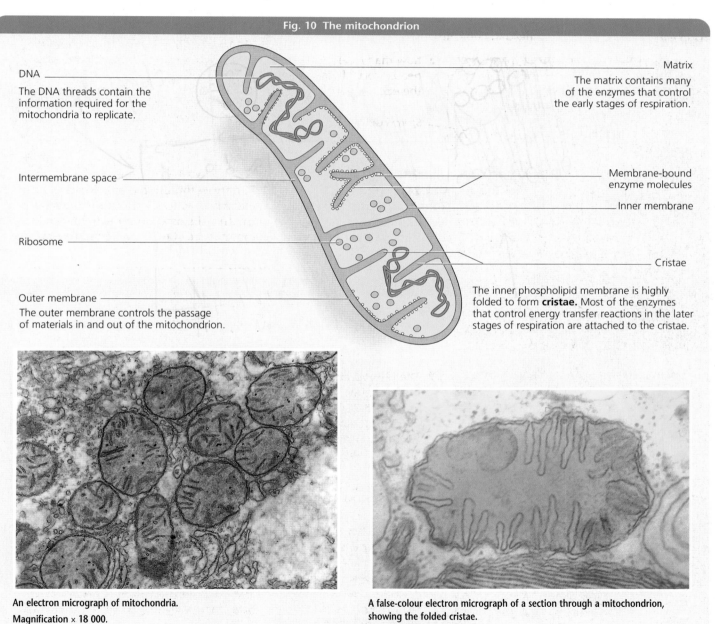

Fig. 10 The mitochondrion

DNA — The DNA threads contain the information required for the mitochondria to replicate.

Intermembrane space

Ribosome

Outer membrane — The outer membrane controls the passage of materials in and out of the mitochondrion.

Matrix — The matrix contains many of the enzymes that control the early stages of respiration.

Membrane-bound enzyme molecules

Inner membrane

Cristae — The inner phospholipid membrane is highly folded to form **cristae**. Most of the enzymes that control energy transfer reactions in the later stages of respiration are attached to the cristae.

An electron micrograph of mitochondria. Magnification × 18 000.

A false-colour electron micrograph of a section through a mitochondrion, showing the folded cristae. Magnification × 144 000.

EXTENSION: Respiration

Aerobic respiration is a complex process involving many stages. The overall process releases 36 ATP molecules by oxidising one molecule of glucose.

Since a complete glucose molecule is too big to enter the mitochondrion, it is first split into two molecules of pyruvate, a smaller 3-carbon compound. Enzymes in the central fluid-filled matrix control the further breakdown of pyruvate to carbon dioxide and hydrogen in a series of reactions known as the Krebs cycle. Finally, enzymes and carrier molecules attached to the cristae combine hydrogen ions with oxygen to produce water, and release ATP. Some ATP is produced as a result of the earlier stages, but the vast majority results from this final stage.

The ATP is made available for all the processes in the cell that need energy. Cells that require a large amount of energy, such as secretory, nerve and muscle cells, contain large numbers of mitochondria.

KEY FACTS

- Mitochondria are the site of aerobic respiration;
- Mitochondria have a double phospholipid membrane;
- The fluid in the inner matrix contains many of the enzymes concerned with the early stages of respiration;
- The highly folded inner membrane forms cristae that bear many of the enzymes that control the later stages that result in the formation of large amounts of ATP;
- ATP is a source of energy for cell processes.

Endoplasmic reticulum

The **endoplasmic reticulum (ER)** is a series of thin, intricate channels. It exists in the space created by folds in the phospholipid membrane that is continuous with the nuclear membrane (see Fig. 9 and Fig. 11). The narrow, fluid-filled space between these membranes acts as a transport network that moves materials through the cell. The folded membrane of the ER creates a large surface area within the cell.

Fig. 11 Endoplasmic reticulum and ribosomes

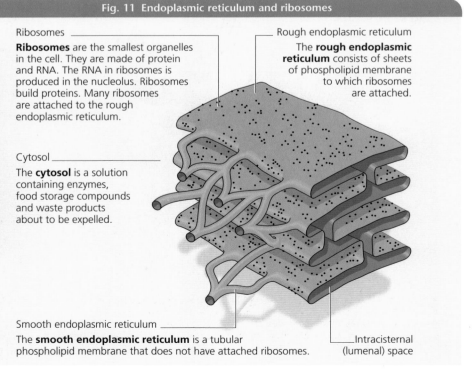

Ribosomes

Ribosomes are the smallest organelles in the cell. They are made of protein and RNA. The RNA in ribosomes is produced in the nucleolus. Ribosomes build proteins. Many ribosomes are attached to the rough endoplasmic reticulum.

Cytosol

The **cytosol** is a solution containing enzymes, food storage compounds and waste products about to be expelled.

Rough endoplasmic reticulum

The **rough endoplasmic reticulum** consists of sheets of phospholipid membrane to which ribosomes are attached.

Smooth endoplasmic reticulum

The **smooth endoplasmic reticulum** is a tubular phospholipid membrane that does not have attached ribosomes.

Intracisternal (lumenal) space

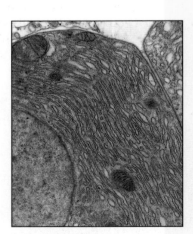

An electron micrograph of endoplasmic reticulum and ribosomes. Magnification × 7500.

Ribosomes and the ER

Much of the outside surface of the ER is dotted with **ribosomes**; this gives the membrane a grainy appearance and also its name, **rough ER**. **Smooth ER**, has no attached ribosomes.

Ribosomes are small, dense organelles, about 20 nm in diameter. They carry out protein synthesis. When a ribosome binds to a length of messenger RNA, it uses information encoded in the RNA to assemble amino acids in the correct order to form a specific protein. Free ribosomes, ones that float around in the cytoplasm, produce proteins for use inside the cell. Ribosomes attached to ER produce proteins destined for export from the cell for use elsewhere.

 8 The electron micrograph in Fig. 11 does not seem to include many of the features of the ER that are shown in the three-dimensional diagram. Suggest a reason for this.

Golgi body and lysosomes

The Golgi body is a group of flattened cavities (Fig. 12). Its function is to take enzymes and other proteins that have been synthesised in the endoplasmic reticulum and to package them into membrane-bound vesicles. The appearance of the Golgi body is constantly changing as material comes in on one side from the ER and is lost from the other as completed vesicles 'bud off'. Such vesicles transport materials to other parts of the cell, or fuse with the cell surface membrane, releasing their contents outside the cell.

 9 Cells in the pancreas produce enzymes. Explain why these cells have large amounts of rough endoplasmic reticulum and Golgi bodies.

Lysosomes (see the photograph below) are vesicles that contain digestive enzymes. These can destroy old or surplus organelles inside the cell or they can be used to break down material that has been taken into the cell by the process of **endocytosis**. Whole cells and tissues that are no longer required can be destroyed if cells nearby allow lysosomes to release their contents at the cell surface. The body uses this process to break down excess muscle in the uterus after birth, and to destroy milk-producing tissue after a baby has been weaned.

 10 Soon after death, cells in the body start to digest themselves. Suggest why.

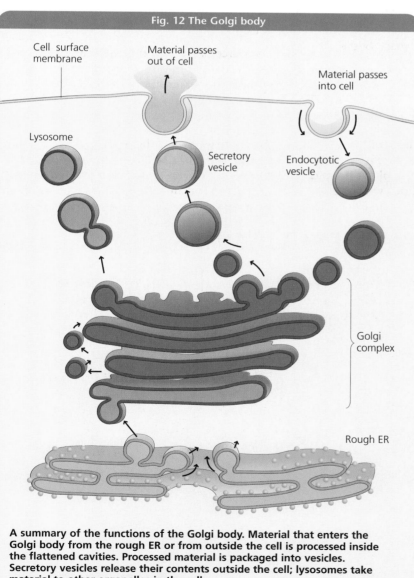

A summary of the functions of the Golgi body. Material that enters the Golgi body from the rough ER or from outside the cell is processed inside the flattened cavities. Processed material is packaged into vesicles. Secretory vesicles release their contents outside the cell; lysosomes take material to other organelles in the cell.

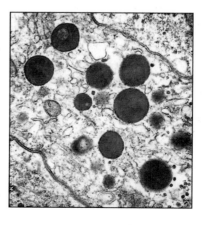

Lysosomes are vesicles that are filled with digestive enzymes. The lysosomes in this photograph have been stained to show up clearly

1.3 Plant cell organelles

Animal cells and plant cells are very similar except for two organelles, the cell wall and the chloroplast, which are found only in plant cells.

The cell wall

The cell wall of a young plant cell is called the primary cell wall and is mainly **cellulose**. Cellulose is a polysaccharide, or carbohydrate polymer. In order to understand the structure and function of the cell wall, we need to look at exactly what this means.

Carbohydrates

All carbohydrates contain the elements carbon, hydrogen and oxygen. The hydrogen and oxygen atoms are always in the ratio 2:1, giving carbohydrates the general formula CH_2O. Plants are rich in several forms of carbohydrate; sugars, starches and cellulose.

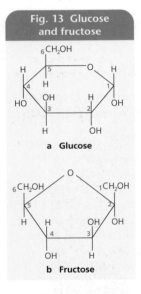

Fig. 13 Glucose and fructose
a Glucose
b Fructose

11 How is the composition of a carbohydrate different from that of a lipid?

There are three basic types of carbohydrate molecule; **monosaccharides**, **disaccharides** and **polysaccharides**. Monosaccharides are single sugars, small molecules that dissolve in water and taste sweet. These are the units from which all larger carbohydrates are made. Glucose is a monosaccharide. It has the formula $C_6H_{12}O_6$ and its structural formula is shown in Fig. 13a. Five of the carbon atoms (numbered 1-5 in the diagram) form a ring. Fructose, another monosaccharide is shown in Fig. 13b. Glucose and fructose are **isomers** – they have the same number of carbon, hydrogen and oxygen atoms but these are arranged differently. This gives the fructose molecule a different shape – and therefore different properties. Honey is a good source of fructose.

Disaccharides form when a condensation reaction joins two monosaccharides. Two glucose molecules combine to form maltose; glucose and fructose combine to form sucrose (Fig. 14). Sucrose is the most common sugar in plants and is also the sugar you probably put in your tea or coffee.

Since sugars are small molecules that are soluble in water, they are easy to transport to different parts of an organism. The energy stored in their molecules can be used to power other essential chemical reactions. Sugars are also converted into substances required for growth and repair of organs (Fig. 15).

Polysaccharides are giant molecules made up from many single sugar molecules joined together by condensation reactions. Starch, glycogen and cellulose are all polysaccharides. Part of a starch molecule is shown in Fig. 16. Starch and glycogen molecules have compact, coiled and branched molecules, making them ideal 'energy' stores. Cellulose molecules have long straight molecules, perfect for forming structural fibres.

12 Look at the shapes of the starch and glycogen molecules. Suggest why animals store carbohydrates as glycogen rather than starch.

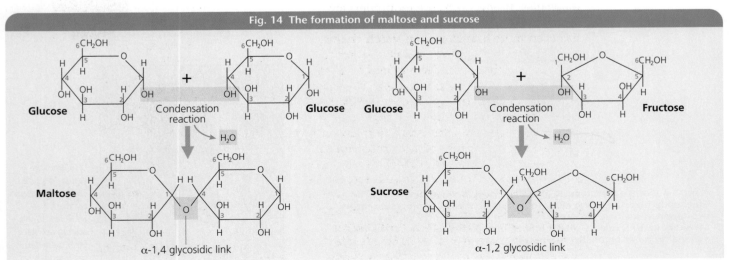

Fig. 14 The formation of maltose and sucrose

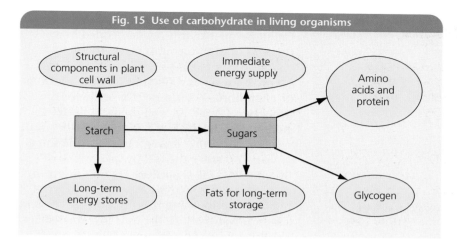

Fig. 15 Use of carbohydrate in living organisms

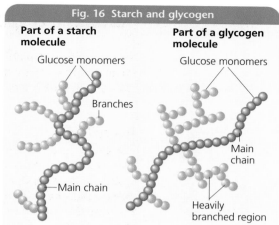

Fig. 16 Starch and glycogen

Monomers and polymers

'Mono' and 'poly' come from Greek words. Mono means one, and poly many. These are used in many biological and chemical names.

Starch and cellulose are both polymers. A **polymer** is a molecule made up of repeating units, rather like links in a chain. The individual units are called **monomers**, and the same monomer can be used to build different kinds of chains – long, short, straight, branched. For example, the same monomer, glucose, is linked in different ways in the polymers starch and cellulose found in plants and the polymer glycogen found in animals.

The glucose monomer in starch is called α-glucose. Cellulose is made from a slightly different glucose monomer β-glucose. Because their monomer molecules differ, the polymer molecules have different shapes. The shape of each polymer determines its function.

The monomers in starch are joined by α-glycosidic bonds. These bonds produce twisted chains of monomers that form branched molecules. In glycogen, the monomers are again linked by α-glycosidic bonds but glycogen has even more branches than starch chains. The coiled and branched chains of starch and glycogen molecules give them a compact shape. This, together with their insolubility in water, makes them ideal 'energy' storage compounds. Starch is the major storage carbohydrate in plants; glycogen does the same job in animals.

The structure of the cellulose molecule, the major component of the cell wall, is shown in Fig. 17. In this polymer, the glucose monomers are joined by β-glycosidic bonds. These bonds result in straight chains. Individual cellulose molecules are therefore long, unbranched chains with many β-glycosidic bonds. The straight molecules lie side by side, forming microfibrils that are strengthened by many hydrogen bonds. This makes cellulose fibres very strong and stable – ideal for a structural material. These fibres are part of most cell walls and give them strength.

 13 A cellulose molecule is a polymer but a phospholipid molecule is not. Explain why.

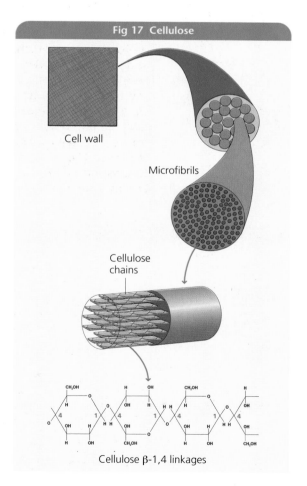

Fig 17 Cellulose

Cellulose β-1,4 linkages

1 CELLS

EXTENSION — Secondary thickening in plant cell walls

Older cell walls often have extra thickening called a secondary wall, which gives a permanent shape and more strength. The trunks of trees contain a large proportion of **xylem cells**. The secondary walls of these cells become impregnated with a tough, waterproof material called **lignin**. This waterproofing layer means that the cells cannot exchange substances with their environment and so the cells die. Cork cells that form the outer layer of tree bark become impregnated with a fatty material called **suberin**, which is also waterproof. It may seem strange when you look at a large oak tree covered in green leaves in the middle of summer that most of what you see consists of dead cells. Only the leaves and thin layers of cells in the trunk and roots are actually alive.

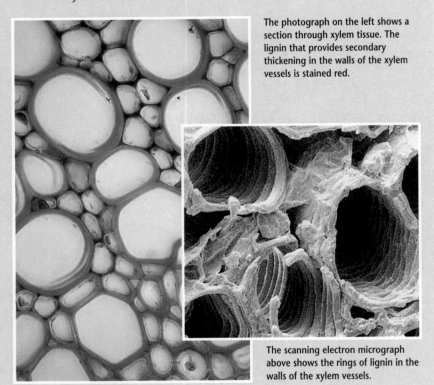

The photograph on the left shows a section through xylem tissue. The lignin that provides secondary thickening in the walls of the xylem vessels is stained red.

The scanning electron micrograph above shows the rings of lignin in the walls of the xylem vessels.

Cellulose and the cell wall

Cellulose microfibrils are embedded in a framework of other substances, the most common being complex molecules called **hemicelluloses** and **pectins**. The arrangement of microfibrils determines the shape of the primary cell wall. As well as being flexible enough to allow the cell to grow, the cell wall needs to be strong enough to resist the force of water entering the cell by osmosis (see Chapter 2). This is particularly important in young plants, where the strength of the cell wall is the only thing that supports stems and leaves. If water is lost, the cells lose their shape. Think how quickly a young plant droops if it loses water.

Chloroplasts

Chloroplasts (Fig. 18) are organelles that occur only in plant cells. They are found in many of the cells in the green parts of a plant (mainly the leaves and young stem). The primary function of the chloroplast is to make the sugars that form the basis of all carbohydrates by **photosynthesis**. Different parts of the chloroplast carry out the different stages of photosynthesis. First, light energy is absorbed by pigments such as **chlorophyll** and transferred to chemical energy, mainly in the form of ATP. Energy from ATP is then used to fix atmospheric carbon dioxide into carbohydrate.

Like the mitochondria, the chloroplasts also have two bilayered phospholipid membranes. The outer one surrounds the organelle; the inner membrane is highly folded, creating **thylakoids**, in between which is a fluid **stroma**. In places, the thylakoid membranes are arranged into structures called **grana**, which look a little like piles of green coins.

KEY FACTS

- **Cellulose**, the major component of the **cell wall** in plant cells is a **carbohydrate**;
- The three classes of carbohydrate are the **monosaccharides**, the **disaccharides** and the **polysaccharides**;
- Monosaccharides (single sugars) combine to form di- or polysaccharides by **condensation reactions**;
- Monosaccharides and disaccharides are sugars – small, soluble, diffusible molecules that are easily transported around organisms. They can be used to release energy and to build other molecules;
- Polysaccharides (carbohydrate polymers) include starch and glycogen, as well as cellulose;
- The properties of these polymers depend on the shape of their molecules;
- Starch and glycogen molecules have compact, coiled and branched molecules, making them ideal 'energy' stores;
- Cellulose molecules have long straight molecules, perfect for forming structural fibres.

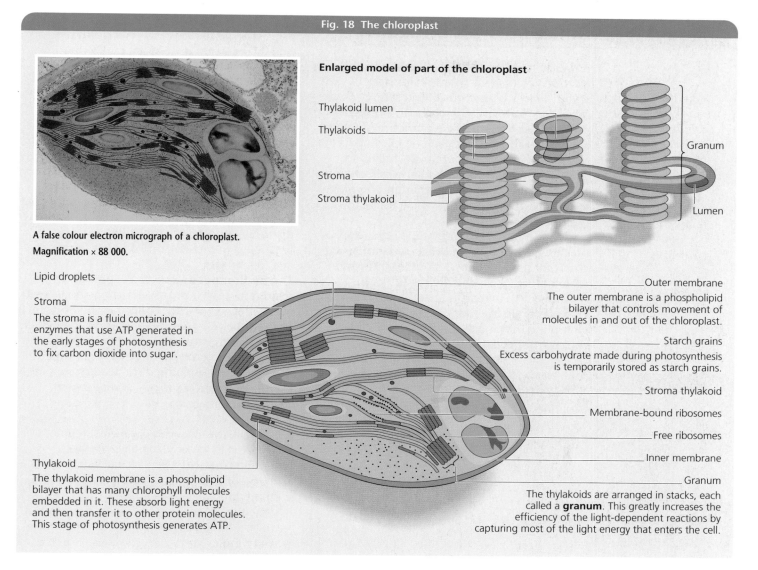

Fig. 18 The chloroplast

The thylakoids contain chlorophyll molecules that capture light energy from sunlight. The arrangement of the thylakoids into grana, and the distribution of the grana, maximises the absorption of light energy. In the first stage of photosynthesis, chemical reactions in the grana split water molecules into hydrogen ions and oxygen. The oxygen is given off as a 'waste product'. The hydrogen ions are used, along with energy from ATP, to reduce carbon dioxide to form sugars. Enzymes in the fluid-filled stroma control this second series of reactions.

14 Describe how chloroplasts and mitochondria are:
a similar in structure and function;
b different in structure and function.

KEY FACTS

- Chloroplasts contain light absorbing pigments such as chlorophyll;
- These pigments are located in double phospholipids membranes called thylakoids;
- In places the thylakoids form stacks called grana;
- In the grana, light energy is transferred into chemical energy;
- In the fluid stroma, enzymes catalyse reactions that use ATP and hydrogen ions produced in the grana to fix carbon dioxide into sugars.

1.4 Differential centrifugation

It is easy to see that the availability of the electron microscope was a major help to the scientists in working out the structure of cell organelles. But, just looking at the organelles in preserved sections of tissue cannot reveal much about their function. In fact, people had found out a great deal about what organelles do, particularly about chloroplasts and mitochondria and the sequence of reactions in photosynthesis and respiration, *before* they had access to electron micrographs.

They investigated individual organelles by breaking down cells and then separating out chloroplasts, mitochondria, ribosomes and membranes. To do this successfully, the organelles had to remain intact, with all internal enzymes and structures in place. The technique that made this possible is called **differential centrifugation**.

As Fig. 19 shows, if a mixture of particles of different sizes is spun at high speed in a **centrifuge**, the larger particles tend to accumulate at the bottom of the tube. Since organelles vary in size, this principle can be used to separate them. Cells are broken up in a **homogeniser**, a device rather like a kitchen blender, that breaks down the outer membrane of the cells but leaves the organelles intact. The liquid used in the homogeniser is ice cold to reduce the rate of enzyme activity, and has the same concentration of solutes as the organelles to prevent shrinkage or bursting due to osmosis.

The **homogenate** is then spun at relatively slow speed. The nuclei, the largest organelles in the cell, collect at the bottom of the tube. The suspension above the cells is known as the **supernatant**. This supernatant is then spun at a higher speed. This time the mitochondria separate out. To separate out the ribosomes the process is repeated at higher centrifuge speed.

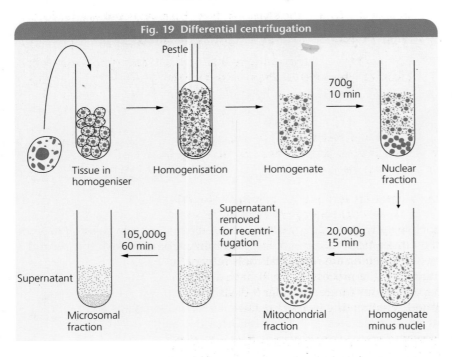

Fig. 19 Differential centrifugation

15
a Place the following organelles in order of size with the largest first: chloroplasts, ribosomes, mitochondria, nucleus.
b Draw a centrifuge tube to show the order of settling of these four organelles when centrifuged.

In 1937 Robin Hill used differential centrifugation to isolate chloroplasts from leaf cells and was able to show that, when illuminated, they could carry out the first light-dependent stage of photosynthesis. In the same year Hans Krebs isolated mitochondria from animal cells and worked out the sequence of reactions in the final stages of respiration. One series of these reactions has been named the **Krebs cycle** in his honour.

1.5 Are all cells the same?

In this chapter, we have looked in detail at the internal structure of plant and animal cells. These are **eukaryotic cells**; they have a nucleus that contains the cell's DNA and they have a collection of complex organelles that have specific functions. There is also another major type of cells; **prokaryotic cells**. Bacteria are prokaryotes (Fig. 20) and these organisms evolved long before the eukaryotes. The term prokaryote means 'before the nucleus' and refers to the fact that prokaryotes contain nucleic acids, but these are not confined inside a definite nucleus with a nuclear membrane. Instead there are strands of DNA in the centre

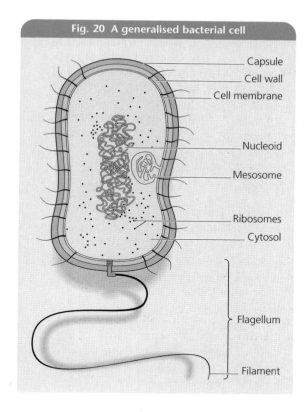

Fig. 20 A generalised bacterial cell

Labels: Capsule, Cell wall, Cell membrane, Nucleoid, Mesosome, Ribosomes, Cytosol, Flagellum, Filament

APPLICATION Bacteria have their good points

Mention bacteria and most people will say 'harmful'. But disease-causing bacteria are in the minority. Indeed, we need bacteria in our intestines to stay healthy and some firms now sell bacteria that are supposed to be good to eat.

Most of the bacteria in the body are in the colon – many different species live there. They are harmless and they live on undigested carbohydrates. Because there is little oxygen in the colon, the bacteria release energy via a special type of anaerobic respiration. The end products of this respiration include short-chain fatty acids such as acetate, propionate and butyrate.

These short chain fatty acids seem to be beneficial to the epithelial cells that line the colon. They also increase blood flow through the walls of the colon and enhance muscular activity in the colon. Butyrate may even lower the risk of colon cancer. These helpful bacteria flourish if the owner of the colon eats plenty of non-starch polysaccharides. Several firms are now marketing 'probiotic bacteria'. The ads claim that if you eat them, they will flourish in the colon and produce cancer-preventing chemicals. However, the evidence about the potential health benefits is not yet conclusive.

1 Why do bacteria in the small intestine respire anaerobically?

2 What is meant by 'non-starch polysaccharide'?

3 How could you obtain evidence that probiotic bacteria do what the firms who advertise them claim they do?

of the cell that are collectively known as the **nucleoid**. There are other differences between prokaryotes and eukaryotes.

16 Look at Table 1 and compare Figs. 3 and 20. Which organelles are present in eukaryotes but absent in prokaryotes?

The cell walls of prokaryotes contain polysaccharides, but not cellulose. They also contain protein and lipids. Many prokaryotes also secrete a protective capsule outside the cell wall. Some prokaryotes have a long whip-like structure called a **flagellum** that allows them to move. Prokaryotes increase the area of their internal membranes by infoldings of the cell membranes called **mesosomes**.

17 Which organelles in higher organisms have internal projections similar to the mesosomes in bacteria?

Prokaryotes are important organisms. Bacteria are involved in the decay of dead organisms and the cycling of nutrients. Some bacteria 'fix' atmospheric nitrogen, adding nutrients to the soil. Bacteria are now vital tools in genetic engineering, a process that will have a big impact on our future. Chapter 9 deals with genetic engineering in more detail.

Table 1 A comparison of prokaryotic and eukaryotic cells			
Feature	Prokaryotic cell	Eukaryotic plant cell	Eukaryotic animal cell
Cell wall	Polysaccharide, but not cellulose	Cellulose	No cell wall
Phospholipid membranes	Present	Present	Present
DNA	Not enclosed in nucleus	Enclosed in nucleus	Enclosed in nucleus
Chloroplasts	Not present	Present	Not present
Mitochondria	Not present	Present	Present
Ribosomes	Present	Present	Present

1 CELLS

EXAMINATION QUESTIONS

1 The drawing shows part of a cell as seen under the electron microscope.

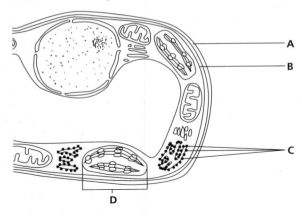

a Identify structures A to D (2)

b Explain why all the detail in this diagram could not be seen using a light microscope. (2)

c Describe how a sample of chloroplasts could be obtained from leaf tissue. (2)

AQA BY01 June 98 Q1

2 The diagram shows the structure of glycerol and a fatty acid.

Glycerol

$$\begin{array}{c} H \\ H-C-OH \\ H-C-OH \\ H-C-OH \\ H \end{array}$$

Fatty acid

$R-COOH$

Draw a diagram to show how glycerol and fatty acids combine to form a triglyceride. (2)

AQA BY01 June 98 Q8

3 The diagram represents a phospholipid molecule.

a Name the parts of the molecule A, B and C. (1)

b Explain how the phospholipid molecules form a double layer in a cell membrane (2)

c Cell membranes also contain protein molecules. Give two functions of these protein molecules. (2)

AQA BY01 March 98 Q1

4 Liver cells were ground to produce an homogenate. The flow chart shows how centrifugation was used to separate organelles from liver cells.

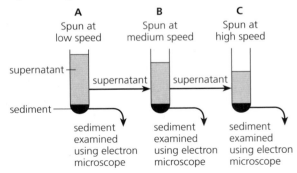

Drawings of electron micrographs of three organelles separated by centrifugation are shown below. The drawings are not to the same scale.

a Copy and then complete the table below. (2)

Electron micrograph	Name of organelle	Centrifuge tube in which the organelle would be the main constituent of the sediment
1		
2		
3		

b Explain why it is possible to separate organelles in this way. (2)

AQA BY01 Feb 95 Q2

5 a Show how the two glucose molecules shown in the diagram may join together. (2)

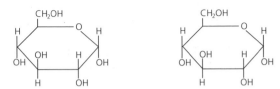

b Explain two ways in which the properties of a cellulose molecule make it suitable for its role in the structure of plant cell walls. (2)

c Give two ways in which the appearance of an epithelial cell would differ from a prokaryotic cell such as a bacterium. (2)

AQA BY01 Jun 98 Q2

KEY SKILLS ASSIGNMENTS

How did cells evolve?

We can never be certain how life on Earth began but many scientists believe that the first cells arose spontaneously about 3.5 billion years ago. Conditions were very different then. The atmosphere contained hydrogen, methane, ammonia and water, with very little or no oxygen. Just right for the formation of the chemicals we find in living organisms. In the early 1950s, Miller and Urey passed electric sparks through a mixture of these gases. The sparks simulated lightning. After several days they detected newly formed organic compounds.

Prokaryotes

The earliest prokaryotic cells probably had no internal organelles. They stored energy in long chain hydrocarbons and perhaps used ATP from their surroundings. One thing is certain – since there was no oxygen in the atmosphere these early cells respired anaerobically. This process released carbon dioxide into the atmosphere and, over the next 500 million years carbon dioxide levels rose steadily. This gas reduced the amount of infra-red and ultra-violet radiation, causing the atmosphere to cool. This reduced the rate of production of organic compounds in the atmosphere. The prokaryotes that depended on these organic molecules for energy (heterotrophs) would probably have died out had it not been for the development of eukaryotes that could produce their own organic compounds (autotrophs). These eukaryotes used the energy from light to convert carbon dioxide into carbohydrate.

One of the earliest groups of Eubacteria to do this was the Cyanobacteria (also called blue-green algae). These organisms still exist in large numbers and their fossil record shows that they have changed very little over the last 3 billion years. They contain molecules of chlorophyll a, the pigment found in all chloroplasts. In fact some Cyanobacteria look remarkably similar to the chloroplasts found inside the green plants of modern cells. Cyanobacteria use energy from sunlight to split water into hydrogen and oxygen. The oxygen and some of the hydrogen is given off, but some of the hydrogen is used to reduce carbon dioxide to carbohydrates.

About 2.5 million years ago, as oxygen levels increased, some prokaryotes evolved the ability to use oxygen to generate ATP using a mechanism that is very similar to that used by mitochondria today. So, by this time there were three types of prokaryote:
- those that could photosynthesise;
- those that could respire aerobically;
- those that could respire anaerobically.

Eukaryotes

The first eukaryotic cells appeared about 1.5 million years ago. The 'Endosymbiotic Theory of Eukaryote Evolution', first proposed in the 1960s, tries to explain how they evolved. This theory states the ancestors of modern eukaryotic cells were 'symbiotic consortiums' of prokaryotic cells. For example, aerobic bacteria might have invaded larger amoeba-like anaerobic bacteria. Both organisms would benefit – the anaerobic bacteria would ingest organic material and aerobic bacteria would oxidise this to provide ATP for both organisms. The embedded aerobic bacteria assumed the role of what we now call mitochondria. These cells were the ancestors of eukaryotic animal cells. Some amoeba-like bacteria also formed symbiotic consortia with cyanobacteria. These cells were the ancestors of eukaryotic plant cells.

At first this theory was ridiculed, but there was an important test. If chloroplasts and mitochondria were prokaryotic symbionts, they would have their own DNA – and this turns out to be the case. The division of mitochondria and chloroplasts is not under the control of the nucleus of the cell, but is controlled by DNA in the organelles.

1 Draw a time line for eukaryotic cell evolution.

2 Do you think that cells could arise on a planet where the atmosphere and climate were similar to those on Earth at the moment. Support your answer with information from the text.

3 Many scientists believe that cells originated spontaneously but many other people believe in Creationism. Creationists believe cells were created in a completely different way.

a Search for a clear explanation of their views with the evidence they use to back them up. You may find the internet useful for this task.

b Using the information in this textbook and any others you find, prepare two documents:
  1 A simple account of cell evolution as described by the endosymbiont theory suitable for the general public.
  2 A more complex article for a scientific journal EITHER to persaude a creationist that the endosymbiont theory is correct OR to convince a scientist that the creationist view is possible.

2 Cells and their environment

Many sports drinks claim to be 'isotonic' and to provide replacement fluids and minerals that 'restore the body's balance'. Many athletes buy these drinks, which cost about 10% more than ordinary soft drinks, because advertisements stress the importance of providing a balanced replacement of salts. But are they worth the extra money?

Isostar Sport is one of several drinks that have become very popular with people who do stamina sports or athletics.

2.1 A balancing act

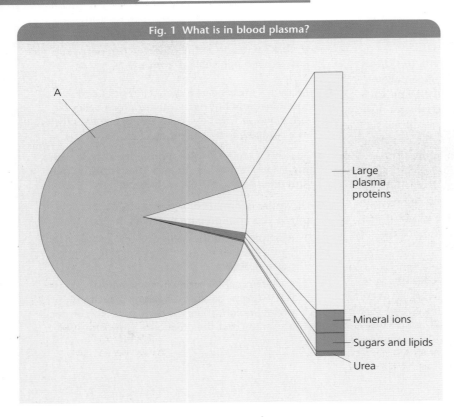

Fig. 1 What is in blood plasma?

To find out whether isotonic drinks are worth buying, we need to find out more about what happens to the body when we do strenuous exercise. As well as getting out of breath, we get hot and we sweat. The fluid lost through the skin evaporates and this has a cooling effect. That is good, but losing large quantities of sweat leads to dehydration if we don't drink fluid to replace what has been lost. Drinking water helps, but is that all we lose in sweat?

Sweat is formed from the fluid part of blood – the **blood plasma**. Plasma contains water, large plasma proteins, mineral ions (Na^+, Cl^-, K^+), sugars, lipids and urea in the proportions shown in Fig. 1.

1 Look at Fig. 1. What is the substance labelled A?

After we have been sweating, our blood plasma has fewer mineral ions such as sodium (Na^+) and chloride (Cl^-), as well as less water. Just as we need to replace the water lost in sweat, these lost minerals need to be replaced too. Sports drinks claim to do this because they contain mineral ions as well as water. The concentration of Na^+ and Cl^- ions in these drinks is very similar to the concentration of the ions in blood plasma (Table 1). Solutions containing the same concentration of mineral ions are said to be **isotonic**.

Controlled studies have shown that athletes who drink small amounts of isotonic drinks when they are doing hard, sustained exercise can exercise for longer than those who just drink water. In one experiment, cyclists given isotonic drinks cycled as hard as they could for about 17 minutes longer than those given water. In a race, this could mean the difference between a gold medal and fourth place. So perhaps sports drinks are worth the price – but how do they work? One theory is that keeping the body's fluid and mineral ion levels constant allows the body to keep its blood plasma in balance.

As long as the plasma and the inside of the red blood cells are isotonic, the cells are fine. They are normal volume and there is no net water gain or water loss (point A in Fig.2). When the body loses a lot of sweat, the mineral ion concentration of the plasma can become slightly higher than the inside of the cell, and water leaves the cell to try to restore the balance. The volume of the blood cells decreases (Zone B in Fig. 2). In extreme cases, this can cause the red blood cells to become star-shaped. If an athlete drinks only water after exercise, the plasma can become less concentrated than the inside of the cells and water enters the cells. This causes them to swell (Zone C in Fig.2).

In both cases, the changes in volume of the red cells, even if they are slight, affect the way the red cells work. To carry oxygen efficiently, the volume of the red blood cells needs to be kept as constant as possible. Drinking isotonic drinks during exercise makes this easier for the body to achieve.

Table 1 Plasma, sweat and an isotonic drink compared			
Mineral ion	Na^+	Cl^-	K^+
Concentration in blood plasma (%)	0.34	0.35	0.02
Concentration in sweat (%)	0.15	0.2	0.02
Concentration in Isostar (%)	0.4	0.39	0.17

2 Look at Fig. 2 and then describe the relationship between the volume of red cells and the concentration of sodium chloride (NaCl) in the external solution.

3 Suggest one way in which the function of red blood cells might be affected if they were swollen (but not burst).

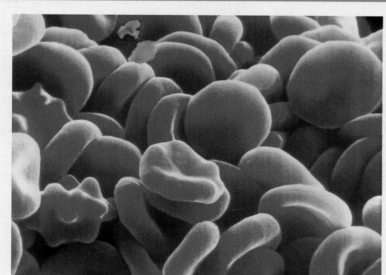

The optimum biconcave shape of the red blood cell provides the maximum surface area for the exchange of molecules. This biconcave shape depends on the concentration of ions in the blood plasma. A high ion concentration causes red blood cells to become star-shaped.
Magnification × 3000.

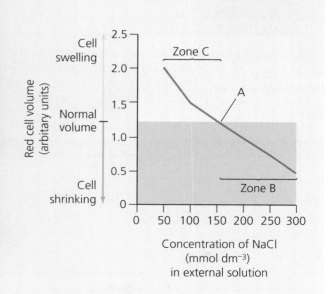

Fig. 2 Red blood cells and salt

2.2 Moving substances in and out of cells

Red cells work better when their environment is ideal and this is also true for all the other cells in the body. Maintaining this ideal environment needs constant effort and depends largely on the way substances pass in and out of individual cells.

Cells are complex factories and they need constantly to import raw materials and get rid of waste. Molecules and ions with different sizes and electrical charges enter and leave all the time. Some of the exchange of materials occurs as a result of passive processes such as diffusion and osmosis, but not all movement of substances is free movement. This would make it impossible for a cell to maintain the high concentrations of valuable substances that it needs to function efficiently. Cells must control the passage of substances through their membranes. The presence of special 'gateways' for individual substances and the use of energy-requiring transport mechanisms such as facilitated diffusion and active transport enables a cell to keep what it needs. The same transport processes also occur inside the cell to segregate particular substances within organelles.

We will look at diffusion, osmosis, facilitated diffusion and active transport in detail in the rest of this chapter.

2.3 Diffusion

All the particles in liquids and gases are in constant random motion. This motion results in a net movement of particles from a region of high concentration to a region of lower concentration. This process is called **diffusion** (Fig. 3).

In a mixture of gases, diffusion causes each gas to spread evenly through the space that the mixture occupies. In the same way, a soluble substance spreads through a liquid until it is evenly dispersed.

4 Look at the photograph of the tea bag in the beaker. Why is the solution darker near the tea bag?

Particles of gas or solute can also diffuse through a membrane, as long as the membrane has pores that are larger than the particles (Fig. 4).

Illustrating diffusion with a tea bag: the flavour and colour from the tea inside the bag diffuse through the water.

Fig. 4 Diffusion through a membrane

Pores in this membrane are wide enough to allow diffusion

Pores in this membrane are too narrow to allow diffusion

Diffusion is a **passive** process. It happens without any energy input from the organism. The rate at which diffusion occurs depends on three factors:

- The difference in concentration between two areas. This is called a **concentration gradient**;
- The distance between the areas;
- The size of the molecules that are diffusing.

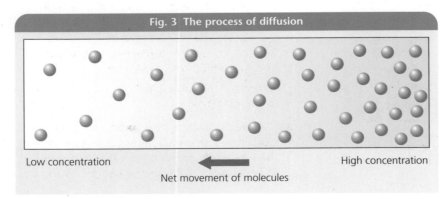

Fig. 3 The process of diffusion

Low concentration ← High concentration

Net movement of molecules

So the greater the concentration gradient and the smaller the particles, the quicker the net movement of molecules from the area of high concentration to the area of low concentration. And diffusion happens faster when molecules need to move only microscopic distances than when they have to travel much larger distances. For example, it would take a small molecule such as oxygen at least 4 minutes to diffuse to the centre of a cell 1 mm in diameter. Some cells, such as those in the human gut and inside plant leaves, have special adaptations to allow diffusion to happen quickly. These adaptations often reduce the distance over which diffusion occurs.

The overall rate at which a substance diffuses through a membrane also depends on the surface area in contact with the substance. The biconcave disc shape of a red blood cell gives it a much greater surface area than if it were spherical, allowing the maximum amount of oxygen to diffuse into it. Look out for other examples of cells that maximise their surface area for efficient diffusion later in this chapter.

You can work out the rate at which a substance diffuses using a formula that takes into account all the factors that affect diffusion:

$$\text{Rate of diffusion} = \frac{\text{Surface area} \times \text{Concentration difference}}{\text{Thickness of membrane}}$$

This relationship is called **Fick's law**.

APPLICATION: Diffusion and the red blood cell

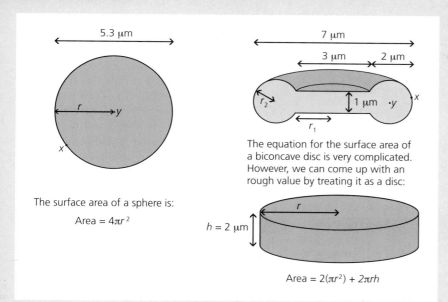

The equation for the surface area of a biconcave disc is very complicated. However, we can come up with an rough value by treating it as a disc:

Area = $2(\pi r^2) + 2\pi rh$

The surface area of a sphere is:

Area = $4\pi r^2$

Distance that molecules need to diffuse	Time required for small molecules to diffuse
1 µm	0.4 milliseconds
10 µm	50 milliseconds
100 µm	5 seconds
1000 µm (1 mm)	8.3 minutes
10 000 µm (1 cm)	4 hours

The diagram on the left shows a cross-section of a red blood cell. The diagram far left shows a section through a spherical cell with the same volume as a red blood cell. In this exercise, we look at the efficiency of diffusion in a biconcave cell compared with a spherical cell.

1. Using the formulae given, estimate the surface area of each of the two cells.

2. Look at the table and calculate the speed of particles diffusing across a distance of:
   a 10 µm;
   b 1 mm.

3. How long would it take for a molecule of oxygen to diffuse from the point marked x to the point marked y on:
   a the red blood cell?
   b the spherical cell?

4. Assume that the concentration difference of oxygen between the inside and outside of the two cells is 1, and both cells have a membrane that is 10 nm thick. Then use Fick's law to calculate the rate of diffusion of oxygen that would occur in:
   a the red blood cell;
   b the spherical cell.

5. Look at the change in volume of the red blood cell in the graph in Fig. 2. Explain how the rate of diffusion of oxygen into the cell would change if the red blood cell swells.

2.4 Osmosis

Not all substances can pass through the cell surface membrane; water molecules do but larger solute molecules do not. This means the membrane is **partially permeable**. The movement of water through a partially permeable membrane from a region of higher concentration of water molecules to a region of lower concentration of water molecules is called **osmosis** (Fig. 5).

Osmosis depends on factors other than the differences in the number of water molecules (see Fig. 6). Solute molecules make weak chemical bonds with water molecules. Solutions with many solute molecules bind most of the water molecules. However, a solution with few solute molecules binds only a few water molecules. Water molecules bound to solute molecules move more slowly than free water molecules. So, as well as there being fewer water molecules, those that are present move more slowly than water molecules in pure water. This enhances osmosis by lowering the concentration of free water in the strong solution.

Fig. 7 explains the change in the volume of the red blood cells shown in Fig. 2 (page 27) in terms of osmosis.

- **Isotonic** solutions have the same concentrations of water molecules, so the rate at which water molecules diffuse into and out of the cell is the same.
- **Hypertonic** solutions have a lower concentration of water molecules compared to the inside of a cell and so there is a net movement of water molecules out of the cell.
- **Hypotonic** solutions have a higher concentration of water molecules compared to the inside of a cell so there is a net movement of water molecules into the cell. This causes the cell to swell; a cell that is full of water but has not burst is said to be **turgid**.

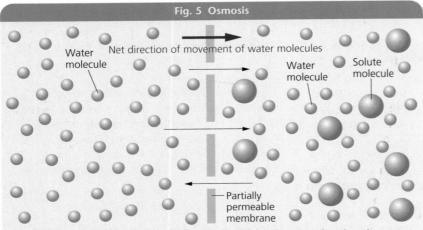

Fig. 5 Osmosis

Water molecules pass through the membrane in both directions, but there is a net movement towards the region of lower concentration. The larger molecules in the cell cannot move outwards through the membrane to balance the inflow of water.

5 What is the main difference between osmosis and diffusion?

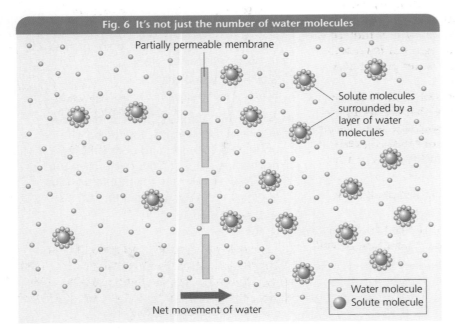

Fig. 6 It's not just the number of water molecules

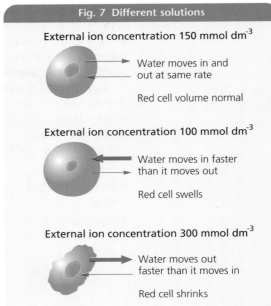

Fig. 7 Different solutions

External ion concentration 150 mmol dm^{-3}
- Water moves in and out at same rate
- Red cell volume normal

External ion concentration 100 mmol dm^{-3}
- Water moves in faster than it moves out
- Red cell swells

External ion concentration 300 mmol dm^{-3}
- Water moves out faster than it moves in
- Red cell shrinks

KEY FACTS

- **Diffusion** is the passive movement of molecules along a concentration gradient from a region of high concentration to a region of low concentration;
- The rate of diffusion can be increased by increasing the concentration gradient, increasing the surface area across which diffusion occurs, increasing the temperature or by decreasing the distance across which the molecules need to travel;
- **Osmosis** is a special type of diffusion;
- In osmosis there is a net diffusion of water through a partially permeable membrane from an area of high concentration of water molecules to an area of lower concentration of water molecules;
- A cell bathed in a solution that is **isotonic** with the cytoplasm can maintain a constant volume since water diffuses into and out of the cell at the same rate;
- A cell bathed in a **hypertonic** solution loses water and shrivels up because water diffuses out of the cell at a faster rate than it diffuses in;
- A cell bathed in a **hypotonic** solution gains water and swells because water diffuses into the cell at a faster rate than it diffuses out.

Water potential

The ability of water molecules to move is known as their **water potential** (Fig. 8). The symbol for water potential is the Greek letter Ψ, pronounced 'psi'. The water potential of pure water at atmospheric pressure is given the value zero. You already know that water molecules that have formed bonds with solute molecules move around more slowly than free water molecules in pure water. Since the water molecules in solutions cannot move as easily as in pure water, solutions always have a water potential value that is *negative* (it is always less than zero). Solutions that have the lowest water potential have the largest negative values. A solution with a water potential of –200 kPa, for example, has a lower water potential than a solution with a water potential of –100 kPa. Water molecules always move towards a region of *lower* water potential – to where the water potential is relatively *more negative*.

The ability of water molecules to move depends on the pressures that are acting on them. The unit of water potential, the kilopascal (kPa), is therefore a unit of physical pressure.

Water potential in a cell depends on two factors:

- the concentration of solutes inside the cell;
- the pressure exerted on the cell contents by the stretched cell surface membrane or cell wall.

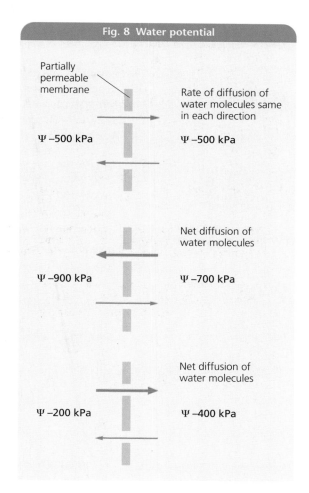

Fig. 8 Water potential

6 Look again at Fig. 6 on page 30. On which side of the membrane do the water molecules have the more negative water potential? Explain your answer in terms of forces between molecules.

2 CELLS AND THEIR ENVIRONMENT

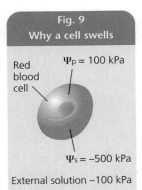

Fig. 9 Why a cell swells

Red blood cell: $\Psi_p = 100$ kPa, $\Psi_s = -500$ kPa

External solution -100 kPa

$\Psi = \Psi_s + \Psi_p$
$= -500$ kPa $+ 100$ kPa
$= -400$ kPa

The water potential of this red blood cell is -400 kPa. It is surrounded by a solution that has a water potential of -100 kPa. In this instance, there is a net movement of water molecules through the cell membrane and into the cell, because water always moves towards a region of lower (more negative) water potential.

Solute and pressure potential

The concentration of solutes inside the cell is called the solute potential and is given the symbol Ψ_s. Solute potential always has a negative value because attractive forces between the solute and water molecules reduce the ability of the water molecules to move. The pressure exerted on cell contents by the cell surface membrane or cell wall is called the pressure potential and is given the symbol Ψ_p. Pressure potential usually has a positive value. This is because the membrane or wall exerts a force on the cell contents to increase the concentration of water molecules in the cell. In effect it squeezes the contents of the cell, trying to push the water molecules out. We can calculate the value of the water potential of a cell by using the water potential equation.

Water potential = Solute + Pressure
of a cell   potential   potential
Ψ Ψ_s Ψ_p

This equation allows us to predict the direction of the net movement of water molecules between systems with different water potentials, for example between a red blood cell and a hypotonic solution (Fig. 9).

7 A cell that has Ψ_s of -500 kPa and a Ψ_p of 100 kPa is placed into a solution that has a Ψ of -600 kPa.
  a In which direction would there be a net movement of water?
  b Explain why red blood cells burst when put into pure water but plant cells do not.

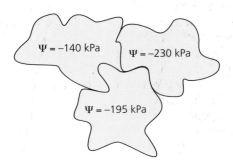

8 The diagram above shows the water potential of three cells that are in contact.
  a Copy the diagram and use arrows to show the net direction of water movement between all three cells.
  b Which cell has the lowest water potential and which cell has the highest water potential?

Osmosis in plant cells

The vacuole of a plant cell usually has a high concentration of ions, so it has very negative solute potential (Ψ_s). This low solute potential causes water to enter the cell by osmosis. The cytoplasm and vacuole both swell and the cell develops a high pressure potential (Ψ_p).

The pressure exerted by the swollen cytoplasm and vacuole on the cell wall of a plant cell that is full of water is known as **turgor pressure**. This pressure keeps the cells **turgid** (rigid) and is the principal method of support in young plants and in plant leaves. If the cells of a young plant lose water, the leaves and stem **wilt**.

9 The diagram below shows the solute potential and the pressure potential of two plant cells. In which direction will water move between the cells? Explain why.

$\Psi_s = -450$ kPa, $\Psi_p = 200$ kPa | $\Psi_s = -400$ kPa, $\Psi_p = 180$ kPa

Water potential in plant cells

If a plant cell is placed in a hypertonic solution (one that has a lower solute potential than the solute potential of its vacuole), three things happen:

- there is a net movement of water out of the cell by osmosis;
- the pressure potential of the cell drops to zero;
- the cytoplasm shrinks and starts to peel away from the cell wall.

The cell is now **plasmolysed** (Fig. 10).

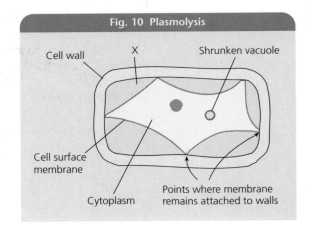

Fig. 10 Plasmolysis

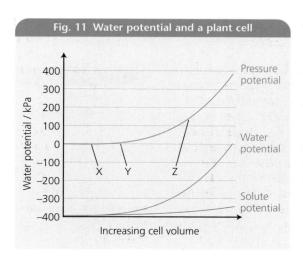

Fig. 11 Water potential and a plant cell

10 Look at Fig. 10. What will occupy the space labelled X? Explain why.

11 The graph in Fig. 11 shows the relationship between water potential, solute potential and pressure potential for a plant cell. The diagram in Fig. 10 shows a plant cell immersed in a solution with a water potential of -300 kPa. Describe the appearance of the cell at the pressure potentials marked X, Y and Z on the graph.

KEY FACTS

- Water molecules that bind to solute molecules in a solution move more slowly than free water molecules in pure water.

- The tendency of water molecules to move from one place to another is called their **water potential**.

- Water molecules always move to a region with a more negative water potential.

- There is no net movement of water molecules between two regions that have an identical water potential. Water molecules move in both directions at the same rate.

- **Osmosis** is the movement of water from a region of higher water potential to a region of lower water potential through a partially permeable membrane.

- You can predict the net movement of water molecules by osmosis if you know the water potentials of the solutions on either side of the partially permeable membrane.

- $\Psi = \Psi_s + \Psi_p$

APPLICATION

Ecstasy

Energetic dancing in a night club or dance club is like any other strenuous exercise – it makes you sweat and it can leave you dehydrated. Ecstasy is an illegal drug that is often used by dancers to 'have a really good time' – but is it worth it? A few people have died after taking Ecstasy in a club after dancing all night. They didn't overdose on the drug; they seem to have overdosed on pure water. Ecstasy seems to induce repetitive behaviour in some people – this has occasionally led to people drinking 20 litres of water or smoking 100 cigarettes in the space of 3 hours.

The urge to drink water constantly may have also been encouraged by the popular belief that water is an antidote to Ecstasy (it isn't) or the drug might affect the way the brain controls the body's water balance. Whatever the reason, people have collapsed because the large amounts of water they have drunk has diluted their blood so much that they have developed a condition called oedema. This happens when cells and tissues in the body absorb too much water and swell; if this happens in the brain this is really bad news. As brain tissue swells it squeezes against the skull and blood vessels and brain cells become squashed. If the centres in the brain that regulate breathing and the beating of the heart are irreversibly damaged, death is usually the result.

1 Use the concept of water potential to explain why the drinking of large amounts of pure water after sweating a lot may lead to oedema in the brain.

2 Would it be a good idea for dance clubs to offer free isotonic drinks instead of water? What would you say to a dance club manager who asked for your advice?

2.5 Facilitated diffusion

In addition to mineral ions, most sports drinks contain small amounts of sugars for 'instant energy'. The market for these drinks is millions of pounds per year so scientists have done a lot of research to find out which sugars are absorbed quickest, and what is the optimum concentration of sugars for the maximum rate of absorption. Some of their findings are surprising. Glucose, for example, is absorbed more quickly:

- than fructose (even though both molecules have the same formula, $C_6H_{12}O_6$)
- from solutions containing sodium chloride than from solutions containing only water and glucose.

In the next section we see how the structure of cell membranes accounts for these observations.

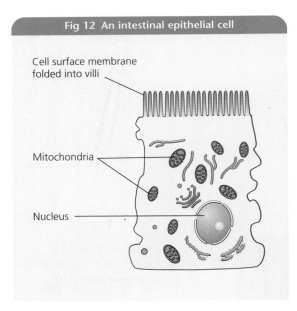

Fig 12 An intestinal epithelial cell

Cell surface membrane folded into villi

Mitochondria

Nucleus

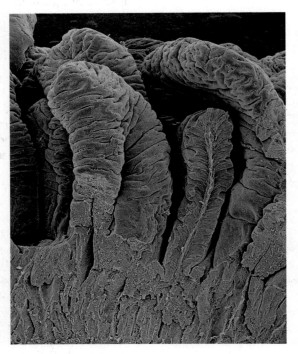

False colour scanning electron micrograph of a section through the wall of the human intestine showing many folded villi. These increase the surface area over which food molecules can be absorbed.

Magnification × 70 000

When we drink an isotonic drink, the water, glucose molecules and mineral ions are absorbed by the cells that line the gut. From here they pass into the blood plasma, and then travel around the body to where they are needed. Most of the initial absorption takes place in the small intestine. The cells that line the small intestine are highly specialised to carry out their function (Fig. 12). The cell surface is highly folded, since this greatly increases the surface area of the cell. A greater surface area means that there is more cell membrane across which diffusion, osmosis, facilitated diffusion and active transport can take place.

12 Explain how the folded cell surface membrane of intestinal epithelial cells helps them to carry out their function.

Each of the three types of particle – water, glucose and mineral ions – is absorbed in a different way through the membrane of the intestine cells. To understand how, you need to revise the structure of the cell surface membrane of these cells.

The bulk of a cell membrane is made up of a double layer of phospholipids (Fig. 8 on page 12). Molecules that do not dissolve in lipids can diffuse across the phospholipid regions of the membrane only if they are small enough to pass between the phospholipid molecules.

13 Water molecules can diffuse between the phospholipid molecules in the cell surface membrane. But they do so at a rate fifteen times slower that the rate that they diffuse in pure water. Suggest why.

Larger lipid-insoluble molecules such as glucose and amino acids need more help. They get it from **intrinsic proteins** within the phospholipid membrane. These protein

2 CELLS AND THEIR ENVIRONMENT

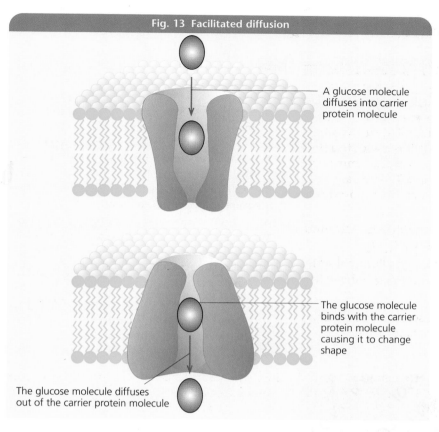

Fig. 13 Facilitated diffusion

- A glucose molecule diffuses into carrier protein molecule
- The glucose molecule binds with the carrier protein molecule causing it to change shape
- The glucose molecule diffuses out of the carrier protein molecule

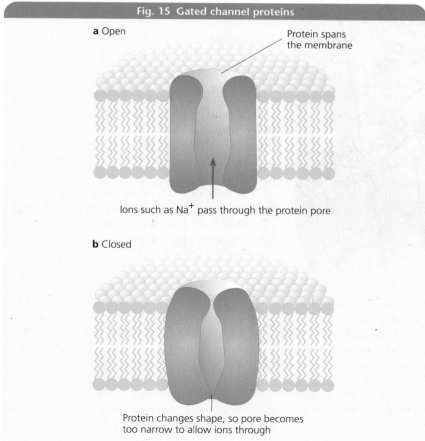

Fig. 15 Gated channel proteins

a Open
- Protein spans the membrane
- Ions such as Na^+ pass through the protein pore

b Closed
- Protein changes shape, so pore becomes too narrow to allow ions through

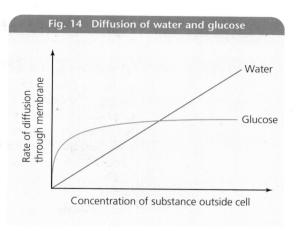

Fig. 14 Diffusion of water and glucose

molecules, called **carrier proteins**, provide transmembrane 'channels' through which small, water-soluble molecules such as glucose can pass (Fig. 13). This process is known as **facilitated diffusion**.

Specific carrier protein molecules transport each substance across the membrane. The proteins recognise the molecules they interact with by their shape, rather like enzymes recognise substrate molecules. Like diffusion, facilitated diffusion only occurs down a concentration gradient and does not need an input of energy from the cell.

14 Fig. 14 shows the rate at which water and glucose move across a phospholipid membrane. Suggest one reason for the difference between the two curves.

Mineral ions also move across membranes by facilitated diffusion but they travel along an **electrochemical gradient**. This is a gradient that occurs between two regions with different charges. So, for example, when there is a high concentration of positively charged sodium ions (Na^+) in one area and a high concentration of negatively charged chloride ions (Cl^-) in another area, the sodium ions will move from positive to negative and the chloride ions will move from negative to positive. Specific proteins called **channel proteins** allow ions to pass through the membrane. Channel proteins are selective – each one has a specific three-dimensional shape with a particular arrangement of electrical charges inside the channel and will only accept one type of ion. Channel proteins can be open or closed and so are called **gated channels** (Fig. 15). The mechanism that enables nerve impulses to be transmitted along nerve cells depends on gated ion channels that open and then close.

2.6 Active transport

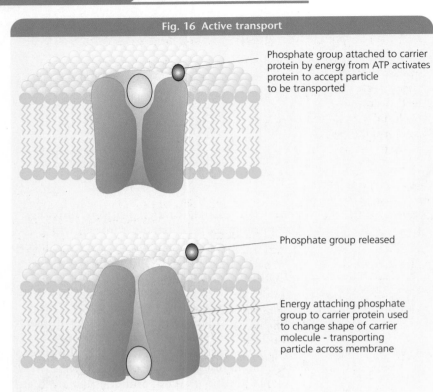

Fig. 16 Active transport

- Phosphate group attached to carrier protein by energy from ATP activates protein to accept particle to be transported
- Phosphate group released
- Energy attaching phosphate group to carrier protein used to change shape of carrier molecule - transporting particle across membrane

Substances can pass in and out of cells along their concentration gradient in several different ways, none of which requires the input of energy. But what happens when the cell needs to move substances against their concentration gradient? In this situation, the cell needs to expend some energy. Some intrinsic protein molecules act as **molecular pumps**. They allow the cell to perform **active transport** to accumulate glucose or ions against their concentration gradient (Fig. 16). Animal and plant cells that specialise in absorption usually have abundant mitochondria to provide the ATP needed to power active transport.

15 Absorbing cells in the intestine have large numbers of mitochondria. Explain how this helps them to function efficiently.

EXTENSION

The power of active transport

Several vital processes depend on active transport including:
- Absorption of amino acids from the gut;
- Absorption of mineral ions by plant roots;
- Excretion of urea and hydrogen ions by the mammalian kidney;
- Exchange of sodium and potassium ions in nerve cells;
- Loading of sugar from the leaf into the phloem in plants.
- Filling of the contractile vacuole in amoeba.

The phloem in the leaves of a plant are rich in sugars because of the high rate of active transport. Lower down the plant, sugars are converted to starch, and the phloem there contains less sugar. This difference in sugar concentration is exploited by people who grow the sugar palm. The sugar is 'tapped' by cutting off the flowers at the top of the plant, just before they are fully formed. When this is done, over 10 litres of rich sugary sap can be collected from the phoem of just one plant. The sap contains 10% sugar and it is boiled to make molasses or refined sugar.

KEY FACTS

- **Facilitated diffusion** is the movement of water-soluble molecules through specific **intrinsic carrier protein molecules** or **channel protein molecules** in the cell surface membrane. It does not require energy.
- **Active transport** moves ions and molecules across a membrane against their concentration gradient. It requires energy.
- Ions are pumped across membranes against the concentration gradient through intrinsic carrier protein molecules called **molecular pumps**.
- Cells that are specialised for absorption have a large surface area and often have many mitochondria.

APPLICATION

Sports drinks and the glucose-sodium symport protein

Cells function normally only when the water potential of the fluids that surround them is kept relatively constant. When we exercise and sweat heavily, we lose water and the blood plasma becomes hypertonic and our performance suffers. The American College of Sports Medicine recommends that runners should take frequent drinks containing water, mineral ions and glucose during long distance runs. Current guidelines for long distance running state:

- It is the responsibility of the race sponsors to provide fluids which contain small amounts of sugar and mineral ions (less than 200 mg of sodium per litre of solution);
- Runners should be encouraged to ingest fluids frequently during competition;
- Race sponsors should provide 'water stations' at 3–4 km intervals for all races of 16 km or more.

1 a Why do they recommend that the maximum concentration of sodium in the drink should be 200 mg per litre?

b Why do the runners need sugar?

As well as replacing the sodium chloride lost in sweat, the sodium ions stimulate the rate of glucose uptake into the blood – so the athlete gets a 'quick glucose fix'. This happens because of **symports** – specialised intrinsic proteins in the cell surface membranes of epithelial cells of the small intestine. The glucose-sodium symport transports glucose and sodium into the cell at the same time but it only works when both substances are present. At the junction between the epithelial cell and a blood capillary, glucose is transported into the blood by facilitated diffusion and sodium ions by active transport.

So, isotonic drinks with a little glucose in them certainly seem to be the best option for someone involved in stamina sports

2 Explain the importance of adding salt to the glucose sports drink.

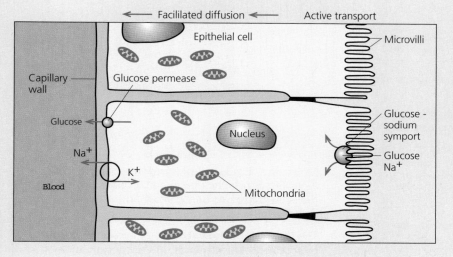

EXAMINATION QUESTIONS

1 The effect of temperature on the permeability of cell membranes can be investigated using fresh carrot. When discs of carrot are placed in water there is a slow release of chloride ions from the vacuoles of the carrot cells. A number of sets of equal-sized discs were cut and placed into water at different temperatures, from 35°C to 70°C. The graph shows the rate of release of chloride ions over this temperature range.

a Suggest why it is necessary to wash the discs before placing them into water at different temperatures. (1)

b Explain the increase in the rate of release of chloride ions between the temperatures of:
   i) 35°C and 45°C;
   ii) 50°C and 60°C. (4)

c What assumption is made about the cell wall of the carrot cells is made in this investigation? (1)

AQA BY01 Feb 97 6

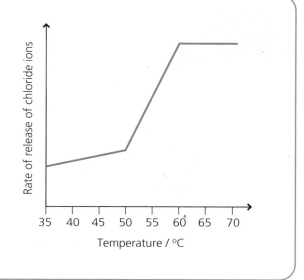

2 CELLS AND THEIR ENVIRONMENT

2 An experiment was carried out to determine the water potential of potato cells. A large number of similar-sized discs was cut from a potato tuber and separated into five sets. Each set was weighed and then placed in a Petri dish containing a glucose solution. The dishes were covered and left for 24 hours. Each set of discs was then removed, reweighed, and the percentage change in mass calculated. The table below shows some of the results.

a Calculate the missing value for the percentage change in mass at a glucose concentration of 0.4 mol dm^{-3}. Show your working. (2)

	Concentration of glucose solution /mol dm^{-3}				
	0.2	0.4	0.6	0.8	1.0
Initial mass /g	8.4	7.9	8.7	8.3	8.4
Mass after soaking /g	8.7	8.0	8.2	7.5	7.3
Change in mass /g	+0.3	+0.1	−0.5	−0.8	−1.1
Percentage change in mass	+3.6		−5.8	−9.6	−13.1

b Explain why it is important to calculate change in mass as a percentage in this experiment. (1)

c Explain in terms of water potential why the mass of the discs increased at lower glucose concentrations. (2)

d At a glucose concentration of 0.43 mol dm^{-3} the percentage change in mass was zero. Explain in terms of water movement why the mass did not change at this concentration. (1)

AQA BY01 June 97 Q5

3 The diagram shows part of a muscle cell membrane.

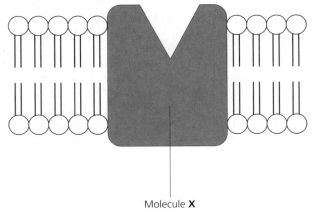

a Name the molecules labelled A and B. (2)

b Explain how glucose enters a muscle cell. (2)

c Suggest the function of the cell labelled B. (1)

d The diagram shows the water potential of three cells in contact with one another.

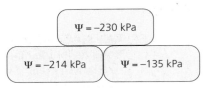

i) Use arrows to show the net direction of water movement between all three cells. (1)

ii) If solute were added to a cell what effect would this have on the water potential within the cell? (1)

AQA BY03 Feb 95 Q5 and AQA BY01 March 99 Q3

4 The diagram shows part of the cell-surface membrane of a red blood cell.

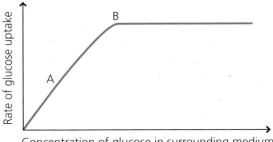

Molecule **X**

a Molecule X acts as a carrier and transports glucose molecules across the membrane. Explain why this molecule transports only glucose and not other carbohydrates. (2)

The graph shows how the rate of uptake of glucose by red blood cells depends on the concentration of glucose in the surrounding medium.

b i) What limits the rate of uptake of glucose from the surrounding medium between points A and B on the graph?. (1)

ii) Explain the evidence from the graph that supports your answer. (1)

c Suggest what is limiting the rate of glucose uptake after point B on the graph. (1)

AQA BY01 June 96 2

Dialysis

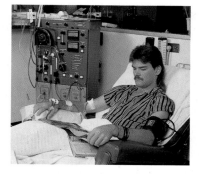

This man is undergoing haemodialysis. He is linked up to a kidney dialysis machine. For several hours, blood flows from his forearm into the machine. The waste from his blood is removed, together with excess salt and water.

The kidneys are the organs in the body that remove waste from the blood. If kidneys fail, these wastes build up in the body. Today, kidney failure can be treated quite successfully using dialysis. Dialysis depends on a membrane that is differentially permeable; it allows small particles to pass pass through but it retains large particles and blood cells. The membrane can be part of an artificial kidney machine (see the photograph above) or the blood can be dialysed using the peritoneal membrane in the abdomen. This method is known as peritoneal dialysis (PD).

Peritoneal dialysis

The peritoneal cavity in the human abdomen is lined by the thin peritoneal membrane, which surrounds the intestines and other internal organs. When someone has PD, their abdomen is filled with dialysis fluid. The fluid is put into the body through a permanent tube opening called a catheter. The body's waste substances leave the blood, pass through the peritoneal membrane and mix with the dialysis fluid. After a time, this fluid is drained from the body and discarded. This 'flushing' process is repeated between three and five times a day. PD enables the person having treatment to live normally whilst it is going on. Dialysis by machine involves many hours sitting in hospital, attached to the large piece of medical apparatus.

There are two types of peritoneal dialysis, continuous and intermittent.

Continuous peritoneal dialysis (CPD)

In CPD about 2 litres of dialysis fluid is in the peritoneal cavity at any one time, so the blood is cleaned constantly. The fluid is changed regularly. There are two ways of doing this. In Continuous Ambulatory Peritoneal Dialysis (CAPD), the dialysis fluid is drained four times a day and the peritoneal cavity is refilled with fresh fluid. This process takes about 45 minutes and most people space the sessions to suit them, perhaps early in the morning, at lunch time, in the late afternoon and just before bed. In Continuous Cycling PD the change of fluid is done by an automatic cycler machine that works overnight. About two litres of dialysis fluid are left in the peritoneal cavity for the day and not changed until the next night.

Intermittent Peritoneal Dialysis (IPD)

IPD usually requires a trip to hospital and most kidney failure patients need between 36 and 44 hours of IPD each week. In IPD, the dialysis fluid is left in the peritoneal cavity for a short time and then drained out. One complete cycle takes about one hour. Because the dialysis fluid in IPD is not changed as frequently as in CPD, the person must take particular care with their diet and fluid intake to avoid a build up of food wastes and water.

1. Dialysis fluid is very similar in concentration to blood plasma in a person with healthy kidneys.

   a. What substances would you expect to find in fresh dialysis fluid?

   b. Would you expect that fresh dialysis fluid would contain urea? Explain the reason for your answer.

   c. Explain why excess water will pass into the dialysis fluid.

2. Describe how dialysis fluid is introduced into the body, then drained from the body.

3. a. Explain the difference between continuous peritoneal dialysis (CPD) and intermittent peritoneal dialysis (IDP).

   b. Summarise the advantages and disadvantages of CPD compared with IPD.

4. Prepare a leaflet for people who have recently been told that they will need dialysis treatment. The leaflet needs to explain why dialysis is important for their health but not frighten them unnecessarily. It should fit onto a single page of A4 paper, printed on both sides.

   Use a DTP package to produce an attractive design if possible. An added benefit of the leaflet is that it will also contain a list of useful websites for patients suffering from kidney disorders. Conduct an internet search to produce this list. You can include up to 10 website addresses but each one must have a brief description saying why it is useful to kidney patients.

3 Exchanges

Why can't an ant grow as big as an elephant or a polar bear? What would life be like for an elephant or polar bear that was as small as an ant?

Insects are an incredibly successful group of organisms that occupy every conceivable habitat. The largest insect that ever lived was probably a prehistoric dragonfly with a wing-span of over 60 cm. Today there are some large insects; the goliath beetle, for example, can grow as large as a man's fist, but it is an exception.

Why has no insect ever grown as big as an elephant? The answer lies in the way organisms exchange materials and heat with their surroundings. Insects 'breathe' through a system of tubes that carry air into the body, allowing oxygen to diffuse the short distance to every individual cell. This system works well for an ant, but imagine using air tubes to get oxygen to every cell of an elephant! To grow this size, mammals like the elephant have evolved a complex blood system and lungs, which provide a large surface area for gas exchange.

Insects do not maintain an internal body temperature; their body temperature changes with the surroundings. So they don't have to worry too much about losing heat. Elephants, like humans, maintain a high, constant body temperature. Their large size means they produce a lot of heat from the food they eat, which they cannot lose very quickly. This can be a problem but if the elephant were the size of an ant, its problems would be far worse. An elephant this small would need to spend 24 hours a day eating to maintain this high constant body temperature – there would be no time for wallowing in mud!

3.1 Heat exchange in mammals

In the first part of this chapter you will find out more about how animals of different sizes deal with the challenges of heat exchange. The rest of the chapter concentrates on the mechanisms that organisms use to exchange gases with the environment. First we will look at the structure and function of lungs in humans and then we take a briefer look at gas exchange in fish and in plants.

The elephant and the polar bear are both large mammals. The larger the mammal, the greater its volume and the more heat it generates by respiration in its tissues. The two animals have opposite problems because of their environment. The polar bear needs to retain heat to keep its body temperature constant despite the cold ice and snow of the Arctic; the elephant needs to keep cool during the hot days on the African savannah.

Mammals transfer heat to the environment by conduction, convection and radiation. The effectiveness of all of these depends on the surface area of the animal. An animal with a very large surface area can lose more heat in a minute than an animal with a smaller surface area. Heat transfer to the environment is also affected by the temperature of the animal's surroundings. On the African savannah, the air is hot during the day and so there is not much of a temperature gradient between the elephant's skin and the air. This makes it nearly impossible for the elephant to lose heat by conduction or convection. But having extremely large ears helps; the diameter of one ear can be as large as 1.5 metres. The earflaps are rich in blood vessels and the elephant flaps them frequently during the hottest part of the day.

In the Arctic winter, the polar bear has the opposite problem. There is a steep temperature gradient between the polar bear's skin and the cold air and the bear loses heat quickly.

1a How does flapping its large ears help an African elephant to reduce its body temperature?

b Explain how the size and shape of the different parts of the polar bear's body help it to conserve heat.

Surface area to volume ratio

An elephant has a much larger surface area than a shrew. But the elephant's volume is also much bigger. If we calculate the ratio of surface area to volume in each animal, we find that the surface area to volume ratio for the elephant is much lower than the value for the shrew. This is a general rule; the larger the animal, the smaller its surface area to volume ratio. The smaller the surface area to volume ratio, the more difficult it is for the animal to gain or lose heat. Large mammals need to compensate for this by having body features that help them to control the way they exchange heat with the environment.

2a A shrew and an elephant in the hot African savannah both need to cool down. Which animal is likely to be more successful and why?

b In Britain, shrews need to eat their own mass of food every day to survive. Suggest why.

Heat exchange in water

Water and air have very different properties with respect to heat transfer. In air, gas molecules are comparatively far apart – so air is a poor conductor of heat. That is why double glazing works. Very little heat is conducted across the air gap between the two glass panes. Molecules in water are very close together so water is a much better heat conductor than air. Animals tend to lose heat to cold water much faster than they do to cold air. A person who falls into the very cold water loses heat so rapidly that their life is put in serious danger. Animals that live in cold water have developed adaptations that help them to survive.

A shrew has a very large surface area to volume ratio. It therefore loses heat very rapidly. It is only able to maintain its body temperature by eating constantly and using large amounts of food in respiration to generate plenty of body heat.

KEY FACTS

- Mammals produce large amounts of heat because of the high rate of respiration in their body cells; much of this heat is lost to the environment.
- Mammals control the rate of heat loss to the environment in order to maintain a constant body temperature.
- As animals increase in size their surface area to volume ratio decreases.
- In cold habitats it is an advantage to a mammal to reduce the surface area of its body to reduce the rate of heat loss.
- In hot habitats it is an advantage to a mammal to increase the surface area of the body to increase the rate of heat loss.

3 EXCHANGES

APPLICATION

The walrus and the manatee

The manatee and the walrus have many features in common but they live in completely different environments. The manatee lives in warm seas near the West Indies, while the walrus lives on the arctic ice shelf off the coast of Alaska. Both must maintain a constant body temperature. Let's look at the way both animals do this. The table shows the relative body sizes of the manatee and the walrus, and a large land mammal, the elephant, for comparison.

1 a Assume the body of each animal has a regular cylindrical shape. Calculate the surface area of each animal.

Surface area of a cylinder = circumference × length

b Animals have approximately the same density as water – 1000 kg m^{-3}. Estimate the volume of each animal.

Density = $\frac{Mass}{Volume}$

c Calculate the surface area to volume ratio of each animal.

Animal	Approximate body length /metres	Approximate body width /metres	Approximate body mass /kilograms
Manatee	4	0.75	1000
Walrus	3.5	1	1700
Elephant	3	2	5000

2 Do mammals lose heat more quickly in air or in water at the same temperature? Explain the reason for your answer.

3 How is the surface area to volume ratio of the two marine mammals related to the temperature of water they can live in?

3.2 Exchanging materials

A coiled beef tapeworm, *Taenia saginata*. This tapeworm is several metres long, as you can see from the ruler. However, its body is thin and flattened.

Living organisms must exchange substances with the environment. The single celled Amoeba exchanges gases through its surface by diffusion. Diffusion is efficient only over small distances, so small, multicellular organisms have become adapted to increase their surface area and to decrease the distance that substances need to diffuse across to maintain their body function. Tapeworms and roundworms have very different habitats. Tapeworms are parasites that live in the guts of animals, roundworms live in the soil. But they both exchange gases with the surroundings through their skin and, as they grow, they both reach their maximum width very quickly. Any further growth is confined to an increase in the length. The flattened shape of the tapeworm ensures that no point in its body is more than 0.1 mm from a supply of oxygen, food and water.

3a What substances does a tapeworm exchange with the gut contents of its host? What mechanism does the tapeworm use to exchange each of these materials?

b How does the body shape of a tapeworm increase the rate at which it can exchange materials? Why is the tapeworm much longer than it is wide?

In larger organisms, diffusion cannot supply the cells at the centre of the body with the oxygen they need and waste such as carbon dioxide cannot escape quickly enough. No adaptations to body shape can make much difference; the problem of a low surface area to volume ratio needs more drastic solutions. Mammals and other animals have overcome this problem because they have evolved special internal exchange surfaces such as the alveoli in lungs and villi in the intestines. Alveoli and villi have a very large surface area that maximises the amount of materials that can be exchanged across them. Animals with internal exchange surfaces usually have a complex blood system to transport substances between the exchange surfaces and the rest of the body tissues.

3.3 Inside human lungs

Larger organisms such as humans have evolved systems that increase the surface area available for gas exchange. This area is called the **respiratory surface**. The respiratory surface in the lungs is large enough to collect enough oxygen to supply all of the body's tissues and to remove carbon dioxide before it builds up to toxic levels.

The lungs have a very large surface area because of the thousands of tiny air sacs they contain. These air sacs are the **alveoli** (Fig. 1). When we breathe in, air enters the alveoli and the oxygen it contains diffuses across the thin walls and into the blood. At the same time, carbon dioxide from the blood diffuses out and into the alveoli to be breathed out.

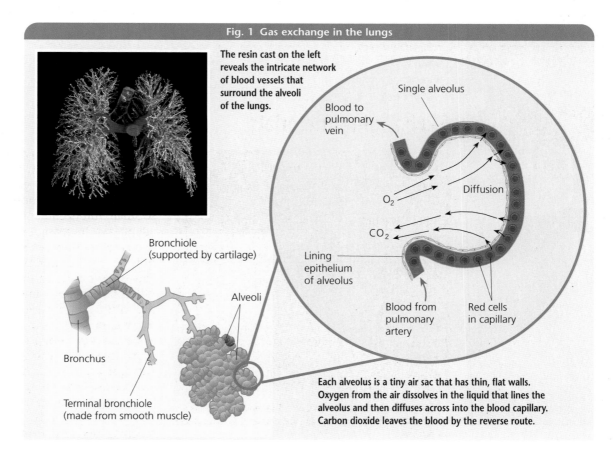

Fig. 1 Gas exchange in the lungs

The resin cast on the left reveals the intricate network of blood vessels that surround the alveoli of the lungs.

Each alveolus is a tiny air sac that has thin, flat walls. Oxygen from the air dissolves in the liquid that lines the alveolus and then diffuses across into the blood capillary. Carbon dioxide leaves the blood by the reverse route.

3 EXCHANGES

APPLICATION

Are women catching up?

Table A shows the winning time for the men's 100 metres sprint in the Olympic games since 1928:

Table A

Year	Winning time / s	Year	Winning time / s
1928	10.80	1968	9.95
1932	10.38	1972	10.14
1936	10.30	1976	10.06
1948	10.30	1980	10.25
1952	10.79	1984	9.99
1956	10.62	1988	9.92
1960	10.32	1992	9.96
1964	10.06	1996	9.84

1. Enter the data into a spreadsheet and use the graph drawing option to draw a graph of the results. You should select the graph type that presents these data in the clearest way.

2. Extend the graph to the year 2012. What would you expect the winning time to be in that year?

Table B at the bottom of this page shows the winning times in men's and women's track events in the Olympic games in 1964 and 1992:

3. **a** Enter the data onto a spreadsheet. Use the spreadsheet to calculate the women's times as a percentage of the men's times for the 100m, 200m, 400m and 800m for both 1964 and 1992.

   b What conclusions can you draw from these percentages about the improvement in women's performances over the 28 years?

4. Explain in terms of body structure why the men's times are faster than the women's times in these events.

5. Suggest why there were no track events longer than 800m for women in the 1964 Olympic games.

Table B

1964 Olympics

Men Event	Time	Women Event	Time*
100 metres	10.0s	100 metres	11.4s
200 metres	20.3s	200 metres	23.0s
400 metres	45.1s	400 metres	52.0s
800 metres	1m 45.1s	800 metres	2m 1.1s
1500 metres	3m 38.1s	No event	–
5000 metres	13m 48.8s	No event	–

1992 Olympics

Men Event	Time	Women Event	Time
100 metres	9.96s	100 metres	10.82s
200 metres	20.01s	200 metres	21.81s
400 metres	43.50s	400 metres	48.83s
800 metres	1m 43.66s	800 metres	1m 55.54s
1500 metres	3m 40.12s	1500 metres	3m 55.30s
5000 metres	13m 12.52s	3000 metres	8m 46.04s
10000 metres	27m 46.70s	10000 metres	31m 06.02s

*Remember that there are 60 seconds in a minute. 2 minutes 6 seconds does not equal 2.6 minutes

Alveoli are well adapted to their function as an exchange surface for gases:

- The epithelial cells that make up the walls of the alveoli support a rich network of blood capillaries. This blood supply is essential to collect the oxygen from the air and to deliver the carbon dioxide to the alveoli.
- The alveolar walls are very thin and the cells of the wall are flattened. This ensures a short diffusion pathway, a large diffusion gradient and so a good rate of diffusion.
- The walls of the alveoli are fully permeable. This allows oxygen to pass into the cells but also lets water out, so the exchange surface is always moist. Oxygen dissolves in the layer of moisture and then diffuses through the wall into the blood.

Having a layer of liquid inside the alveoli presents a problem. If the liquid was just water, surface tension would cause the walls of the alveolus to stick to each other and the air sac would tend to collapse in on itself, reducing the surface area for gas exchange. However, in healthy lungs this does not happen because the liquid in the alveoli contains a **surfactant**, a chemical that reduces the surface tension of the liquid. This surfactant is produced in a baby's lungs from about the seventh month of pregnancy onwards. Very premature babies are born with lungs that do not produce surfactant and many have breathing difficulties. Treatment with artificial surfactant now allows more premature babies to survive and helps to prevent brain damage from lack of oxygen in the first few vital weeks.

KEY FACTS

- The **alveoli** of the lungs form the respiratory surface in humans and other animals.
- The alveoli and the lung capillaries that surround them have a large surface area for gas exchange.
- Alveolar walls are very thin. This makes the diffusion pathway between the air in the lungs and the blood very short.
- The walls of the alveoli that are in contact with air are moist. Oxygen dissolves in this liquid before diffusing across the alveolar wall and into the blood.
- The liquid contains **surfactant** that reduces surface tension and prevents the air sacs collapsing.

Breathing

Breathing draws fresh air into the lungs and forces stale air out again. This process is also called **ventilation**. Ventilation ensures there is always a good supply of 'fresh' air inside the lungs (Fig. 2). This maintains large diffusion gradients for oxygen and carbon dioxide between air and blood. Large diffusion gradients mean efficient gas exchange. Notice the difference between *ventilation* and *respiration*. Respiration is a series of oxidation reactions that occur in cells.

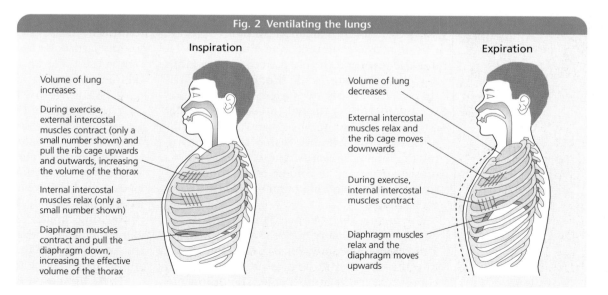

Fig. 2 Ventilating the lungs

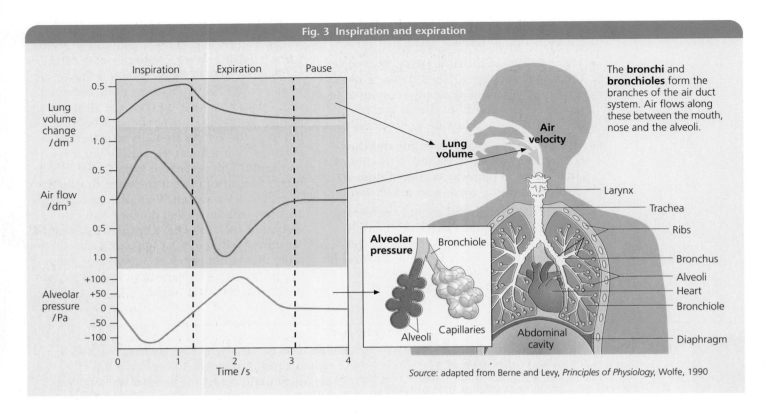

Fig. 3 Inspiration and expiration

Source: adapted from Berne and Levy, *Principles of Physiology*, Wolfe, 1990

Breathing in is known as **inspiration**, breathing out is called **expiration** (Fig. 3). Because the flow of air through the lungs occurs in both directions – in and out – we say that the ventilation is **tidal**. Air passes into and out of the alveoli via the **trachea**, **bronchi** and **bronchioles**. Except for the narrowest bronchioles, the walls of these three air tubes contain rings of **cartilage**. Cartilage is a strong, slightly flexible tissue. It helps to keep the tubes open during the pressure changes that take place as we breathe in and out.

Inspiration is always an active process. When the body is at rest, expiration is a passive process. During gentle expiration, the **diaphragm** muscles relax and the elastic recoil of the lungs and chest wall returns the **thorax** to its original shape. When the body is exercising, expiration is boosted as the **internal intercostal muscles** contract, pulling the ribs downwards, and the muscles in the abdomen wall contract, forcing the diaphragm upwards. These contracting muscles decrease the volume of the thorax and so increase the volume of air exhaled with each breath.

Increasing the volume of the thorax decreases air pressure in the alveoli to below atmospheric pressure (Fig. 3). Air is forced into the lungs until the pressure in the alveoli equals that of the atmosphere. The inflow of air inflates the lungs. Decreasing the volume of the thorax has the reverse effect. As air pressure in the alveoli rises above atmospheric air pressure, air is squeezed out until the alveolar pressure equals atmospheric pressure.

Breathing rate and exercise

It is very important that your breathing rate is matched to your activity. Moving muscles use up energy quickly and the faster you move, the quicker your cells need to burn fuel. This means they need extra oxygen and the body makes the lungs take in more air to supply that need. Air is only 20 per cent oxygen, and so only a fifth of the air that you breathe in can be used. The air in the alveoli contains even less oxygen; approximately 14.5 per cent. This is because breathing replenishes only 30 per cent of the air in the lungs when we are not active. Even when we are exercising and breathing deeply, only 50 per cent of the air in the lungs is changed with each breath. This low concentration of oxygen in the alveoli gives quite a shallow diffusion gradient for oxygen between air in the alveoli and blood in the lung capillaries. Consequently, we only extract about 25 per cent of the oxygen from the air we breathe in. The air we breathe out still contains about 15 per cent oxygen.

These limitations of the breathing system

mean that the body often passes the point at which its need for oxygen is satisfied. At this point, the muscles must respire without oxygen. You get muscle fatigue when a muscle cannot get enough oxygen to produce all the ATP it needs, despite all the puffing and panting. A substance called lactic acid can build up in the muscles. Hydrogen ions from the dissociation of lactic acid may damage the muscle tissue (see Chapter 12).

Training can improve the ability to exercise for long periods without getting muscle fatigue. One of the main aims of training for long distance running is to improve the oxygen supply to the muscles. Training increases muscle mass, it increases the blood supply to the muscles and it improves the rate at which oxygen gets into the blood in the lungs.

The effect of training

The amount of oxygen that a person can take in per minute depends to some extent on their size; a large person has larger lungs and can breathe in more air than a smaller person. A reasonably fit and active adult can take in 12 litres of air in a minute, which delivers three litres of oxygen per minute to the blood. With the right training, long distance runners can deliver over six litres of oxygen to the blood per minute. This increase in rate is due to an increase in the capacity of the lungs and an increased rate of blood flow through the lungs (for more information about blood flow in the body, see Chapter 11).

The increase in capacity of the lungs depends on three factors:

- the **vital capacity** of the lungs. This is the maximum volume of air that can be exhaled from the lungs in each expiration;
- the **breathing rate**. This is the number of inspirations and expirations in a given time;
- the **maximum breathing capacity**. This is the maximum amount of air that can move in and out of the lungs in a given time.

All three are determined by the volume of the lungs and also by the strength of the muscles that move air in and out of the lungs. Three sets of muscles are actively concerned with breathing (Fig. 4):

- the muscles of the diaphragm;
- the external intercostal muscles;
- the internal intercostal muscles.

Training can increase the strength of these muscles, raising vital capacity and improving maximum breathing capacity.

Fig. 4 Muscles involved in breathing

Intercostal muscles are connected to the ribs.

The **diaphragm** is the main muscle for inspiration and is solely responsible for inspiration when the body is at rest.

The lungs and chest wall are elastic. When the muscles relax the lungs and chest fall, or 'recoil', back to shape.

The **thorax** is the chest area.

The flow of air in and out of the lungs is **tidal**.

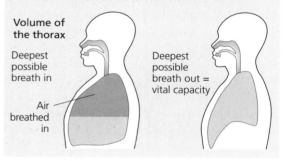

4 The rate of oxygen uptake into an athlete's blood during a training programme is measured in litres of oxygen per minute. Can you think of fairer units in which to measure improvements due to training?

5 In three months of training, an athlete increased her maximum breathing capacity from 130 litres min^{-1} to 180 litres min^{-1}. Calculate the percentage improvement in her maximum breathing capacity.

KEY FACTS

- Breathing maintains large diffusion gradients for oxygen and carbon dioxide between air and blood.
- The **diaphragm muscles**, the **intercostal muscles** and the **abdominal muscles** work together to ventilate the lungs.
- These muscles change the volume of the thorax and hence the pressure of air inside the lungs.
- The flow of air in human lungs is **tidal**.

3.4 Gas exchange in fish

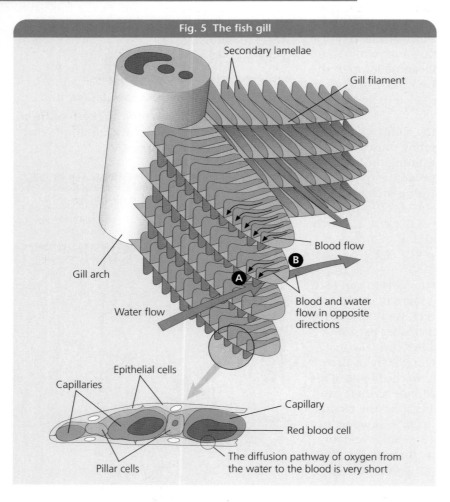

Fig. 5 The fish gill

In fish, water is pumped over the specialised respiratory surface in the gills. Inside the gills, projections called secondary lamellae (Fig. 5), provide a large surface area. The diffusion path for oxygen is short, because the blood that flows within the lamellae is separated from the water outside them by a very thin layer of cells.

However, obtaining oxygen from water rather than air presents particular problems for aquatic animals. Oxygen is not very soluble in water and so water contains only one thirtieth as much oxygen per unit volume as air. Water also has a higher density than air, and is therefore harder to move over a respiratory surface during ventilation. In order to obtain maximum oxygen uptake from the water, fish have evolved the strategy of **countercurrent flow** (Fig. 6). The fish gill acts as a **countercurrent multiplier**.

Countercurrent multiplier

The water that flows over the secondary lamellae and the blood that flows through them travel in opposite directions. This is the countercurrent flow. It increases the efficiency with which oxygen can diffuse into the blood because:

- at point X, the water that has just entered the fish's gills is saturated with oxygen, so there is a large diffusion gradient for oxygen between the water and the blood;

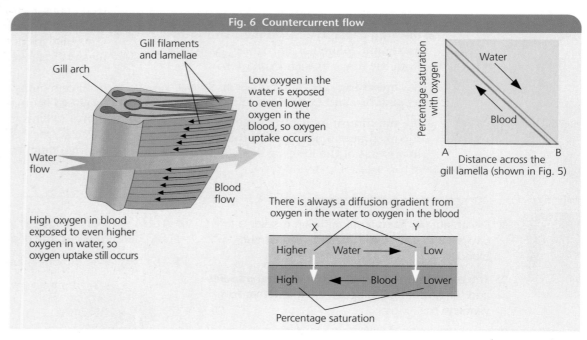

Fig. 6 Countercurrent flow

- at point Y, although much of the oxygen has already diffused from the water into the blood, the blood here has just entered the gills. It therefore has a very low concentration of oxygen and so there is still a large diffusion gradient that favours the movement of oxygen from the water to the blood.

This system is very effective: up to 80% of the oxygen dissolved in water is extracted by the gills. In comparison, our lungs can only extract a maximum of 25% of the oxygen from the air we breathe.

The diffusion of carbon dioxide from the blood into the water is also helped by the countercurrent flow. The countercurrent mechanism works only when the ventilation current moves in one direction only; notice that ventilation in fish is not tidal.

Ventilation in fish

The countercurrent flow system can work well only if there is a regular and unidirectional flow of water over the gills. The fish achieves this by using the muscles in the floor of its mouth to pump water into its mouth (Fig. 7). It then uses its mouth and **opercula** as valves to maintain the unidirectional water flow.

 List 3 differences and 3 similarities in the methods used by humans and fish to obtain oxygen.

Fig. 7 Fish ventilation

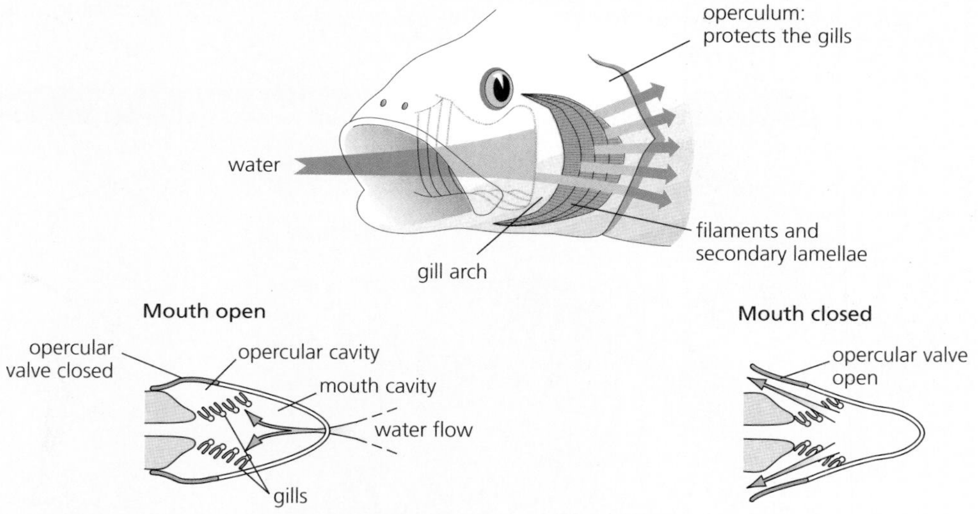

As the fish opens its mouth, it lowers the floor of its mouth cavity. Pressure in the mouth cavity drops and water rushes in.

As the fish closes its mouth, it raises the floor of its mouth cavity. This forces water over the gills.

KEY FACTS

- **Gill lamellae** comprise the respiratory surface in fish. The total surface area of the lamellae is enormous.
- The length of the diffusion path for oxygen is very short, because the layer of cells that separate the blood from the outside water is very thin.
- Water contains only one thirtieth as much oxygen per unit volume as air and water is more difficult to push over the respiratory surface.
- Fish use a **countercurrent system** to maximise the rate of gaseous exchange across the respiratory surface.
- A **countercurrent multiplier** ensures the oxygenating medium flows in the opposite direction to the blood. This means that a diffusion gradient exists across the whole length of the exchange membrane.
- Muscles in the floor of the mouth and in the **operculum** are used to force water over the gill filaments.
- The mouth and operculum act as valves to ensure that water flows only in one direction.

3.5 Gas exchange in plants

Plants do not move from place to place and so do not need nearly so much oxygen as active animals. But they do need surfaces for gas exchange as the processes of cell respiration and photosynthesis both involve an exchange of gases with the atmosphere.

The main gas exchange surface in dicotyledonous plants is inside the leaves (Fig. 8). Root and stem cells obtain most of their oxygen as dissolved oxygen in the water that comes in through the roots.

The leaf has a large internal surface area in relation to its volume. Oxygen and carbon dioxide enter and leave a leaf mainly via microscopic pores called stomata (singular stoma). There are huge numbers of stomata – several thousand per cm^2. Gases that enter the stomata diffuse through the intercellular spaces in the mesophyll. The surface of the mesophyll cells acts as the gas exchange surface. The cell walls are thin and permeable to gases. As in the lungs, water escapes, so the surface is moist. However, excessive loss of water into the atmosphere is avoided because the exchange surface is inside the leaf.

Each stoma is surrounded by two guard cells that can vary the width of the stoma, allowing more or less gas to enter or leave. The action of stomata is covered in more detail in Chapter 13.

7 What are the main gases that the leaf exchanges with the atmosphere:
a at noon on a bright summer day?
b at midnight in summer?

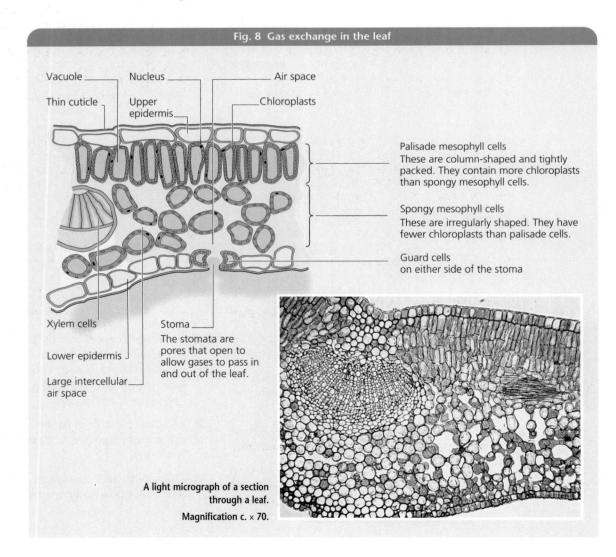

Fig. 8 Gas exchange in the leaf

A light micrograph of a section through a leaf. Magnification c. × 70.

EXAMINATION QUESTIONS

1

Polar bears are found in the extremely cold climate of the Arctic. Rabbits are found in warmer climates. Despite the differences in the temperatures they are exposed to, these animals both have fur with similar insulating properties. Suggest two ways, visible in the drawing, in which the polar bear is better adapted to low temperatures than the rabbit. In each case explain your answer. (4)

BY04 Feb 97 Q6a

2 Very small organisms, such as those consisting of a single cell, have no special tissues, organs, or systems for gaseous exchange. Mammals are large, multicellular organisms and they have a complex system for gaseous exchange. Explain why the animal needs such a system, when the single-celled organism does not. (6)

BY03 Feb 96 Q10 a

3 A thin surface and a diffusion gradient are both features of gas exchange surfaces. Describe how these are achieved at the gas exchange surfaces of:

a a mammal; (3)

b a leaf. (3)

BY03 Mar 99 Q1a

4 Explain how oxygen in the air outside the body reaches the blood in the capillaries surrounding the alveoli of the lungs. (6)

BY03 Jun 96 Q8a

5 The drawing shows a young fish which has just hatched from an egg.

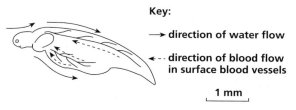

Key:
→ direction of water flow
--→ direction of blood flow in surface blood vessels

1 mm

a Suggest how very small fish like this one can get enough oxygen without gills, although they cannot do so when they grow larger. (2)

b Explain why the direction of blood flow shown in the diagram results in efficient gas exchange. (2)

The table shows some features of three species of fish.

Species	Swimming speed	Water or air breathing	Relative surface area of gills/cm^2 per gram of body mass
A	Fast	Water	13.5
B	Slow	Water	1.9
C	Slow	Air	0.6

c What is the purpose of giving the surface area of the gills in cm^2 per unit area of body mass? (1)

d The area of the gills is related to the way of life of the fish. Suggest an explanation for the fact that the relative surface area of the gills of:
   i) species **A** is larger than the relative surface area of the gills of species **B**;
   ii) species **B** is larger than the relative surface area of the gills of species **C**.

BY03 Feb 97 Q1

6 The diagram shows the way in which water flows over the gills of a fish. The graph shows the changes in pressure in the buccal cavity and in the opercular cavity during a ventilation cycle.

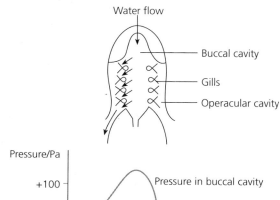

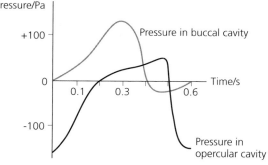

3 EXCHANGES

EXAMINATION QUESTIONS

a Use the graph to calculate the rate of ventilation in cycles per minute. (1)

b For most of this ventilation cycle, water will be flowing in one direction over the gills. Explain the evidence from the graph that supports this. (2)

c Explain how the fish increases pressure in the buccal cavity. (2)

BY03 Jun 97 Q4

7

a Tench are freshwater fish found in water with a low oxygen content. The structure of their gills and the direction of flow over the gills makes them particularly efficient at removing oxygen from the water.

Describe and explain how efficient uptake of oxygen by gills is achieved in a fish such as a tench. (6)

b The diagram shows the arrangement of the respiratory surface in the lungs of birds.

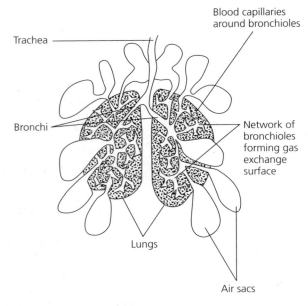

i) Give one similarity and one difference between the gas exchange surface of a bird and that of a mammal. (2)

ii) During inhalation in birds, some air passes through the bronchial system into the air sacs. During exhalation the air may pass out by the same route. Suggest how this increases the efficiency of oxygen uptake. (1)

BY03 Jun 99 Q8

8 The drawing below shows an African elephant.

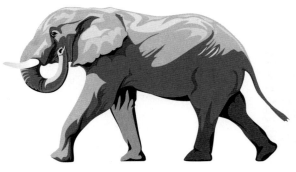

Explain why a large animal such as an elephant might be expected to experience difficulties in regulating its body temperature in a hot climate. (3)

BY05 June 99 7(a)(i)

9

a Very small organisms, such as those consisting of a single cell, have no special tissues, organs, or systems for gaseous exchange. Mammals are large multicellular organisms and they have a complex system for gaseous exchange. Explain why the mammal needs such a system, when the single-celled organism does not. (6)

b How does a molecule of carbon dioxide in the atmosphere reach and then enter the cells inside the leaf where photosynthesis tales place? (6)

BY03 Feb 96 Q10

10 The drawing shows part of a fish's gill.

Use information from the drawing to explain how the direction of flow of blood and water increases the efficiency of the gill filament as an exchange surface. (2)

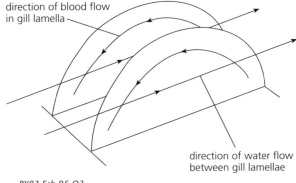

BY03 Feb 95 Q3

KEY SKILLS ASSIGNMENTS

Altitude sickness

Climbing a mountain is hard work. As you climb, the concentration of oxygen in the air remains the same (about 21%) but the number of oxygen molecules that you can take in with each breath decreases as the pressure of air decreases. At sea level air pressure is about 100 kPa, but at 3 500 metres (about 12 000 ft) this has reduced to about 63 kPa. Your body compensates by increasing your breathing rate but if you ascend too quickly, you can start to develop altitude sickness. Climbing above 3 000 metres causes the symptoms of mild altitude sickness in many people. These symptoms include headache, dizziness, shortage of breath, nausea and disturbed sleep. They arise because, when the air pressure falls to a level below the pressure in the lung capillaries, fluid leaks from the capillaries into the alveoli. Within a couple of days the body begins to compensate and the symptoms start to disappear after about three days. This process is known as acclimatisation.

If you ascend very quickly to a very high altitude (4 000 – 5 000 m), the symptoms are much worse – shortness of breath even when resting, mental confusion and the inability to walk. This is acute altitude sickness and is caused by fluid leaking from the capillaries into the cavities in the brain. Someone with these symptoms should be returned to a lower altitude immediately. If this is not possible then they should be placed in a Gamow bag. When the bag is inflated with a foot pump, conditions inside simulate lower altitudes and the patient quickly recovers. Most high altitude climbing expeditions now carry at least one of these bags.

A Gamow bag.

Altitude sickness can be prevented by acclimatising to the change in altitude in stages. This involves spending 2 – 3 days at a particular altitude before trying anything strenuous like rock climbing or skiing and before moving to higher levels. Acclimatisation occurs because, when the body is at high altitudes, it produces a hormone called erythropoietin. This hormone stimulates the body to produce more red blood cells to compensate for the lowered capacity to take in oxygen from the air. Many athletes train at high altitude. The increased number of red blood cells lasts about two weeks and enhances performance, particularly in middle and long-distance running events.

It is possible to inject erythropoietin to obtain the same effect, but this is forbidden by athletics authorities. Athletes who want to simulate the effects of going to high altitudes without having to climb a mountain can however use a high altitude sleeping chamber quite legitimately. The Gamow bag also has a chamber that enables an athlete training at sea level to gain the advantage of training at high altitude while they sleep.

1 Explain why the body obtains less oxygen per breath at high altitude than at sea level.

2 Explain how fluid leaking from capillaries into the alveoli affects gas exchange.

3 Explain what is meant by acclimatisation.

4 Explain how you would recognise if a companion was suffering from acute altitude sickness.

5 Explain how the Gamow bag helps a person suffering from acute altitude sickness.

Discuss the legality of training at altitude, injecting erythropoietin and sleeping in an altitude chamber as methods of training for an athletics event. Work with a group or a partner so that you can work with at least one argument for and one argument against.

a Prepare a list of points in favour of the idea that training at altitude is fair.

b Now do the same for the practice of injecting erythropoietin and sleeping in a Gamow bag.

c Carry out a discussion. Note down the points made by your opponents.

d Modify your ideas, taking into account the views of your opponents.

e Decide on a statement that the whole group can support.

4 Proteins and enzymes

A hydrogen-powered bus built by the Southeastern Technology Center in Augusta, Georgia, USA; cleaner and greener than its diesel-powered brothers.

The dream of using hydrogen as an economical pollution-free fuel moved closer to reality with the discovery of two bacterial enzymes. Working together, the two enzymes use glucose to produce hydrogen gas. During the process, which produces no potentially harmful by-products, they convert glucose to gluconic acid, a substance used in detergents and some pharmaceuticals.

Thermoplasta acidophilum, a type of bacterium first found in a coal tip, produces the enzyme glucose dehydrogenase. This enzyme can pluck two hydrogen atoms from each glucose molecule it interacts with. One of the hydrogen atoms combines with a carrier molecule while the other goes into solution as a hydrogen ion. The reaction is completed when the second enzyme, hydrogenase, catalyses the reaction between two hydrogen ions to form a molecule of hydrogen gas. Hydrogenase is produced by *Pyrococcus furiosus*, a bacterium that flourishes around volcanic vents deep in the Pacific Ocean.

Both the enzymes are unusual in that they can resist quite high temperatures. Glucose dehydrogenase works at up to 60°C, while hydrogenase can cope with temperatures up to 100°C. Being able to operate at these temperatures makes the conversion of glucose very rapid. Contamination by other bacteria is also unlikely because hardly any other bacteria can survive at these temperatures.

4.1 Why cells need enzymes

A huge number of chemical reactions take place inside every living cell. Some release energy, some synthesise new substances and others break down waste products. **Enzymes** enable these reactions to happen and, most importantly, enzymes make it possible for the reactions to happen quickly enough at body temperature. Most chemical reactions need an input of energy to set them going and normally this is supplied as heat. Enzymes catalyse reactions; they reduce the amount of energy that is needed to get a reaction going but they are not themselves changed at the end of the reaction. This makes them very valuable substances, not only in living organisms, but also for human use. For centuries, enzymes from yeast have been used to brew alcoholic drinks. Today, adding enzymes to detergents to digest food stains is commonplace and in the future, many more industrial processes are likely to make use of enzymes.

Enzymes are proteins. The properties of proteins help to explain how enzymes work and how they supply the energy needed to get a reaction going. The structure of enzyme molecules does, however, make them sensitive to environmental conditions. Changes in temperature or pH can disrupt enzyme action. In humans, severe over-heating or getting so cold that core body temperature falls can upset the delicate balance of enzyme-controlled reactions in the body's cells and can be fatal.

4.2 Enzymes and chemical reactions

Sugar left in a bowl does not react with oxygen in the air. Sugar only reacts if heated strongly, and even then it is difficult to start it burning. In living cells, however, sugar is continuously broken down by respiration to release energy. This is just one of many chemical processes going on in a cell and this constant chemical activity is known as the cell's **metabolism**.

In respiration, glucose is broken down in a series of reactions called a **metabolic pathway**. This is quite different from burning, where the energy release is rapid and uncontrolled. A metabolic pathway releases energy in small steps. Enzymes enable metabolic reactions to occur in cells at normal environmental temperatures. They also determine when, where and how fast those reactions take place.

The substances that react together in any reaction catalysed by an enzyme are said to be the **substrates** of that reaction.

Activation energy

Why does sugar react with oxygen only when heated? Before it can react, the bonds linking the atoms in the molecules must be broken, and this requires energy. The heat from a match or a bunsen burner can provide this. The energy gives the molecules of the substrates more kinetic energy. As a result, the molecules of glucose vibrate more vigorously in the solid sugar and the molecules of oxygen move faster in the air nearby. Some of these molecules collide, but does that always lead to a reaction? Normally the electrons that orbit an atom or molecule repel other electrons because both are negatively charged, but by making the molecules move faster, this repulsion can be overcome and the orbits can overlap. The bonds that hold together the molecules of the substrates break and new bonds form. The minimum amount of energy needed to do this and so set a reaction going is called the **activation energy** (Fig. 1).

Once the bonds have been broken, new bonds can be formed. In the case of burning sugar, molecules of carbon dioxide and water are made. These have much less energy stored in their molecules, so the excess energy is released as heat. This keeps the reaction going since it provides energy to activate other substrate molecules. However, this sudden release of heat energy would be bad news inside a living organism; reactions here need to be controllable.

The energy in food can be released by combustion but a large input of energy is needed to get the reaction going. Once the sugar starts burning, the combustion reaction occurs very rapidly.

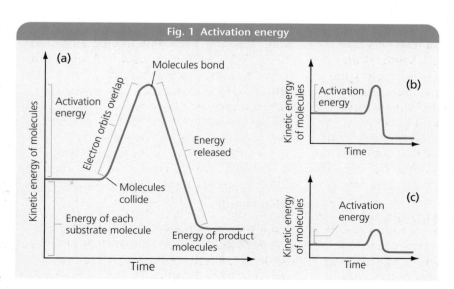

Fig. 1 Activation energy

1a Would the substrate molecules in Fig. 1b react? Explain your answer.

b The substrate molecules in Fig. 1c have less energy than the substrate molecules in Fig. 1b, but they still react. Explain why.

Lowering activation energy

The activation energy of a reaction cannot be changed. However, enzymes *seem* to lower the activation energy of a reaction because they split the reaction into stages, each with a lower activation energy (Fig 2). The cell can

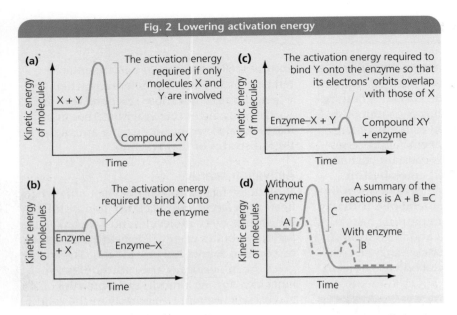

Fig. 2 Lowering activation energy

supply this much smaller amount of energy from ATP produced in respiration. This allows many reactions to take place easily at the temperatures that normally occur in the body of the organism.

2. The graph in Fig. 3 shows the activation energies for the same reaction with and without an enzyme. Using the letters on the graph, indicate:

   a the curve that shows the enzyme-controlled reaction;

   b the activation energy required when no enzyme is present;

   c the difference in activation energy with and without the enzyme;

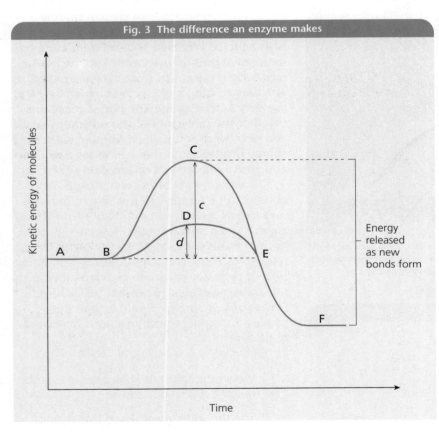

Fig. 3 The difference an enzyme makes

Speedy enzymes

It is hard to appreciate how fast enzymes can work. Hydrogen peroxide is produced as a waste product in some cells but is highly toxic. Rapid removal is essential. The enzyme catalase breaks down hydrogen peroxide into water and oxygen. A single molecule of catalase can break down about 100 000 molecules of hydrogen peroxide per second, that's in less time than it takes to say "one hundred thousand".

Table 1 shows the **turnover number** of catalase and some other enzymes. This is the number of substrate molecules that can be acted upon by a single molecule of the enzyme in one minute.

Table 1 Enzymes in action	
Enzyme	**Turnover number**
carbonic anhydrase	36 000 000
catalase	5 600 000
β-galactosidase	12 000
chymotrypsin	6 000
lysosyme	60

KEY FACTS

- Enzymes are **catalysts**. They speed up the rate of chemical reactions.
- Enzymes regulate the metabolic processes that occur in living cells.
- The energy needed to start a reaction is the **activation energy**.
- Enzymes lower the activation energy, so reactions can take place at acceptable temperatures inside living organisms.

4.3 The nature of enzymes

Enzymes are proteins. Proteins are important compounds in living organisms – not just as enzymes but in many other ways. You saw in Chapter 1 the role of proteins in cell membranes. Proteins are also major structural components in tissues such as muscle, cartilage and bone, making them vital for growth. Haemoglobin, insulin and antibodies are all proteins and, since there are over 10 000 different proteins in the human body, you will come across many other examples.

The primary structure of proteins

How can proteins have such a range of different functions? Like polysaccharides, they are polymers. Protein monomers are amino acids and a protein molecule can contain hundreds or thousands of amino acid units. Unlike starch, which is made from only one type of monomer (glucose), proteins are made from 20 different naturally occurring amino acid monomer units. These can be assembled in any order, making an endless variety of protein structures possible. It is like having an alphabet of twenty letters to make up words hundreds of letters long.

All amino acids have an **amino group**, which is chemically **basic**. This means it reacts with and neutralises acids. They also have an acid group called the **carboxyl group**. Different amino acids have a different group attached to the central carbon atom. This group is referred to as 'R' (Table 2).

Fig. 4 shows how two amino acids molecules join by a condensation reaction. This is the same type of reaction that joins the sugar monomers in polysaccharides. Water is removed and the two amino acids link together by a **peptide bond**. The substance produced is a **dipeptide**. Notice how the acid group links up with the amino group. The result is that the dipeptide molecule has an acid group at one end and a basic amino group at the other. The long chains that eventually form the protein are made by adding more amino acids. Each protein has its own unique sequence of amino acids; this is known as the **primary structure** of the protein.

Table 2 Some amino acids	
Amino acid	**R group**
Glycine	H
Alanine	CH_3
Cysteine	CH_2SH

Fig. 4 Amino acids and dipeptides

A molecule of amino acid can be represented like this.

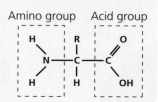

Two amino acid molecules join by a condensation reaction to form a dipeptide.

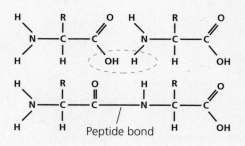

3 Look at Fig. 4 and answer the following questions:

a Which four elements are present in every amino acid?

b Which of these elements is not present in a carbohydrate or lipid?

c What is the formula of the amino group?

d What is the formula of the acid group?

e When the carboxyl group ionises in solution which two ions are formed?

4a Draw the structural formula of a molecule of each of the three amino acids in Table 2.

b Write an equation to show the formation of a dipeptide by condensation of alanine and glycine.

c Draw the backbone chain of carbon and nitrogen atoms formed when three amino acid molecules condense. Label the peptide bonds.

4 PROTEINS AND ENZYMES

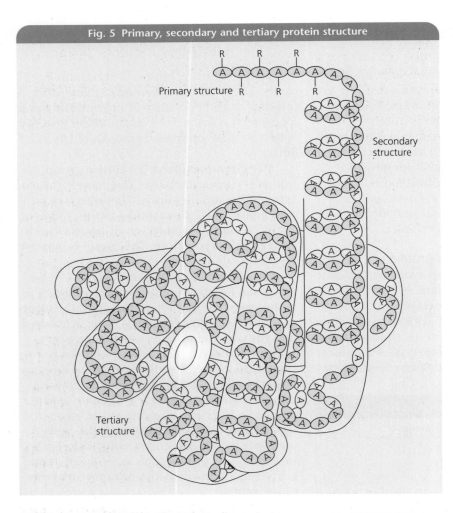

Fig. 5 Primary, secondary and tertiary protein structure

Secondary, tertiary and quaternary structure

Many proteins consist of several short chains of amino acids. These chains are **polypeptides** and they can form coils or pleats. The coiling or pleating of a polypeptide chain is known as the **secondary structure** of a protein. The polypeptide chains often fold and are joined by weak chemical bonds that give a complex three-dimensional shape. This is the **tertiary structure** of a protein (Fig. 5). Some proteins, such as haemoglobin, have a fourth level of organisation, in which polypeptide chains and non-protein groups combine to give a **quaternary structure** (Fig. 6).

The final shape of a protein molecule depends ultimately on its primary structure. The order of amino acids in a polypeptide chain determines how the molecule folds into its secondary and tertiary structure. Many of the connections that maintain the folds are hydrogen bonds that occur where a hydrogen atom on one part of the polypeptide chain is attracted to an oxygen atom at another position on the chain (Fig. 7). Other connections depend on the R groups. Two amino acids contain sulphur atoms, and these link together to form bonds that also determine the three-dimensional shape of a protein molecule.

There are two basic types of protein molecule; fibrous and globular.

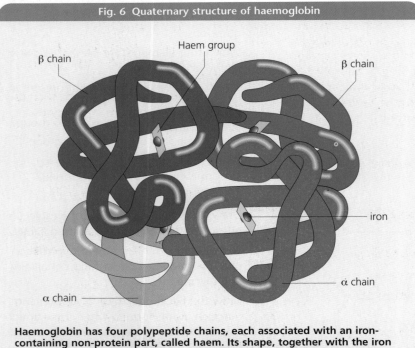

Fig. 6 Quaternary structure of haemoglobin

Haemoglobin has four polypeptide chains, each associated with an iron-containing non-protein part, called haem. Its shape, together with the iron in the haem, gives haemoglobin its special oxygen-absorbing properties.

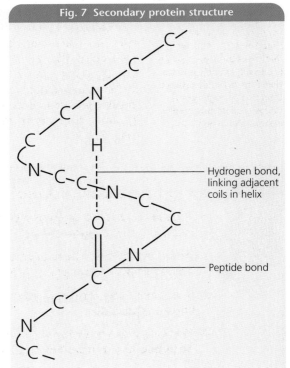

Fig. 7 Secondary protein structure

4 PROTEINS AND ENZYMES

Fibrous proteins

The structure of proteins relates to their function. In **fibrous proteins** such as keratin, which occurs in hair and nails, the chains of amino acids have a regular pattern of hydrogen bonds that cause them to coil into long helices that resemble thin springs. Three of these coils twist round each other, as in a rope (Fig. 8a). This secondary structure makes the keratin strong and flexible and so well suited to its role.

Collagen, another fibrous structural protein, gives bone and cartilage their strength. Collagen molecules also have three twisted polypeptide chains, but these are more tightly bound than in keratin. This means that collagen can form more rigid structures. Often the molecules are grouped together to make almost rigid rods (Fig. 8b).

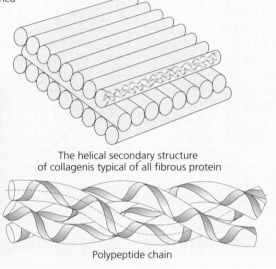

Fig. 8 Three-dimensional structure of fibrous proteins

Collagen fibres in the cornea of the eye are stacked in neat crosswise piles.

The helical secondary structure of collagen is typical of all fibrous protein

Polypeptide chain

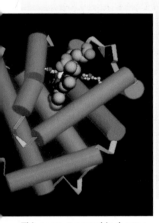

This computer graphic shows a myoglobin molecule. Myoglobin is a globular protein that stores oxygen inside muscles. The shape is important because the haem that carries the oxygen must fit into exactly the right position. Notice the difference between the regular arrangement of collagen and the myoglobulin molecule shown here.

5a In cartilage the collagen rods are arranged in many different directions. Suggest the advantage of this.

b In the cornea collagen rods are stacked in very neat piles. Suggest the advantage of this, bearing in mind that light enters the eye through the cornea.

Globular proteins

Enzymes are **globular proteins**, and their ability to function depends largely on their tertiary structure. Bonds between R groups at different points on the polypeptide chain make the molecule fold up into an almost spherical shape. This makes it easy for the molecules to move around inside cells or to be secreted through membranes. The exact three-dimensional shape of an enzyme molecule is very important. This is determined by the sequence of amino acids in the polypeptide chains that make up the enzyme (see Fig. 5 in this chapter) and so any change in the sequence can affect its function. If even one amino acid in the chain is changed, the 'correct' pattern of hydrogen bonds might not form to give the protein its proper shape and the enzyme may then not interact with its substrates. The importance of the shape of the active site in an enzyme molecule is explained on page 60.

6 (To answer this you may need to look back at the sections on cell membranes in Chapters 1 and 2)

a Suggest why the proteins in cell membranes are globular rather than fibrous.

b Explain why the carrier proteins used for facilitated diffusion have a variety of different shapes.

KEY FACTS

- Proteins are polymers of amino acids.
- Proteins contain nitrogen as well as carbon, hydrogen and oxygen.
- Amino acids have a basic **amino group** ($-NH_2$) and an acid **carboxyl** group (-COOH).
- Two amino acids combine by condensation to form a **dipeptide**.
- The long chain of amino acids forms the **primary structure** of a polypeptide or protein.
- Coiling or pleating of this chain produces the **secondary structure**.
- Folding produces the **tertiary structure**, and addition of other groups makes the **quaternary structure**.
- The precise shape of the protein molecules determines their function. Structural **fibrous proteins** have long twisted molecules; enzymes and carrier molecules are **globular proteins** with roughly spherical molecules.

4.4 How do enzymes work?

Enzyme molecules have a complex tertiary structure. The substrate molecules of the enzyme must be precisely the right shape to fit into part of the molecule called the **active site**. The substrate molecules are attracted to the active site and form an **enzyme-substrate complex**. This complex exists for only a fraction of a second, during which time the products of the reaction form.

The traditional way of explaining how an enzyme and its substrates fit together was to use the analogy of a lock and key. The enzyme can be thought of as a lock into which the substrates, but no other molecules, fit. However biochemists now know that the active site changes shape as the substrate fits into it (Fig. 9). They have modified the **lock and key hypothesis** into the **induced fit hypothesis**. Only the substrates for a reaction catalysed by a particular enzyme cause the changes in shape necessary for the enzyme to function.

When two substrate molecules are attracted to adjacent positions in the active site, the forces of attraction between them causes new bonds to form. Because the enzyme brings the two substrate molecules very close together, only small amounts of energy are needed for the two molecules to react, and the reaction needs much less energy than when uncatalysed. Once the product is formed it is immediately released, so the enzyme molecule is available for reuse.

7a Enzymes are specific. A particular enzyme can catalyse only one chemical reaction. For example, catalase will only break down hydrogen peroxide. Some enzymes break only one type of bond – peptidase, for example, breaks a peptide bond. Use your knowledge of how enzymes work to explain why enzymes are specific.

b Explain why the 'induced fit hypothesis' of enzyme action is better than the 'lock and key' hypothesis.

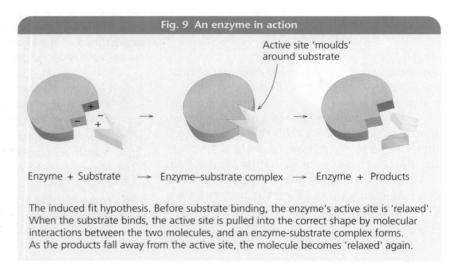

Fig. 9 An enzyme in action

Enzyme + Substrate → Enzyme–substrate complex → Enzyme + Products

The induced fit hypothesis. Before substrate binding, the enzyme's active site is 'relaxed'. When the substrate binds, the active site is pulled into the correct shape by molecular interactions between the two molecules, and an enzyme-substrate complex forms. As the products fall away from the active site, the molecule becomes 'relaxed' again.

KEY FACTS

- Enzyme molecules have a very specific shape, due to the tertiary structure of the protein.
- The **substrates**, the molecules of the reacting substances, are attracted to a particular part of the enzyme molecule, called the **active site**. An **enzyme-substrate complex** is formed, the enzyme molecule is distorted and the substrates react.
- Activation energy is low in a reaction controlled by an enzyme because little energy is needed to bring the two substrate molecules together.
- The products of the reaction are released rapidly, so the enzyme can be reused immediately.

Temperature and enzyme activity

Enzyme activity is affected by temperature (Fig. 10). As the temperature rises, the rate of a chemical reaction normally increases. This is because the molecules gain kinetic energy, so they move faster, collide more often and the collisions are more likely to lead to a reaction. This also holds true for reactions controlled by enzymes because there is an increased chance of substrate molecules colliding with the active site. However, as temperatures rise the atoms within the enzyme molecules also gain energy and vibrate so rapidly that the weak bonds that maintain the tertiary structure of the protein molecule break and the molecule can unravel. Since the shape of the active site in an enzyme molecule is crucial for it to work, any change in shape will inactivate the enzyme. Once broken, the hydrogen bonds do not re-form in their original positions. So, even when the temperature falls, the enzyme cannot regain its functional shape. We say the enzyme is **denatured**. Notice that you should never say that enzymes are 'killed', since they were never 'alive' in the first place.

Most enzymes in the human body are denatured by temperatures above about 45°C. They work fastest just below this, at about 40°C; this is therefore their **optimum temperature**.

8a Explain why the rate of reaction is so low below 10°C.

b From Fig. 10, what are the rates of reaction at 10°C, 20°C, and 30°C?

c What do you notice about the effect of increasing the temperature by 10°C?

d What would be the rate of reaction if some of the enzyme and its substrates - were frozen and then warmed to 40°C?

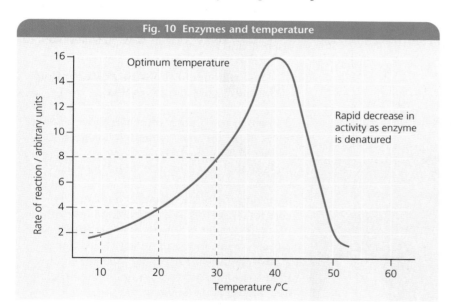

Fig. 10 Enzymes and temperature

Human body temperature is about 37°C, but do not think that all the body's enzymes have an optimum at that temperature. Many work fastest at about 40°C. However, if our temperature does rise to this level, we feel ill because our metabolic pathways are no longer co-ordinated properly.

Many organisms have enzymes that work perfectly at much higher or lower temperatures. Some fish live in the Antarctic sea, where the water temperature never rises above 2°C. They are active at this temperature, but die if the water warms by more than a few degrees. Worms and bacteria can live near volcanic vents in the ocean and close to hot springs in Iceland, where the temperatures are close to boiling point. Many of the enzymes from these specialised organisms are proving useful in manufacturing industry (see page 54).

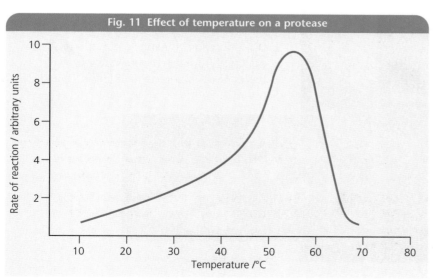

Fig. 11 Effect of temperature on a protease

9 The graph in Fig. 11 shows the effect of temperature on a protease used in washing powders.

a What is the optimum temperature of this protease?

b Suggest the advantage of using a protease with this optimum temperature in a washing powder.

c Proteases break down protein molecules into amino acids. Explain how proteases in washing powders help to remove protein stains, such as blood or egg, from clothes.

pH and enzyme activity

As well as having an optimum temperature, enzymes also have an optimum pH. Most enzymes are denatured in solutions that are strongly acidic or strongly alkaline. This is because the hydrogen ions (H^+) in an acid or the hydroxyl ions (OH^-) in an alkali are attracted to the charges on the amino acids in the polypeptide chains that make up the enzyme. They interact with the amino acids and disrupt the bonds that are maintaining the molecule's three-dimensional shape, destroying the active site of the enzyme. Once the original structure is lost, it cannot reform and the enzyme no longer binds to its substrates.

APPLICATION — Enzymes and extreme pH

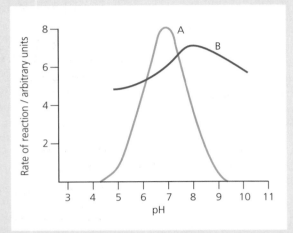

Extremes of pH are rare inside the cells of living organisms. Most enzymes have an activity curve similar to curve A in the graph shown on the left, with an optimum somewhere near neutral pH. However, some enzymes work best at extreme pH values. For example, proteases in the acid environment of the stomach have an optimum pH of about 2 and are denatured above pH 6.

1 Sketch a graph to show the effect of pH on the activity of stomach protease.

Curve B in the graph on the left shows the effect of pH on an enzyme that has been extracted from the marine shipworm, a mollusc that can bore into woodwork. This enzyme is a protease and it is very good at removing the protein that accumulates on contact lenses. It has an advantage over other proteases used for this purpose because it remains active in the presence of the slightly alkaline solution of hydrogen peroxide that is used to disinfect lenses. A single solution that can cleanse and sterilise at the same time can thus be used in place of the conventional two solutions.

2 Use the graph to explain why the protease obtained from the marine shipworm is better for cleaning contact lenses than a protease with a curve like curve A.

The marine shipworm *Spirorbis borgalis*

Concentration and enzyme activity

Enzymes work by forming enzyme-substrate complexes. The rate of reaction depends on the number of substrate molecules that bind with enzyme molecules in any given time. Clearly, the more enzyme molecules in the solution, the greater the chance of a substrate molecule finding an active site, and the faster the rate of reaction. Similarly, the more substrate molecules in a solution, the greater the chance that an enzyme-substrate complex will form, and the greater the rate of reaction.

Fig. 12 Enzyme and substrate concentration

a Enzyme concentration constant

b Substrate concentration constant

10 These questions refer to Fig. 12.
- **a** Explain in terms of active sites why the curve in Fig. 12a flattens out.
- **b** What do the letters X and Y in Fig. 12b show?
- **c** What might increase the rate of the reaction at position Z in graph b?
- **d** Explain one set of conditions in which no increase would occur.

KEY FACTS

- Temperature, pH, enzyme concentration and substrate concentration are all factors that affect the rate of enzyme-controlled reactions.
- Rising temperature increases the rate of reaction up to an **optimum**.
- High temperatures **denature** enzymes, thus destroying the active site.
- Most enzymes are denatured by strongly acid or alkaline conditions.
- Increasing the concentration of enzyme or substrate increases rate of reaction up to a maximum at which all active sites are fully used.

Fig. 13 Competitive and non-competitive inhibition

a Competitive inhibition

The usual substrate molecule can form a complex with the enzyme

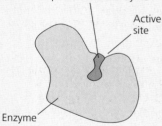

A competitive inhibitor molecule can also form a complex with the enzyme, preventing the usual reaction from taking place. The substrate molecule and the inhibitor molecule compete for the active site.

b Non-competitive inhibition

The substrate can form a complex with the enzyme

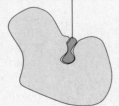

The substrate molecule is unable to form a complex with the enzyme

The inhibitor site is usually unoccupied

An inhibitor molecule attaches to the enzyme, changing the enzyme's shape

Competitive inhibition

Many enzymes have a high **substrate specificity**. This means that only molecules of one particular substrate can attach to the active site. However, sometimes molecules with a very similar shape to the usual substrate attach to the active site. If such molecules are present in the same solution as the normal substrate they *compete* for the active site. They are called **competitive inhibitors** (Fig. 13 a).

The molecules of the competitive inhibitor block the active site, so fewer molecules of the enzyme are available for the normal reaction and the reaction rate is reduced. The degree of inhibition depends on the relative concentrations of the substrate and the competitive inhibitor. The inhibitor does not attach permanently to the active site, nor does it damage the site. Therefore, if the concentration of the normal substrate is increased, its molecules compete more successfully for a place in the active site of the enzyme and the rate of the reaction increases.

APPLICATION: Using enzymes

Enzymes are widely used in food manufacture, for example in cheese-making and in extracting juice from fruit. One use is to produce fructose syrup from starch. Fructose is a monosaccharide sugar that is twice as sweet as glucose and so is useful in slimming foods as a sweetener. The flow chart shows how corn starch is converted to fructose.

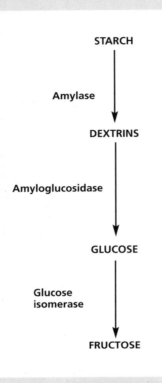

Starch is obtained from corn (maize), a very cheap source. It is boiled briefly in water to make a paste.

Amylase is extracted from bacteria, and is active above 90 °C. It breaks down starch into short glucose chains called dextrins.

Amyloglucosidase is extracted from a fungus. This enzyme breaks down dextrins to form glucose. It works best at about 55 °C.

Glucose syrup is refined to remove any residual enzymes.

Extracted from bacteria, this enzyme converts glucose into its isomer, fructose. This reaction occurs at 60 °C.

Syrup containing about 42% fructose and 55% glucose is formed at the end of the process.

1. What is the advantage of using fructose syrup in slimming foods instead of glucose?
2. What is the economic advantage of using corn starch instead of cane sugar?
3. What is the economic advantage of using enzymes that work at high temperatures?
4. Look back to Fig. 13 in Chapter 1 that shows glucose and fructose. Explain what glucose isomerase does.
5. Suggest why the glucose syrup is refined to remove the enzyme.
6. Glucose isomerase is expensive to extract. For use it is immobilised; it is fixed to inert granules in a column through which the glucose syrup slowly drains. Suggest the advantage of doing this.
7. Suggest why manufacturers use fructose syrup that contains a high percentage of glucose, instead of obtaining pure fructose.

Non-competitive inhibition

Some substances inhibit enzyme reactions in a different way. Their molecules do not attach to the active site, but to a different part of the enzyme. This alters the shape of the enzyme molecule so that the active site can no longer bind with the substrate. These substances are called **non-competitive inhibitors** (Fig 13b). The shape of a non-competitive inhibitor molecule may be completely different from that of the usual substrate. The inhibitor may remain permanently attached to the enzyme, so the enzyme is effectively destroyed. Increasing the amount of substrate present does not help. Some non-competitive inhibitors can affect many enzymes. This is why heavy metal ions such as mercury, lead and arsenic are so poisonous; they prevent many of the body's metabolic reactions taking place. Others are more specific and they can be used as insecticides and drugs.

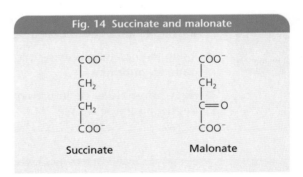

Fig. 14 Succinate and malonate

11. Fig. 14 gives the structure of succinate and malonate. Succinate dehydrogenase is an enzyme involved in the metabolic pathway of respiration. It removes hydrogen from succinate.

   a. Explain why malonate is a competitive inhibitor of succinate dehydrogenase.
   b. Explain why molecules of non-competitive inhibitors are quite different from the molecules of the enzyme's substrate.
   c. Some non-competitive inhibitors can be used as pesticides. What factors would have to be considered before using one as a pesticide for general use?

4 PROTEINS AND ENZYMES

EXAMINATION QUESTIONS

1

a The diagram below represents a small polypeptide consisting of eight amino acids s to z.

i) Copy the diagram and write the formula (in the box provided), of the chemical group which would appear at the end of amino acid z. (1)
ii) Name the type of reaction by which amino acids are joined together. (1)
iii) Name the chemical element that occurs in all amino acids but that is not found in monosaccharides. (1)

b Describe how the secondary structure of a fibrous protein is related to its function. (3)

BY01 Mar 97 Q1

2 A student carried out an investigation to find the effect of temperature on the activity of the enzyme amylase and plotted the graph shown below.

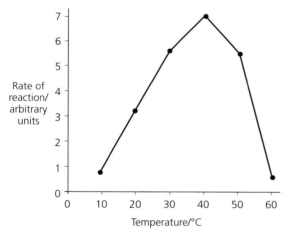

a Explain the results obtained
i) between 10° C and 20° C;
ii) above 50° C. (5)

b The student concluded that the optimum temperature for amylase activity was 40° C. Why may this conclusion not be valid? (1)

BY01 Mar 98 Q6

3 Succinate dehydrogenase is an enzyme which catalyses the conversion of succinate to fumarate.

a Use your knowledge of enzyme structure to explain why succinate dehydrogenase catalyses this reaction only. (2)

b Malonate is an inhibitor of succinate dehydrogenase. The structural formulae of succinate and malonate are shown in the diagram below.

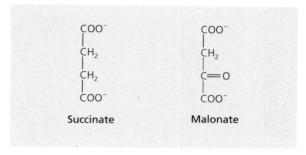

Use the information in the diagram to explain how malonate inhibits the enzyme. (2)

BY01 Mar 98 Q8

4 The graph shows the results of an investigation into the effect of a competitive inhibitor on an enzyme-controlled reaction over a range of substrate concentrations.

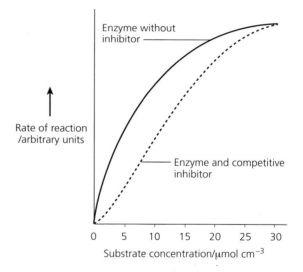

a Give one factor which would need to be kept constant in this investigation. (1)

b i) Explain the difference in the rates of reaction at the substrate concentration of 10 $\mu mol\ cm^{-3}$. (2)

ii) Explain why the rates of reaction are similar at the substrate concentration of 30 $\mu mol\ cm^{-3}$. (1)

BY01 Mar 97 Q3

5

a Diagram **A** shows an enzyme, and diagram **B** shows the substrate of this enzyme.

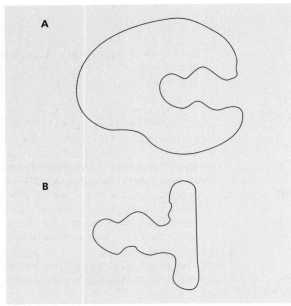

Copy this diagram and then draw onto your copy to show how a competitive inhibitor would affect the activity of the enzyme. (2)

b The graph below shows how the rate of an enzyme controlled reaction is affected by changing the subsrate concentration. Explain why increasing substrate concentration above the value shown (**X**) fails to increase the rate of reaction further. (2)

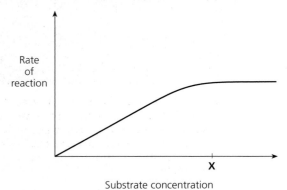

c Explain how adding excess substrate could overcome the effect of a competitive inhibitor. (1)

d Explain what is meant when we say that an enzyme molecule has been denatured. (2)

BY01 Mar 99 Q5

6

a Explain how the following are related to the protein structure of enzymes:
   i) the effect of high temperature on an enzyme-catalysed reaction;
   ii) substrate specificity.
   iii) the effect of inhibitors. (10)

b Suggest a simple method by which you could find out whether an enzyme catalysed reaction is being inhibited by a competitive or a non-competitive inhibitor. (2)

BY01 Jun 97 Q9

KEY SKILLS ASSIGNMENTS

Enzymes for the future

Use the passage at the start of the chapter to help you answer these questions.

1. After removing the hydrogen from glucose, the gluconic acid is made. Which group in gluconic acid makes it acid?

2. Sketch graphs to show the likely rates of reaction for glucose dehydrogenase and hydrogenase at different temperatures.

3. If both processes go on at the same time in the same container, what would be the optimum temperature to use? Explain your answer.

4. One possible source of glucose for this process is the cellulose in waste paper. How could glucose be obtained from the cellulose?

5. In 1999 glucose obtained from cellulose cost about 18 cents per kilogram in the United States. Gluconic acid sells for $2.65 per kilogram.

   a Assuming that each kilogram of glucose yields approximately a kilogram of gluconic acid, what percentage profit would be produced?

   b About 7 million tonnes of cellulose from waste paper could be available. What could be the value of the gluconic acid produced?

   c What other source of profit would there be from the process?

   d The demand for gluconic acid would be limited and would be less than the amount produced. Explain how this could affect the economics of the process.

   e What other factors would have to be taken into account in assessing profitability?

6. Research and explain the possible benefits to the environment of using hydrogen instead of hydrocarbon fuels like petrol.

7. Enzymes now appear in a wide range of products including shampoos, cosmetics, washing powders and even certain types of foods. All of these products need to be tested to check that they are safe to use. Many people are concerned about the amount of testing on animals and would like to see the tests stopped. Some are opposed to all animal testing. Some are opposed to testing of cosmetics but agree that medicines must be tested on animals before they are used on humans.

   a Construct a questionnaire to investigate attitudes of people to animal testing of products containing enzymes. Use your questionnaire with at least 50 people and then present the results as a five minute talk to your teaching group. Your talk needs to:

   b Include a number of graphs and charts to shown the range of people's responses

   c Explain how you analysed the results, state your understanding of the most common opinion amongst the people you questioned, and give your own opinion on this issue.

5 Food and digestion

Feel bloated after breakfast? ... Lethargic after lunch? ... Stuffed after supper? Chances are, you're experiencing the effects of poor or inadequate digestion. Besides causing a lot of discomfort, improper digestion can cheat your body out of the good nutrition your food and supplements are supposed to supply. Because good health depends on good digestion, our company offers a wide assortment of highly effective supplements created specifically to assist and improve the digestive process and help you feel better. Our enzyme complex was designed exclusively for senior citizens. Tablets contain protease, amylase, cellulase, lipase, invertase, lactase and maltase...

This type of advert appears frequently on the internet and in health food magazines. The manufacturers claim that taking digestive enzymes as a food supplement solves problems that are associated with over-eating. But will adding extra enzymes make our digestive systems more efficient? Or is that just advertising hype?

5.1 The food we eat

Every day in wealthy countries such as Britain, people shovel large quantities of food into their mouths. The food that we swallow is not in a form which it can be used directly by the body; steak, which is muscle, cannot simply be added to our biceps to make them grow. So what happens to it?

Chewing breaks up food into smaller pieces and the mechanical churning action of the stomach continues this process. However, most food molecules are too large to be absorbed into the bloodstream. They have to be **digested** into smaller molecules that are then able to pass into blood capillaries in the wall of the gut and then from the blood into the cells where they are to be used. Digestion depends on enzymes, which catalyse the reactions that break food molecules down, allowing the reactions to take place rapidly and at body temperature. In this chapter we look at the foods in our diet and investigate the different digestive enzymes that work on food as it passes through the gut.

What is food?
Humans, like all other animals, must consume food substances that are derived from other living organisms. Unlike plants we can not use sunlight to synthesise organic compounds from inorganic ones. Our diet must provide the carbohydrate, fat and protein building blocks that we use to grow and repair cells and tissues. In addition we need small amounts of vitamins and mineral ions, as well as water, but these do not have to be digested before they can be absorbed and used by the body.

Food is a complex mixture. Even something as apparently simple as bread and butter is a bewildering mixture of ingredients. The bread contains flour, water, oil, salt and the yeast that was used to make it rise. Butter is just fat and water. However, this relatively short list masks another level of complexity. Flour is a mixture of food chemicals; carbohydrates, proteins, a small amount of fat, a selection of vitamins and minerals, some roughage and a tiny amount of water.

In this chapter, we will use the word *food* to mean the ingredients of meals; things like bread, beef and apple pie. For the chemical components of these foods, such as carbohydrates, proteins and fats, we will use the term *food chemicals*.

Food has two main functions in the body – to act as a source of energy and to provide the raw materials for cell metabolism and growth. In humans much the greater proportion is used as a fuel; as much as a half may be needed to provide the energy just to maintain body temperature. Energy is also needed for muscular activity, building new cells, active transport and other metabolic processes. The amount of energy used for each type of process varies according to the age of the person, their activity levels and the environmental temperature. Different individuals therefore need different proportions of carbohydrate, fat and protein. A balanced diet should contain all the nutrients that the body needs, including vitamins, minerals, fibre and water. This does not mean that you should eat equal amounts of each nutrient but that you should tailor your diet to your needs.

1. Why do vitamins and minerals not need to be digested before they are absorbed?

APPLICATION

What is food used for?

The carbohydrate, fats and proteins in the human diet have a range of functions in the body. Their major functions are given in the table below.

Nutrient	Function in body	Main storage form
Carbohydrates	As a source of energy. Carbohydrate is the main substrate in respiration	Glycogen
Lipids	As an energy store. Lipids can be converted into substrate for respiration. Lipids, especially the subcutaneous fat stored under the skin, are also important for insulation. They are also used as the raw materials for the synthesis of membrane phospholipids.	Triglycerides (fats)
Proteins	As raw materials for the synthesis of proteins needed by the body. These include the structural fibrous proteins in bone, cartilage and muscle fibres, the carrier and receptor proteins in membranes, enzymes, some hormones, haemoglobin and blood plasma proteins.	No long-term store. Unused protein broken down in liver.

1. Suggest why adults are advised to take in less energy in their diet now than they were 50 years ago.

2. Why do:
   a. babies require a higher proportion of proteins in their diet than adults?
   b. marathon runners need a high proportion of carbohydrates?
   c. people living in the Arctic eat a high proportion of dietary lipids?

3. Most men have thinner layers of subcutaneous fat than most women. If a man and a woman who are equally active were compared, which of them would need more energy-containing food in their diet and why?

4. a. Why do adults, who have stopped growing, still need protein in their diet?
   b. An adult man only needs about 50–60 g of protein per day. Why does he gain no benefit if he eats much larger amounts?
   c. Why does someone having kidney dialysis treatment need to restrict their protein intake?

5.2 Investigating food

Table 1 Some common biochemical tests

Nutrient	Reagent used	How test is carried out	Positive result
Reducing sugar	Benedict's solution	Add Benedict's solution to sample in a test-tube. Heat in a water-bath.	Orange-red precipitate
Non-reducing sugar	Hydrochloric acid, and Benedict's solution	Boil sample with dilute acid. Add sodium hydrogencarbonate to neutralise. Carry out reducing sugar test, as described above.	Orange-red precipitate
Starch	Iodine solution	Add a few drops of iodine solution to the sample.	Blue-black staining
Lipid	Ethanol	Shake the sample with ethanol in a test-tube. Allow to settle. Pour clear liquid into water in another test-tube.	Cloudy-white emulsion
Protein	Biuret solution	Add biuret solution to sample in a test-tube. Warm very gently.	Lilac / mauve colour

Most foods consist of mixtures of carbohydrates, lipids and proteins. You cannot tell which food chemicals are in a particular food just by looking at it. Even if you know the ingredients you may not be able to tell which food chemicals are present. One way is to use a biochemical test; another is to use chromatography.

Biochemical tests

Biochemical tests are often used to detect the presence of sugars, starch, lipids and proteins (Table 1).

Tests for sugars can distinguish two groups of sugars; the **reducing sugars** and the **non-reducing sugars**. Reducing sugars such as the monosaccharides glucose and fructose are readily oxidised. When this reaction occurs the sugars lose electrons to another substance, which is said to be *reduced*. Some disaccharides, notably sucrose, are not readily oxidised and so do not reduce other substances. For this reason they are termed non-reducing sugars.

You must take care when handling food test reagents. Copper(II) sulphate used in Benedict's and biuret solutions is toxic, even at quite low concentrations. Biuret solution includes concentrated sodium hydroxide solution, which is corrosive. Ethanol gives off flammable vapour and should not be used near naked flames.

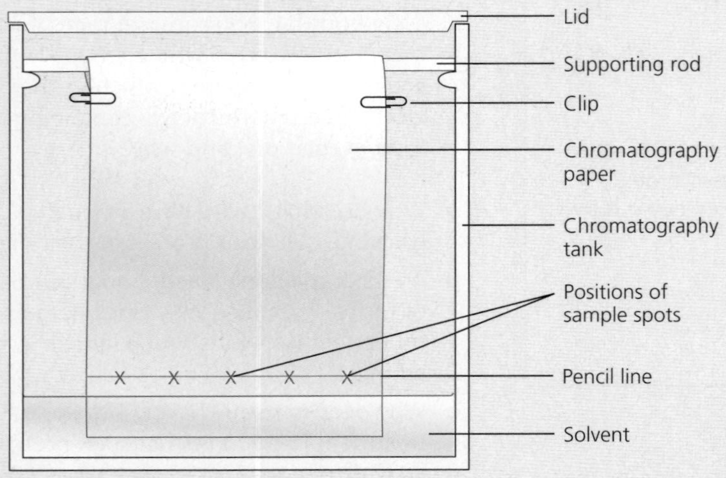

Chromatography

You have probably used blotting paper or filter paper to separate the pigments in inks or food colourings. This technique is much used by biologists as a method of separating and identifying the compounds in a mixture, especially where only small quantities of each may be present. Chromatography is also useful to distinguish between substances when biochemical tests are not suitable; Benedict's test, for example, cannot distinguish glucose from fructose because both are reducing sugars and give a positive result.

Chromatography in action

Fig. 1 shows a chromatography tank. The paper is prepared by drawing a pencil line about 2 cm from one end. Drops of the samples are added to the marked points on the line. To reduce spreading, each drop of each sample is allowed to dry before adding the next. Several drops are needed to make the spot sufficiently

Fig. 1 A chromatography tank

(Labels: Lid, Supporting rod, Clip, Chromatography paper, Chromatography tank, Positions of sample spots, Pencil line, Solvent)

5 FOOD AND DIGESTION

EXTENSION What's the difference between reducing and non-reducing sugars?

A redox reaction

Glucose → (intermediate) → Gluconic acid (Oxygen from oxidising agent)

Look at the structure of glucose. You will see that the carbon atoms 1 and 5 are linked by an oxygen atom. In the presence of an oxidising agent, a substance that readily releases oxygen or takes up electrons, this link is broken. The —CHO group is oxidised to form the carboxyl acid group, –COOH. Reactions like this are called **redox reactions** (see the diagram above). In any redox reaction, one substance is oxidised and loses electrons while another is reduced and gains electrons. The sugar acts as a **reducing agent** because it takes oxygen away from the **oxidising agent**. Such reactions are common in respiration and photosynthesis.

When two glucose molecules combine by a condensation reaction to form maltose, the glycosidic bond C–O–C, is formed between carbon atoms 1 and 4, as shown below. As you can see, there is still a carbon atom at one end of the molecule, which is free to form an exposed –CHO group. However, when glucose combines with fructose to make sucrose, carbon atom 1 of the glucose molecule becomes 'buried' inside the sucrose molecule, and is no longer free to produce a –CHO group. Sucrose, therefore, is a non-reducing sugar, whereas maltose, although a disaccharide, is still able to act as a reducing sugar.

Benedict's solution contains copper sulphate. In solution, copper sulphate has copper(II) ions, Cu^{2+}. Because glucose is a reducing sugar, it releases electrons to the copper(II) ions. These gain electrons, and copper(I) ions, Cu^+, are formed.

$$Cu^{2+} + e^- \rightarrow Cu^+$$

Copper(I) oxide is insoluble and produces a red precipitate. In the blue solution of copper sulphate, small quantities of precipitate can look yellow, or even green.

1. Explain why the Benedict's test works with maltose but not with sucrose.

2. Boiling a disaccharide with acid breaks it down into monosaccharides. Explain why Benedict's test gives a positive result after sucrose has been treated with acid.

3. For low concentrations of glucose, a positive Benedict's test ranges from green through yellow to orange and brick red. By comparing the colour of the sample solution with the colour of a standard solution, the glucose concentration of the sample can be estimated. Explain how you would prepare standard solutions for comparison.

(a) Glucose + Glucose → Maltose + H_2O

(b) Glucose + Fructose → Sucrose + H_2O

concentrated. The paper is then lowered into the tank so that the end, but not the spots, dips into the solvent. While the chromatography is in progress the tank must be sealed with a lid to keep the atmosphere saturated with solvent; otherwise the solvent evaporates from the paper as it is soaked up. Different solvents are selected for the separation of different substances. The distance a substance moves up the paper depends on its solubility in the solvent used.

If the substances to be separated are pigments, such as the photosynthetic pigments in a leaf, their position on the paper can be seen easily.

Colourless substances, such as sugars or amino acids are invisible and must be stained to show up; amino acids can be located by spraying the paper with ninhydrin, which stains them purple. Ninhydrin spray is toxic and must be used with care in a fume cupboard.

When a known solvent is used under standard conditions, a particular substance always travels the same distance up the paper. One way to confirm the presence of a substance in a mixture is to run a known sample of that substance alongside the mixture on the same chromatogram (Fig 2a). If the known substance and the mystery substance behave in the same way, your suspicions are confirmed. Alternatively, a simple calculation can be done to work out how far the substance has moved compared with the solvent. This figure is called the R_f **value** (Fig 2b). Your experimental R_f value can be compared with R_f values from known substances to help you identify it.

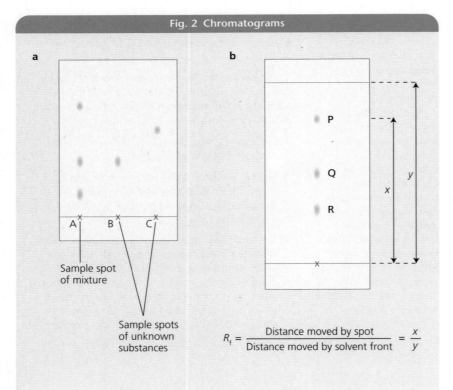

Fig. 2 Chromatograms

$$R_f = \frac{\text{Distance moved by spot}}{\text{Distance moved by solvent front}} = \frac{x}{y}$$

2a Fig. 2a shows that mixture A contains substance B. How many substances does mixture A contain? Does the mixture contain substance C? Give the reason for your answer.

b In Fig. 2b the R_f value of substance P is 41/50, which is 0.82. Calculate the R_f values of substances Q and R. Which substance is least soluble in the solvent used?

KEY FACTS

- Food is needed to provide the body with energy and raw materials for growth and metabolism.
- The main organic constituents of the diet are carbohydrates, lipids and proteins.
- The molecules of most foods are too large to pass through cell membranes or into the bloodstream. They must be broken down by digestion before they can be absorbed.
- Biochemical tests and chromatography can be used to identify the substances in foods.
- Chromatography separates substances in a mixture. A **solvent** is used to carry the components of the mixture through the chromatography paper. The more soluble the substance, the further it is carried by the solvent.
- The distance moved by the substance divided by the distance moved by the solvent is the R_f **value**. Two-way chromatography is used to separate mixtures containing several substances with similar R_f values.

5 FOOD AND DIGESTION

APPLICATION

Using chromatography

Chromatography can be used to find which amino acids are present in a protein or in a small section of a protein, such as a polypeptide. First, enzymes are used to digest the polypeptide into its amino acids. The chromatography is then carried out using paper, or a specially coated glass or plastic sheet. The principle is the same in both cases; the solvent and the substances in solution move up the chromatogram by capillary action.

After running, the dried chromatogram is sprayed with ninhydrin. The table (right) shows the R_f values of some amino acids using a particular solvent.

Amio acid	R_f value
Alanine	0.70
Arginine	0.72
Glutamic acid	0.38
Glycine	0.50
Tyrosine	0.66
Leucine	0.91
Proline	0.95

Amino acid chromatography

1 The diagram (left) shows a chromatogram of the amino acids in one polypeptide. Which amino acids are present?

2 Look at the diagram below, which shows two-way chromatography.

   a How many different substances were detected in the mixture?

   b Which spots in chromatogram 1 contained more than one substance?

   c Which substances in chromatogram 2 had R_f values of about 0.32 in the first solvent?

   d Which substance was insoluble in the second solvent?

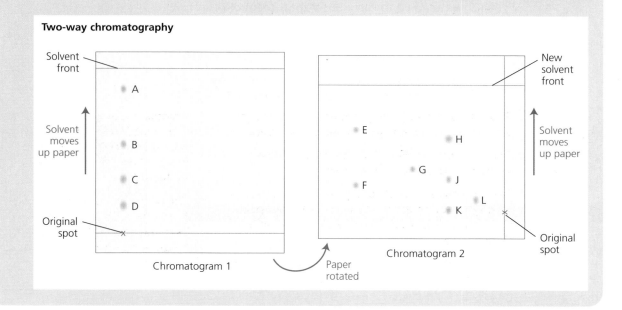

5.3 What is digestion?

Food compounds such as starch, proteins and lipids are large, insoluble molecules that cannot be absorbed directly into the blood. They must first be digested into smaller molecules. Digestion breaks down polymers into the monomers or smaller molecules of which they are made. The reaction that splits polymers is the reverse of the condensation reaction that joined them together. Condensation involves the removal of water; digestion involves the addition of water. The reaction involved in digestion is therefore termed **hydrolysis**, which means 'water splitting'. The hydrolysis reaction can be illustrated by looking at how the peptide bonds between the amino acids are broken when a protein is digested (Fig. 3). Compare this with the process of condensation shown in Fig. 4 on page 57 of Chapter 4.

> **3** Draw a diagram to show the hydrolysis of the disaccharide, maltose. Refer to the diagram of condensation in Fig. 14 on page 18 of Chapter 1 to help you.

The human gut

Digestion is catalysed by enzymes in all living organisms. Humans, like most other animals, secrete enzymes into the **lumen**, the central cavity of the gut where digestion takes place.

As food travels along the gut it passes through several different regions. Each region is adapted to carry out a specific function:

- to break large lumps of food into smaller lumps, a process called **mechanical digestion**;
- to break down the food compounds by hydrolysis, a process called **chemical digestion**;
- to absorb the soluble products of digestion;
- to get rid of undigested materials and other waste products.

Food enters the gut by the mouth. In the mouth cavity, chewing starts the process of mechanical digestion. Some chemical digestion also begins here since the saliva contains an enzyme that breaks down starch. Most of the time though, people don't chew their food for long enough for this to have much effect. The saliva softens and lubricates the food so that it can be swallowed easily. From the mouth, the food passes down the oesophagus into the stomach. The rest of the gut is in effect a long tube with a variable diameter. The muscles of the tube contract and relax in a regular and constant succession of waves. This process, called **peristalsis**, moves food steadily through the gut (Fig. 4)

Fig. 4 Peristalsis

Peristalsis involves the rhythmical contraction and relaxation of circular muscle. Peristaltic waves of contraction move along the gut

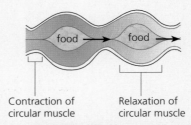

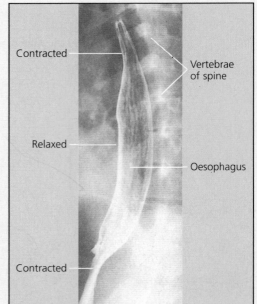

Fig. 3 Hydrolysis of a dipeptide

5 FOOD AND DIGESTION

The gut wall

The gut wall is divided into three main layers:

- an outer muscle layer, protected by a thin coating of fibres;
- a middle layer, called the **submucosa**;
- an inner layer, called the **mucosa**.

The structure of the wall is not the same all along the gut (Fig. 5). The layers have special features in different regions that allow that part of the gut to carry out specific functions. In an early stage of evolution the gut was probably just a simple tube; over millions of years changes have occurred and individual regions have developed with specific features.

These special features are called **adaptations**. You can find out more about adaptation on page 83. The main functions and adaptations of the regions of the gut are summarised in Table 2.

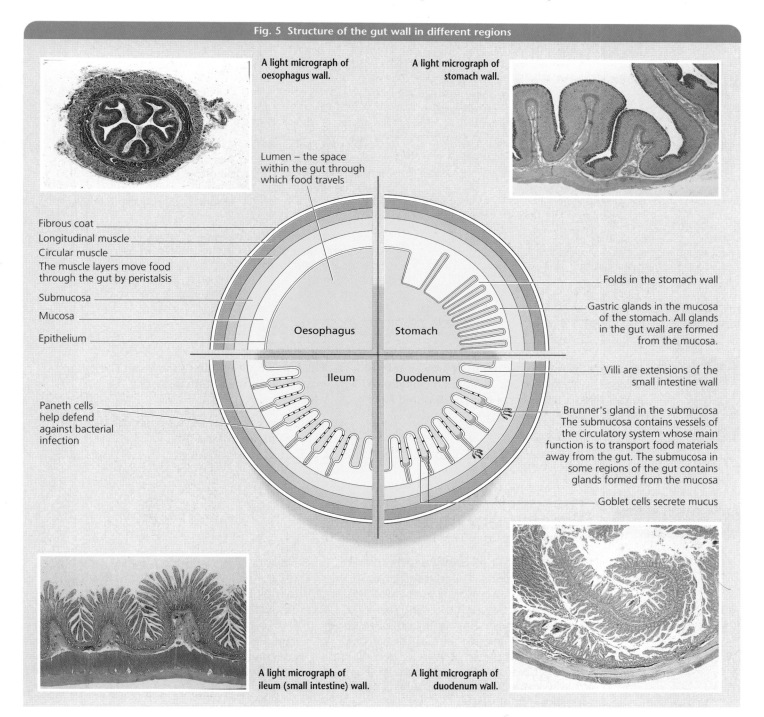

Fig. 5 Structure of the gut wall in different regions

5 FOOD AND DIGESTION

Table 2 Functions and adaptations of the gut

Region of gut	Main function	Adaptations of muscle layer	Adaptations of submucosa	Adaptations of mucosa
Oesophagus	To push food to the stomach.	Two thick layers force solid food along by peristalsis.	Elastic to allow expansion as food passes. Glands secrete mucus to lubricate passage of food.	Lining has several layers of flattened cells; outer layers can be rubbed off as food passes without causing damage to cells underneath. Folds allow expansion as food passes.
Stomach	Temporary food store. Muscular churning mixes and breaks up food. Hydrochloric acid produced kills microorganisms in food. Some digestion.	Three layers run in different directions. As the layers contract and relax, this creates an effective churning action.	Separates muscular and glandular layers.	Layer is thick with deep pits. These contain many glands that secrete mucus, enzymes and acid.
Duodenum (first 25 cm of small intestine)	Neutralisation of stomach acid. Point of entry for pancreatic juice and bile. Digestion and some absorption	Two layers for peristalsis.	Contains Brunner's glands that secrete alkaline mucus. This helps to neutralise stomach acid.	Contains many glands that secrete mucus and enzymes. Folded into numerous projections called villi. These increase the surface area for absorption of digested food.
Ileum (lower part of small intestine)	Completion of digestion. Absorption of products of digestion.	Two layers for peristalsis.	Contains many blood and lymph vessels that take up absorbed foods and transport them around the body.	Similar to duodenum, but fewer glands. Patches of cells called Paneth cells, at base of glands, which help defend against bacterial infection. Some enzyme production.

KEY FACTS

- Digestion involves the **hydrolysis** of large, insoluble molecules of carbohydrates, lipids and proteins to smaller soluble ones. These can be absorbed from the gut.
- The human gut wall has three main layers, the muscle layer, the **submucosa** and the **mucosa**.
- These layers are adapted for particular functions in different regions of the gut.

5.4 Digestive enzymes

Digestive enzymes hydrolyse carbohydrates, lipids and proteins. The gastric glands of the stomach and other glands in the wall of the stomach and the small intestine secrete digestive enzymes directly into the lumen of the gut. Other glands, such as the salivary glands and the pancreas secrete their digestive enzymes through tubes called **ducts** (Fig. 6). Whatever method is used to deliver the enzymes, the digestion is *extracellular*, because it happens in the lumen, outside the cells of the gut wall. In the duodenum and ileum, many of the enzymes secreted by cells in the gut wall stay attached to the cell membranes. Here digestion occurs very close to the surface of the gut wall rather than in the lumen. The enzymes are released into the lumen only when cells in the mucosa are broken down by the scouring effect of food passing along the intestine.

5 FOOD AND DIGESTION

Fig. 6 Digestive glands

Stomach
Endopeptidase ● and lipase ○ are secreted into the gastric glands and pass into the lumen of the stomach.

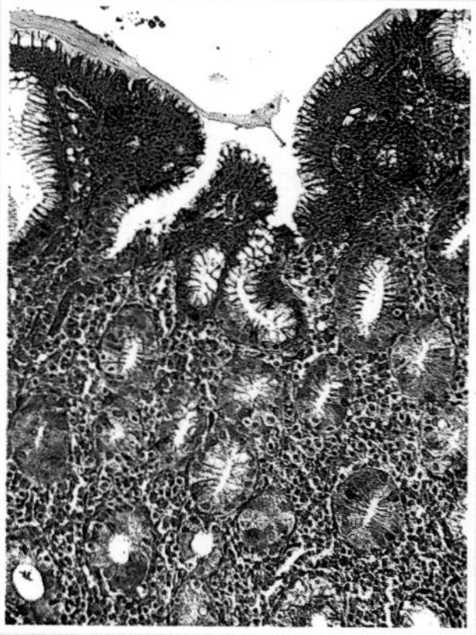

A light micrograph of a section through the mucosa of the stomach, showing gastric glands

- ● Carbohydrate digestion
- ● Protein digestion
- ○ Fat digestion

Salivary glands
Amylase ● is secreted into the salivary duct and passes into the mouth.

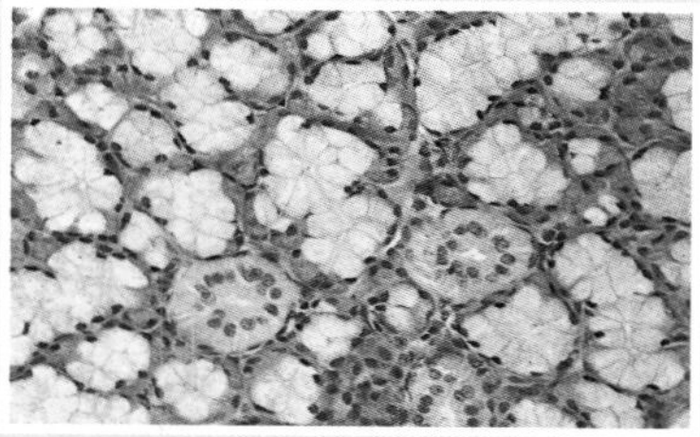

A light micrograph of parotid salivary glands

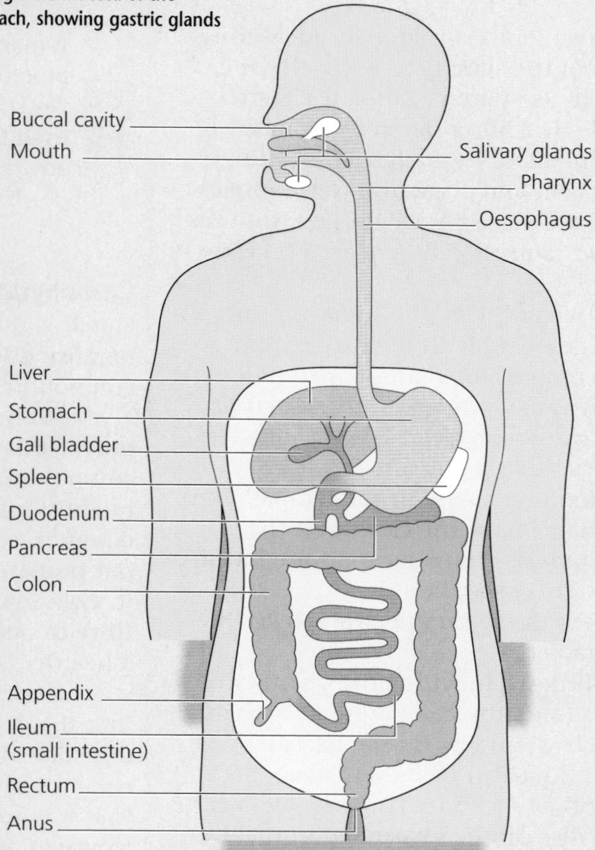

Liver
Bile ○ is secreted into the bile duct and passes into the duodenum.

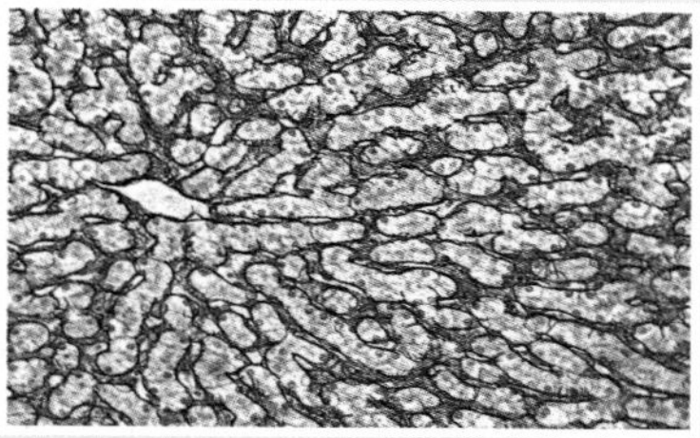

A light micrograph of liver cells

Small intestine
Maltase ● and exopeptidases ● in the outer membrane of the surface cells of the microvilli.

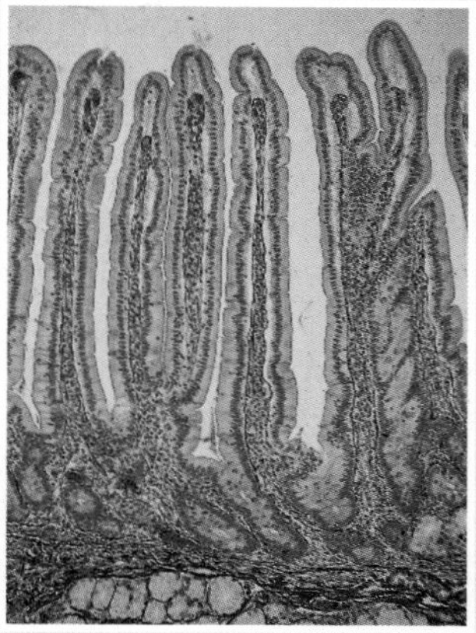

Pancreas
Lipase ○ amylase ● and endopeptidases ● are secreted into the pancreatic duct and pass into the duodenum.

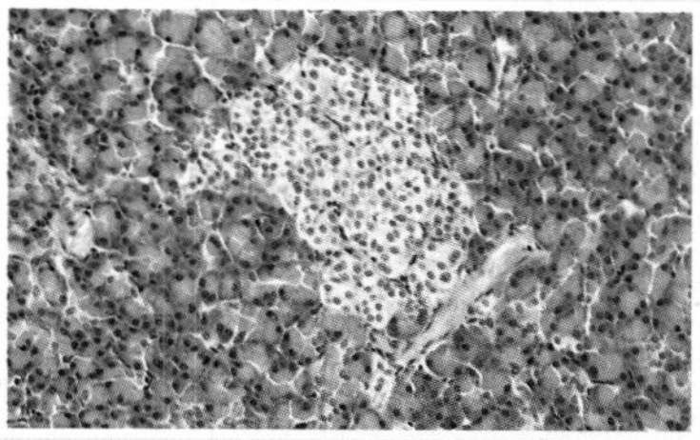

A high power light micrograph of islet cells in pancreas

A light micrograph of a section of the epithelial lining of the small intestine, showing many villi.

Labels on diagram: Buccal cavity, Mouth, Salivary glands, Pharynx, Oesophagus, Liver, Stomach, Gall bladder, Spleen, Duodenum, Pancreas, Colon, Appendix, Ileum (small intestine), Rectum, Anus

Table. 3 Sources of digestive enzymes

Digestive gland	Enzymes produced	Substance digested	Product of digestion
Salivary glands	Amylase	Starch	Maltose
Stomach (gastric glands)	Endopeptidase	Protein	Polypeptides
Pancreas	Amylase	Starch	Maltose
	Lipase	Fats/lipids	Fatty acids + glycerol
	Endopeptidases	Protein	Polypeptides
Small intestine (Duodenum/ileum)	Exopeptidases	Polypeptides	Dipeptides, amino acids
	Maltase	Maltose	Glucose

  What substances are produced from hydrolysis of the following food compounds: starch; maltose; sucrose; triglyceride; protein; dipeptide?

5 FOOD AND DIGESTION

Measuring enzyme activity

The activity of a digestive enzyme can be determined in two ways:

- by measuring the quantity of products;
- by measuring how much substrate is used up in a given time.

One convenient technique for measuring the activity of the starch-digesting enzyme, amylase, is to use starch agar plates. Starch agar is made by adding starch to liquefied agar jelly. The molten starch agar is poured into a petri dish and allowed to set. Samples to be tested are placed in cavities cut into the agar, or solid samples can be placed on the surface (Fig. 7).

After a several hours, the surface of the agar plate is covered with iodine solution. Areas that contain starch are stained blue-black, whereas areas in which the starch has been digested remain clear. The size of a clear area can be used as a measure of the concentration of amylase in the sample. The larger the diameter of the clear area, the further the amylase must have diffused from the sample. This is because the higher the concentration in the sample, the steeper the diffusion gradient.

This technique, in which the quantity of a substance is found by comparing its activity with a standard sample, is called an **assay**. A similar technique can be used to assay the activity of other enzymes. For example, white protein powder can be suspended in the agar. Protein-digesting enzymes make the milky-white agar turn clear.

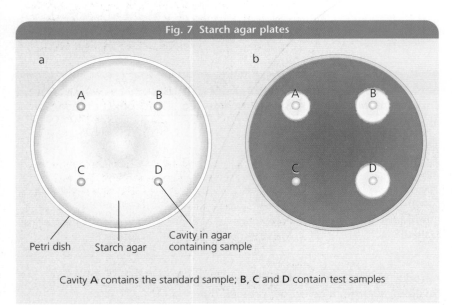

Fig. 7 Starch agar plates

Cavity **A** contains the standard sample; **B**, **C** and **D** contain test samples

5
a Look at the clear areas in Fig. 7. What can you conclude about the samples?
b What substance would you expect to find in the clear areas of agar?

6 A manufacturer wants to test three strains of a fungus as possible sources of amylase. Describe how starch agar plates could be used to find the strain that produces most amylase.

Carbohydrate digestion

Starch is the main carbohydrate in our diet because it is such a common storage compound in plants. We eat small amounts of glycogen because although animals use this polymer as their main carbohydrate storage compound, they store only relatively small quantities in their liver and muscles. Another common carbohydrate polymer is cellulose, but humans do not have an enzyme to digest it. Cellulose therefore forms the bulk of the **fibre** in our diet that is necessary to maintain a healthy colon and bowel.

The enzyme amylase breaks down starch into the disaccharide, maltose. Some digestion of starch occurs in the stomach as a result of the continuing action of the amylase that was added to food in the mouth. However, this tends to be short-lived as amylase is inactivated rapidly by the acid in the stomach. No further digestion of starch takes place until the food reaches the duodenum, where pancreatic amylase completes the process.

7 Explain how amylase is inactivated in the stomach.

The epithelial cells lining the small intestine have huge numbers of very thin, finger-like projections on their surface, called **microvilli** (Fig. 8). The membranes of these microvilli contain the enzymes that break down disaccharides into monosaccharides. Maltase, for example, digests maltose into glucose, which passes immediately into the cytoplasm of the nearby epithelial cells (Fig. 9). Carbohydrate digestion is complete by the time the gut contents have passed through the small intestine.

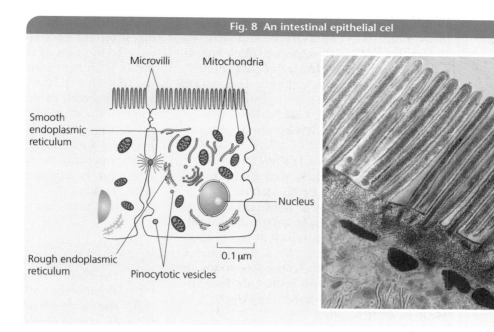

Fig. 8 An intestinal epithelial cell

EXTENSION

Digestion in other animals

Rabbits and cows consume large quantities of grass, which contains a high proportion of cellulose. Yet, like other mammals, neither is able to make the enzyme, cellulase to hydrolyse the cellulose. Instead they both have adaptations that enable them to maintain a store of bacteria that do produce cellulase and that can break it down to glucose.

A cow has four chambers to its stomach, one of which is a large bag-like **rumen**. This does not produce digestive enzymes but does contain large quantities of bacteria and other microorganisms secrete cellulase. The microorganisms use some of the glucose produced for their own growth, and they also digest proteins and other nutrients in the grass. The microorganisms are eventually regurgitated and passed into the rest of the gut. They themselves are then digested by the action of the cow's enzymes, releasing more nutrients. A rabbit also has microorganisms that produce cellulase. They grow in the rabbit's large appendix, situated at the junction between the small intestine and the colon. Some of the products of digestion are absorbed in the appendix, but most of the contents pass out as faeces. Rather than lose valuable nutrients as waste, rabbits have an additional adaptation; they eat their own faeces.

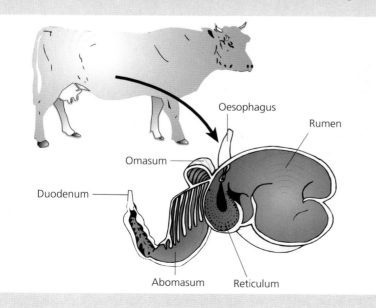

1. Both the rumen of the cow and the appendix of the rabbit are side branches, off the main route of the gut. Suggest the advantage of this.

2. The relationship between the microorganisms and the cows is called 'symbiotic', because both organisms benefit from it. Explain the advantages to:
   a the cows;
   b the microorganisms.

3. Explain why it is an advantage to rabbits to consume their own faeces.

5 FOOD AND DIGESTION

Protein digestion

Protein digestion begins in the stomach. The stomach produces an **endopeptidase** that is often called **pepsin**. Pepsin breaks proteins into shorter polypeptide chains. This enzyme is unusual in that it has an optimum between pH 2 and pH 3. This allows it to work in the very acid environment of the stomach lumen.

The pancreas secretes other endopeptidases and these pass along the pancreatic duct into the duodenum. The resulting polypeptides in the small intestine are then hydrolysed by **exopeptidases** that break only the peptide bonds at the ends of the polypeptide chains. This action releases amino acids, or in some cases dipeptides and amino acids. The amino acids produced are absorbed into the epithelial cells by facilitated diffusion.

Some exopeptidases and dipeptidases are attached to the cell membranes of the microvilli. Protein digestion, like the digestion of disaccharides, therefore occurs close to the wall of the small intestine, rather than in the lumen. In fact, because the digestive enzymes are so intimately associated with the cell surface membrane, some of the enzymic activity happens just inside the cell (Fig. 9).

There are actually whole families of endo- and exopeptidases. Each individual enzyme works on the bonds between specific amino acids, making the whole process of protein digestion quite complex.

8 Endopeptidases break long chains of amino acids in proteins into shorter ones. The exopeptidases then complete digestion by cutting off amino acids at the ends of the chains. Explain the advantage of starting digestion with endopeptidases.

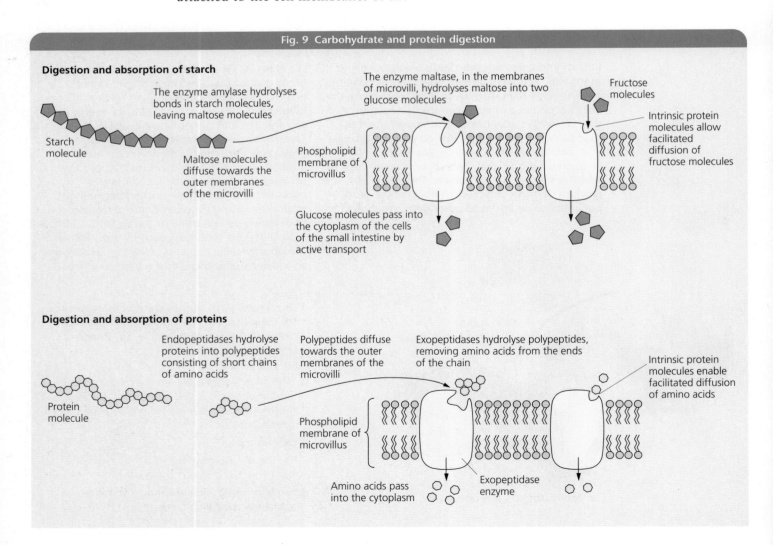

Fig. 9 Carbohydrate and protein digestion

Lipid digestion

Most lipid digestion occurs in the small intestine. The warmth in the stomach and the churning action turns solid fats to liquid. However, because they do not dissolve in the watery contents of the stomach, the fats form a suspension of large droplets. Bile, which is secreted by the liver, does not contain digestive enzymes. However, it aids digestion because it creates an alkaline environment that breaks down the large fat droplets into much smaller ones. This process is called **emulsification** (Fig. 10). Bile is produced constantly but is stored in the gall bladder, a small sac just under the liver. Bile is released down the bile duct and into the duodenum when food enters the small intestine from the stomach.

Lipase produced by the pancreas hydrolyses triglycerides into fatty acids and glycerol. Each glycerol molecule remains attached to one fatty acid, forming a monoglyceride. Monoglycerides, together with the phospholipids and cholesterol in the food diffuse rapidly through the membranes of the gut epithelial cells.

> **9** Suggest why monoglyceride molecules pass quickly through the membrane of the gut epithelial cells, whereas glucose and amino acids are absorbed only by active transport or facilitated diffusion.

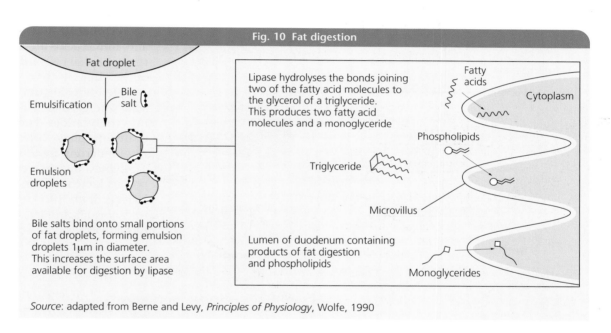

Fig. 10 Fat digestion

Source: adapted from Berne and Levy, *Principles of Physiology*, Wolfe, 1990

KEY FACTS

- Amylases break down starch into the disaccharide, maltose. Maltase then breaks maltose into glucose.

- Endopeptidases and exopeptidases break down proteins and polypeptides to give amino acids.

- Lipase digests lipids to fatty acids and glycerol. Each glycerol molecule remains attached to one fatty acid, forming a monoglyceride.

- Protein digestion takes place in the stomach and small intestine. The enzymes responsible are produced in the gastric glands and pancreas.

- The amylase produced by the salivary glands starts the process of starch digestion in the mouth. However, most of the starch in food is digested in the small intestine.

- Lipids are digested in the small intestine. The liver produces bile, an alkaline mixture that emulsifies large fat droplets into smaller ones. This provides a larger surface area for the fat-digesting enzyme lipase.

- Enzymes in the microvilli of the intestinal epithelium complete the digestion of disaccharides and dipeptides to monosaccharides and amino acids.

5.5 Absorbing the products of digestion

As digestion is completed, the products are absorbed from the small intestine into the blood or lymph systems. The duodenum and ileum are well adapted for absorption. The surface area through which the products of digestion can pass out of the lumen is huge – it covers about 350 m², over three times the floor area of the typical living room or bedroom. The large surface area is due to:

- The length of the gut; it is about 6 metres long and is coiled up inside the abdomen;
- The large folds in the inner wall (see Fig. 5);
- The huge numbers of projecting villi that cover these folds (Fig. 5);
- The extra surface area of the microvilli on the surface of the gut epithelial cells (Fig. 8).

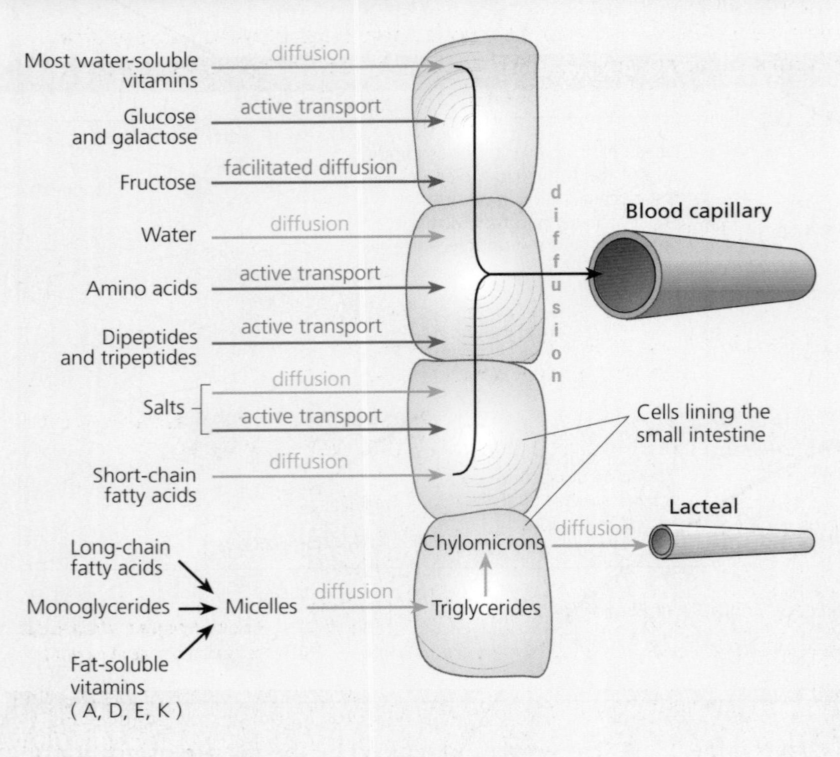

Fig. 11 summarises how different food chemicals are absorbed in the intestine. Simple sugars and amino acids enter the epithelial cells by active transport and facilitated diffusion (see Chapter 1). The carrier protein that transports glucose through the membrane also carries sodium ions. The passage of glucose is much more rapid when the concentration of sodium ions in the lumen is relatively high. The presence of enzymes in the membrane also assists the uptake of both sugars and amino acids. Sugars and amino acids are actively transported from the epithelial cells into the blood capillaries of the villi.

After diffusing into the epithelial cells, fatty acids and glycerol recombine by condensation into triglycerides. These form tiny protein-coated droplets, called **chylomicrons**. Chylomicrons are too large to enter the blood capillaries, but they can enter the much more porous **lacteal** in the centre of a villus. The lacteals drain into larger lymph vessels that transport fat droplets to the blood system. The lymph vessels connect to a large vein near the base of the neck.

Fig. 11 Absorption in the intestine

Substance	Mechanism
Most water-soluble vitamins	diffusion
Glucose and galactose	active transport
Fructose	facilitated diffusion
Water	diffusion
Amino acids	active transport
Dipeptides and tripeptides	active transport
Salts	diffusion / active transport
Short-chain fatty acids	diffusion
Long-chain fatty acids, Monoglycerides	→ Micelles → diffusion → Triglycerides → Chylomicrons → diffusion → Lacteal
Fat-soluble vitamins (A, D, E, K)	

All enter epithelial cells; from there diffusion into Blood capillary, except chylomicrons which enter the Lacteal.

10
a The epithelial cells of the small intestine contain large numbers of mitochondria. Explain why these mitochondria are needed.
b Vitamins A and D are fat-soluble. Suggest, with reasons, how they are absorbed into the gut wall.

KEY FACTS

- Monosaccharides and amino acids are absorbed into the blood capillaries in the villi of the small intestine.
- Fatty acids and glycerol recombine into fats after absorption by the epithelial cells, and tiny droplets are absorbed into the **lacteals** of the lymph system.
- Monosaccharides and amino acids enter the epithelial cells by active transport and facilitated diffusion. They are transferred to the blood capillaries by active transport.
- Fatty acids and glycerol diffuse into the epithelial cells.
- The surface area of the gut is large due to the length of the gut, the folded inner wall and the presence of villi and microvilli.

APPLICATION

Milk intolerance

Milk contains a disaccharide called lactose, or 'milk sugar'. Human milk is about 7% lactose and so is quite sweet. Babies are able to produce an enzyme called lactase in their intestine. Lactase digests lactose into two monosaccharides, glucose and galactose. Like maltase, lactase is bound to the membrane of the microvilli in the epithelial cells. In most mammals, the lactase degenerates when the baby is old enough to start eating solid food. In many humans, however, the enzyme persists and is still present in adults, although the amount produced often declines with age. In some people, most of the lactase disappears from the intestine at an early age. As a result they experience lactose intolerance. This condition is particularly common in people of oriental descent. They cannot digest lactose and this milk sugar passes intact from the ileum into the colon. Here, the excess sugar reduces the uptake of water by the body and tends to cause diarrhoea. Also bacteria feed on the lactose, producing acids which irritate the bowel, causing excessive wind and abdominal pain. People who are lactose intolerant need to avoid milk and many dairy products, although natural yoghurt can usually be tolerated because the lactase produced by the bacteria in it pre-digest the lactose before it is eaten.

1 Explain why milk sugar is not digested by sucrase, which breaks down cane sugar.

2 What evidence suggests that lactose intolerance is inherited?

3 Use your knowledge of osmosis to explain why excess lactose in the colon causes diarrhoea.

5.6 Adaptation and digestion

Adaptation is a very important idea in biology. Every living organism has features that suit it for survival in its environment. Consider the adaptations of each the animals shown in the photographs below.

Left: Cheetah in hot pursuit of an impala.

Below: Peregrine falcon about to tear into a dead pigeon.

Above: Giant anteater probing for his next meal.

Right: Red admiral butterfly feeding on Buddleia.

5 FOOD AND DIGESTION

Internal adaptations

Animals also have adaptations in the internal structures and enzyme systems that they use to process the food chemically. It seems fair to assume that, at early stages in evolution, the gut wall was simply a muscular tube similar to that seen in earthworms. The earthworm gut is a more or less simple tube that extends from one end of the body to the other. The wall of the intestine has muscular layers to push the food material from the mouth to the anus but it has no need of anything more elaborate.

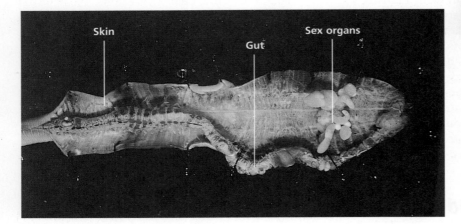

The body of an earthworm is mostly taken up by its gut, as this dissection shows.

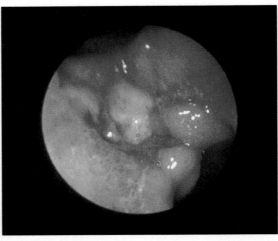

This photograph (above) shows a stomach ulcer, as seen through an endoscope. This is a typical ulcer; it is a yellowish crater with an even, raised rim. The crater is caused when the body's own enzymes digest and destroy a small patch of stomach lining.

Gut adaptations in humans

In humans, the oesophagus is a simple muscular tube, as shown in Fig. 5. Other parts of the human digestive system have developed adaptations. The stomach wall does not just have circular and longitudinal muscle, it also has layers of muscle fibres that run at angles to each other. When they contract and relax, food is churned, squeezed and broken down mechanically.

The production of acid in the cells of the gastric glands may have evolved as a means of protecting the necessarily thin absorptive surfaces of the intestine from harmful bacteria. This acid acts as an effective 'antiseptic'. However, it is also capable of damaging the cells in the stomach lining, and other adaptations have developed to prevent this. Some cells secrete an alkaline mucus, a mixture of proteins and sugars that form a watery gel that coats the stomach surface and protects the mucosa. Failure to produce enough mucus leaves gaps in the protective layer and can allow acid to attack the stomach lining, causing ulcers that bleed into the stomach.

The endopeptidase that digests protein in the stomach is an example of 'chemical adaptation'; this enzyme has evolved to be able to work at low pH. In fact, it needs to come into contact with stomach acid in order to work properly. The enzyme only becomes active and able to break down protein when the hydrochloric acid in the stomach cavity removes a section and exposes the active site. This is an advantage for two reasons:

- It prevents the protein-digesting enzyme digesting stomach tissue. Endopeptidase is secreted by cells deep in the gastric glands; if the enzyme was active here, the secretory cells and their neighbours would be harmed by digestion of membrane proteins.
- It focuses all the protein-digesting activity in the lumen of the stomach, where the food is.

A similar trick is used to protect the pancreas from the endopeptidase produced there. An inactive form of the enzyme is secreted in the pancreas and this only starts working when it is activated by an enzyme secreted by cells in the wall of the duodenum.

Feeding in fungi

Like animals, fungi need to use organic compounds as a source of energy. Many fungi feed on the organic compounds in the dead bodies of plants and animals. Organisms that feed on dead organic matter are called **saprophytes**, and they are important agents in the process of decay. The bodies of dead plants and animals are soon colonised by fungi, which start to break down their tissues. One of the principal groups of saprophytic fungi are the **moulds**.

The body of a mould consists of extremely thin threads called **hyphae** (Fig. 12). The hyphae have an outer wall made of a polymer similar to cellulose, but they do not contain separate cells. The organelles, including nuclei and mitochondria, are spread through the cytoplasm. The extensive network of hyphae forms a **mycelium**, which spreads through the food source. This mycelium may be vast. The largest single organism known is a soil fungus in a North American forest that covers about 6 square kilometres.

Digestion in fungi is extracellular since it takes place outside the body of the fungus. Hyphae secrete enzymes that diffuse through the cell wall and onto the food. This is comparable to the stages in the human gut in which the enzymes are secreted from glands into the lumen of the stomach and intestine. The enzymes hydrolyse the organic compounds into soluble monomers. These monomers are then absorbed into the hyphae, probably by facilitated diffusion and active transport. As they feed, the hyphae branch and grow through the decaying food material. The thin hyphae and large number of branches ensure that the mould has a large surface area to volume ratio. Thus it is well adapted for secreting enzymes over a large area and absorbing the products of digestion.

11 The photo shows a mould growing on a papaya fruit, which has a thick juicy flesh.

a Explain why the mould secretes enzymes only at the tips of the feeding hyphae and not in the older parts of the mycelium.

b Most of the mitochondria are concentrated at the hyphal tips. Suggest the advantages of this.

c The branching hyphae spread across the surface of the fruit; they do not penetrate deep into the flesh. Suggest why they need to stay near the surface.

d The fruiting bodies that produce spores grow upright away from the surface of the fruit. Suggest the advantage of this.

The mould Aspergillus growing on papaya fruit.

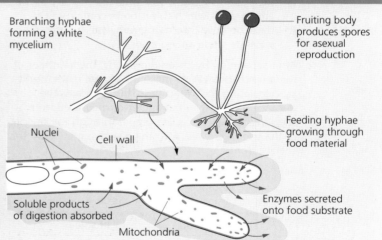

Fig. 12 Saprophytic digestion

Source: adapted from Green et al., Biological Science 1 & 2, Cambridge, 1990.

KEY FACTS

- An adaptation is a feature of an organism that suits it for a particular function and helps the organism to survive in its environment.
- Animals have adaptations that help them to obtain food. Their gut is adapted for digesting that food and for absorbing the products of digestion.
- Specific adaptions in the human stomach prevent the wall being damaged by stomach acid or by protein-digesting enzymes.
- **Saprophytic fungi** feed on dead organic matter. They secrete enzymes that digest food extracellularly.
- Thin, branching **hyphae** provide a large surface area for absorption of products of digestion.

5 FOOD AND DIGESTION

EXAMINATION QUESTIONS

1

a Copy and complete the table below, which gives the action of two of the digestive enzymes that are produced in the small intestine. (2)

Enzyme	Substrate	Product(s)
exopeptidase		
maltase		

b The goblet cells in the epithelium of the small intestine produce mucus. Suggest **one** function of this mucus. (1)

c Explain how the following features of the cells which line the small intestine help the efficient absorption of the products of digestion.
   i) microvilli; (1)
   ii) large numbers of mitochondria. (2)

BY03 Feb 95 Q1

2

a Describe a chemical test you would perform to confirm that albumen, a protein, is present in egg white. (2)

b Starting from a pure solution of albumen,
   i) describe how you would obtain a mixture of the amino acids which this protein contains; (1)
   ii) describe how chromatography can be used to separate these amino acids. (3)

c After separation by chromatography, the identities of the amino acids which make up a protein are found by calculating their Rf values. This compares the distance moved by an amino acid, measured from the starting point to a line drawn through the middle of the spot it creates, with the distance moved by the solvent front.

$$Rf = \frac{\text{distance moved by amino acid}}{\text{distance moved by solvent front}}$$

Amino acid	Rf value
Aspartic acid	0.24
Glycine	0.26
Serine	0.27
Glutamic acid	0.30
Threonine	0.35
Proline	0.43
Tyrosine	0.45
Methionine	0.55
Valine	0.60
Leucine	0.73

The diagram shows the results of separating the two different mixtures of amino acids using this technique.

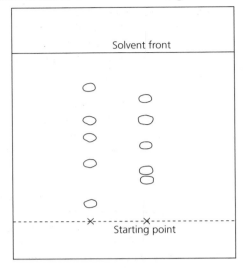

Use the table of Rf values to name the amino acid found in both mixtures. Show your working. (2)

BY01 Mar 99 Q1

3 The drawing shows the human digestive system.

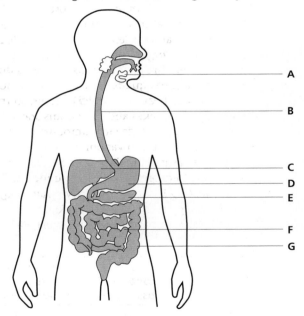

a Give the letter of an organ that produces.
   i) endopeptidase
   ii) maltase

b Name the compounds produced by the digestion of triglyceride. (1)

c Describe **one** role of bile in digestion of triglycerides. (2)

BY03 Jun98 Q1

The importance of water

Water is a major constituent of the tissues and is an important constituent of the diet. Typically about 60% of the water we take in comes from drinks, 30% comes from food and the remaining 10% is metabolic water, a product of respiration. Water from food and drink can be absorbed by all parts of the gut. However, since large amounts of water enter the gut from the glands that produce enzymes and other secretions, net water absorption takes place mainly in the lower parts of the small intestine and the colon.

Most substances in the body dissolve in water; exceptions are fats and other large polymers such as fibrous proteins. Molecules in solution can move around in water and take part in reactions in the presence of enzymes; water is therefore the medium in which metabolic reactions take place. Water is also the medium in which substances are transported within cells and around the body.

Water is also important for body temperature regulation in most organisms. Water has a high specific heat capacity, which means that it needs a lot of heat energy to raise its temperature and it retains heat better than most substances. This means that its temperature does not fluctuate as rapidly as, for example, air temperature. This is particularly important for organisms that do not control their body temperature as mammals and birds do. It is also an advantage to aquatic organisms; water does not cool down or heat up as quickly as air, so the temperature in lakes and the oceans stays relatively constant. Water needs large amounts of energy to turn it from liquid to vapour. It, therefore, does not evaporate easily, but when it does, it absorbs large amounts of energy from its surroundings. This is why sweating is such an effective way to speed up loss of energy as heat from the skin.

This table shows the water content of human tissues.

Tissue	Water content of tissue / %	Total water content of tissue as % of total body mass
Muscle	75	32
Blood	83	5
Brain	75	1.5
Skeleton	22	1
Adipose (fat storage)	10	0.01

1 An adult man weighs 70 kg. 40% of his mass is muscle tissue. Calculate:

a The percentage of his body mass that is water in his muscle tissue;

b The mass of water in his muscle tissue.

2 A man's brain weighs 1400 g. How much water does it contain?

3 An average man's body is about 65% water. In a very fat man the percentage is only 60%. Use information in the table to explain why.

4 The percentage of water in a baby's body is higher than in an adult. Suggest why.

5 Explain how 'metabolic water' is produced.

Percentage of normal water level	Symptoms
100	Happy and healthy!
99	You start to feel thirsty
98	You feel very thirsty and uncomfortable. Any sort of activity is difficult.
96	You feel tired, sick and moody.
94	You start to look pale and ill. You become irritable.
90	You stop sweating to save water, but this means your body temperature starts to rise.
89	You need medical help now!
80	You are dead by now.

6 a Calculate how many litres of water an average man can loose before he is so dehydrated that he stops sweating.

b An explorer with a body mass of 70kg lost 4.5 litres of water during a trek across a dry plain. How serious is this water loss? Explain how you worked out your answer.

6 The genetic code

For hundreds of thousands of years, human genes have passed from generation to generation, their presence shown only by the features that they have conferred on the human beings. At the start of the 21st century, our knowledge of genes is changing rapidly. The recent massive expansion of communications technology and the advances made in genetics mean that genes are now visible as sequences of letters (ACTG) recorded on CD ROM and published on the internet. The consequences of this knowledge is yet to be explored but runs way beyond any thoughts that Crick and Watson could even dream about in the 1950s, when they first deciphered the structure of DNA. In the first few years of the 21st century, the human genome project will have decoded all of the genes within the 23 pairs of human chromosomes, giving most researchers in the field direct access to the complete code.

Whilst this is good news in some respects – medical progress may well accelerate – it also raises some ethical issues that we have never had to face before. Does someone who has a mutation that gives them a very 'unusual' gene have any rights over that sequence? Can they say what the sequence can be used for? Do they 'own' it? Should they be paid for it? Can a pharmaceutical company or a biotechnology company apply for a patent on genes? And should companies refuse to employ someone because they know the person has a particular gene? Will our knowledge of genetics create a new genetically-challenged underclass?

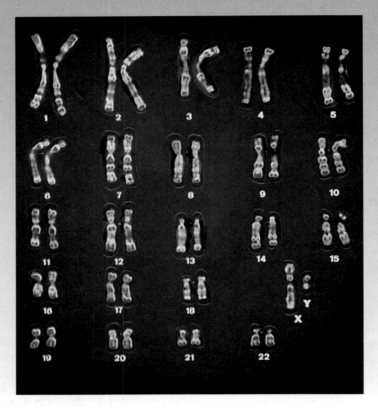

The full complement of human chromosomes (male) arranged in numbered homologous pairs. The female karyotype is the same, but has two X chromosomes instead of an X and a Y.

6.1 The genetic code

The collection of all genetic information within an organism has two remarkable properties:

- it carries the information that codes for the characteristics of that organism;
- it can copy itself exactly and pass a complete copy on to every new cell.

Every human being starts life as a fertilised egg. This single cell contains two sets of coded information, one set from the mother and one set from the father. All the information is copied every time a cell divides, so the nucleus of every one of the billions of cells in the body has a full complement of **genes**. We do not know for sure exactly how many human genes there are – latest estimates say between fifty and a hundred thousand. The term **genome** is used to refer to all the different genes in a single individual and the Human Genome Project aims to discover the sequence of bases that makes up every one of them.

In this chapter, we see how genes determine the nature and development of humans and other organisms.

What are genes?

A **gene** is a chemical code that contains the instructions for making a complete protein, or, more usually, a polypeptide. Often two or more polypeptide chains must be joined together to produce a functional protein. For example, a haemoglobin molecule contains two copies of two different types of polypeptide (see Fig. 6, Chapter 4). Genes are important because the proteins they code for determine the characteristics of an organism. As we saw in Chapter 4, the huge variety of different proteins act as enzymes, structural components, carriers and hormones. Our genes contribute to the development of every human feature that we recognise, such as the colour of our eyes or the shape of our nose. Usually many genes are involved in shaping a particular feature; human beings are so complex that is rare to find a single gene that has one clear-cut effect. Of course, the environment also has an effect, it modifies the action of individual genes and groups of genes. For example, a serious illness during childhood in a person whose genes code for a tall, athletic build could lead to a shorter, less muscular adult. This means that discovering the sequence of all human genes will be just the beginning - the huge task of finding out what they all do and how they interact with each other and the environment will still lie ahead.

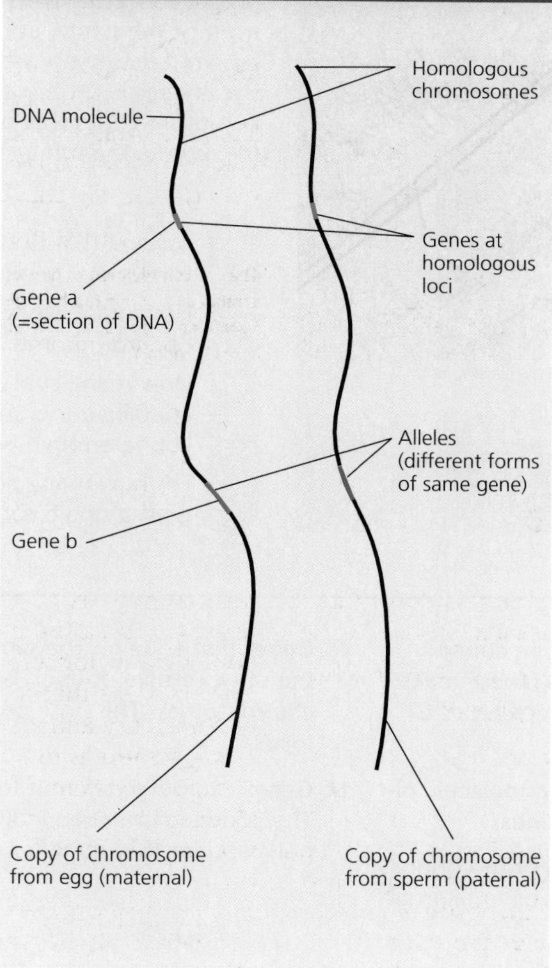

Deoxyribonucleic acid

Deoxyribonucleic acid (DNA) is the **nucleic acid** that carries the genetic code. DNA is a remarkable substance. Its properties make it the key to all life on Earth and it has essentially the same structure in bacteria, plants or mammals. It has survived throughout evolution as the one substance that can store blueprints for each of the millions of species that have existed. DNA molecules:

- are huge, and able to store vast amounts of information in a small volume;
- have small variations in structure that act as a simple code;
- are stable, so that the information is not easily corrupted;
- can reproduce themselves and so copy the information.

Just before cells divide, their DNA is copied so that the information it contains can be passed on. The DNA then contracts into chromosomes and remains in this condensed form until cell division is complete. When the cell is not dividing, the DNA exists in its uncondensed form as chromatin and is used as a guide for making proteins that the cell needs. In humans, DNA is organised into 46 **chromosomes**. Each chromosome consists of a single, very long DNA molecule surrounded by proteins. The DNA contains sections that code for particular proteins – these are the genes. There are also sections in between the genes that do not code for proteins.

As the photograph on page 88 shows, the 46 chromosomes in a body cell consist of 23 pairs; one of each pair is a copy of a chromosome from the egg and the other is a copy of a chromosome from the sperm. The members of a pair are not identical. The genes they carry occur at the same position on each chromosome in the pair. This position is called the **locus** (plural; **loci**) of the gene (Fig. 1) Each cell therefore has two copies of the gene that codes for every protein. The chromosomes

Fig. 1 The gene locus

- DNA molecule
- Homologous chromosomes
- Gene a (=section of DNA)
- Genes at homologous loci
- Alleles (different forms of same gene)
- Gene b
- Copy of chromosome from egg (maternal)
- Copy of chromosome from sperm (paternal)

6 THE GENETIC CODE

In blue eyes the iris has a patchy white layer on a black background. In brown eyes there is an extra black layer in front of the white one.

themselves are called **homologous chromosomes**. The genes that code for a particular polypeptide are normally found at same position in the DNA molecule of their chromosome in every cell of the body.

You might expect the two genes that code for a particular polypeptide to have exactly the same DNA codes. In some cases this is true; both copies of the gene are identical. In other cases, the two genes have slightly different codes, although they still occur at the matching locus on their chromosome. These different forms of the same gene are called **alleles**. A variant allele can produce a polypeptide with a different structure and function to the one produced by the normal allele. Sometimes the variant polypeptide does not function at all. Although each individual person can have only two alleles of the gene, a wide variety of alleles for that gene can occur within a population. These different alleles are one of the sources of variation between individuals of the same species.

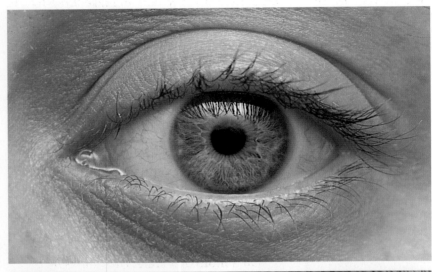

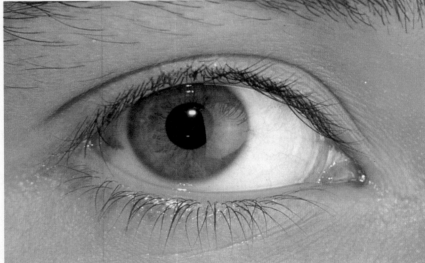

The photographs show a blue and a brown eye in close-up. It is difficult to see from photographs but the iris of an eye has no blue or brown pigment. Its colour is due to reflections from black and white patches. There are two alleles of the gene for eye colour. One allele causes extra black pigment to be produced in the iris. This forms a black layer in front of the white, and makes the eye look brown. Usually the white and black layers are streaky because they are incomplete. There are also other genes that affect the structure of the iris, producing other shades of colour.

1 Assume that the human chromosome 22 carries 450 genes.

a How many DNA molecules does one copy of chromosome 22 contain?

b How many copies of chromosome 22 are contained in a sperm, a fertilised egg, a young embryo with 16 cells?

c For how many polypeptides or proteins does a single chromosome 22 code?

KEY FACTS

- Genes are sections of DNA that contain coded information for making polypeptides. These make the proteins that determine the characteristics of organisms.

- **Chromosomes** contain one very long molecule of DNA. Each molecule carries many **genes**.

- In body cells, chromosomes occur in **homologous pairs**. Each pair consists of a copy of one maternal and one paternal chromosome.

- Genes that code for the same polypeptide occupy the same relative position on homologous chromosomes. This position is called the **gene locus**.

- Genes can have different forms, called **alleles**. The coded information in alleles differs, so the polypeptides they code for also differ.

6.2 The structure of DNA

People have selected for favourable features in animals and plants for roughly ten thousand years – really since the start of agriculture. For most of that time they were unaware of the rules of genetics which controlled the inheritance of those features. Genetics as a science began roughly a hundred years ago. Molecular genetics, which allows us to explain the reactions of the chemicals that control the inheritance of genes, is even younger. It all started in 1953 with Francis Crick and James Watson, at Cambridge University when they worked out the molecular structure of DNA. Crick and Watson unlocked the puzzle of how the components of DNA fit together into a complex three-dimensional structure.

DNA is made up of monomers called **nucleotides**. Each nucleotide has three parts:

- a sugar;
- a phosphate group;
- a base.

The sugar in DNA is called **deoxyribose**. There are four different bases, all of which contain nitrogen. These four bases are called **adenine**, **thymine**, **cytosine** and **guanine**, and are often referred to by their initial letter. This means that there are four types of nucleotide in a DNA molecule. The nucleotide monomers link together to make long strands, forming a polymer called a **polynucleotide** (Fig. 2).

A DNA molecule consists of two polynucleotide strands joined together to make a structure rather like a twisted ladder. Weak hydrogen bonds form between the bases to produce the 'rungs' of the ladder. The hydrogen atoms on one base are attracted to oxygen and nitrogen atoms on another. The shapes and sizes of the bases mean that the correct distance between the two sugar-phosphate backbones can only be maintained by adenine-thymine and cytosine-guanine bonding. This produces a regular and stable DNA molecule with two sugar/phosphate sides joined by pairs of bases (see Fig. 2).

On a molecular scale, DNA molecules are huge. This is an advantage to an organism because it means that a vast number of different genes are confined to a fairly small number of DNA molecules. This makes it more likely that all the information is passed on during cell division and from one generation to the next. It would be disastrous for an

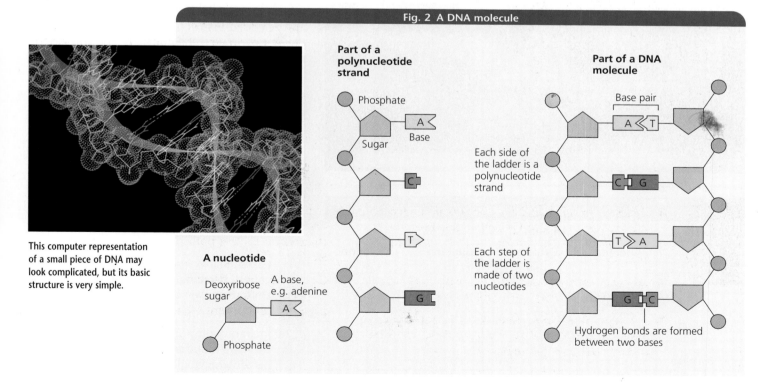

Fig. 2 A DNA molecule

This computer representation of a small piece of DNA may look complicated, but its basic structure is very simple.

6 THE GENETIC CODE

organism if some of its cells did not have a full complement of genes, since it would have no instructions for synthesising some of its proteins.

The twisting of the two strands, like two long springs plaited together, earns DNA its famous nickname – the **double helix**. Twisting the strands into a helical structure ensures that the weak hydrogen bonds linking the bases are protected in the centre of the molecule, which prevents the code being corrupted by other chemicals present in the nucleus. DNA is a very stable molecule and can withstand relatively high temperatures. Samples of intact DNA have been found in centuries old woolly mammoths frozen in the arctic permafrost, and even from 20 million year old fossils of insects preserved in amber – the basis of the book *Jurassic Park*.

The way in which the bases pair up means stored information can be copied quickly and accurately. When the DNA molecules untwist, the hydrogen bonds break so that the strands can be separated like the sides of a zip. Exact copies can then be produced, since each exposed base will only combine with one of the four types of nucleotide.

The only difference between different molecules of DNA is the number and order of the pairs of bases that join the two sugar-phosphate backbones. Different genes have different sequences of base pairs. Only the four base pair combinations shown in Fig. 2 are possible. In effect the instructions in genes are written in an alphabet with only four letters. But, by having long sequences, DNA can code vast amounts of information. After all, computers can store massive amounts of information with only a two-letter `binary' code. A human fertilised egg has about a billion (10^9) pairs of nucleotides. These carry all the coded instructions for the development of the adult body and for the maintenance of all the metabolic processes.

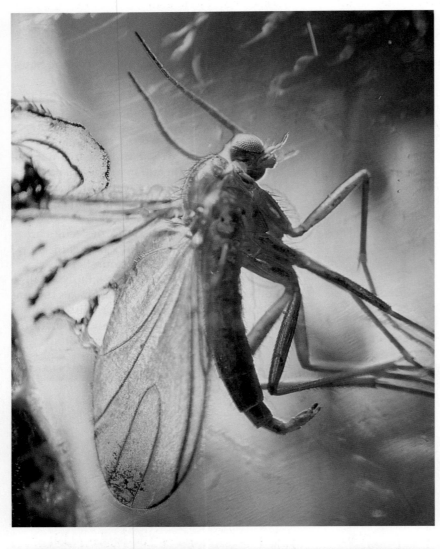

A photograph of a fossilised midge insect embedded in Baltic amber. This specimen is approximately 40 million years old.

2a Draw a diagram of a polynucleotide strand with the bases in the following order: thymine, thymine, adenine, guanine, cytosine, adenine, using the shapes shown in Fig. 2.

b Complete the other strand of the DNA molecule you have just drawn.

3a Table 1 shows the proportions of the four bases in DNA from four organisms. Use your understanding of the structure of DNA to explain the pattern in these proportions.

b In an organism 26% of the bases in the DNA are found to be adenine. What percentage would be cytosine?

Table 1 Bases in DNA

Organism	Amount of each base (%)			
	Adenine	Cytosine	Guanine	Thymine
Human	31	19	19	31
Locust	29	21	21	29
Yeast	32	18	18	32
Tuberculosis bacterium	15	35	35	15

Source: Herskovitz, *Principles of Genetics*, Collier Macmillan, 1977

Copying the DNA

Every time a cell divides, it makes a complete copy of its DNA; it copies every single one of its genes. The copy must be exactly the same as the original to preserve the information. But how does a molecule with such a complex structure make a perfect copy of itself? Fig. 3 shows what happens. The hydrogen bonds that connect the bases are broken by an enzyme. The two strands separate easily, exposing the bases. The bonds between the sugar and phosphate groups in the polynucleotide strands are relatively strong, and they keep the separate strands intact.

Another enzyme, called **DNA polymerase**, attaches free nucleotides to the exposed bases on each strand. Only the complementary bases will fit together. A new strand is built on each of the original strands, so that the two new DNA molecules are exactly the same as the original. This process of making perfect copies of DNA is called **replication**. As you can see from the diagram, each of the two new molecules of DNA has one of the original polynucleotide strands and one new one made from the supply of nucleotides in the cell. The system is therefore called **semi-conservative replication**, because one strand in each molecule is conserved (Fig. 4).

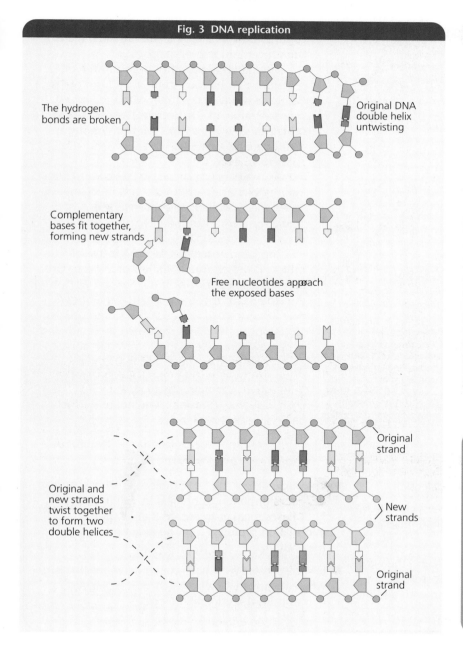

Fig. 3 DNA replication

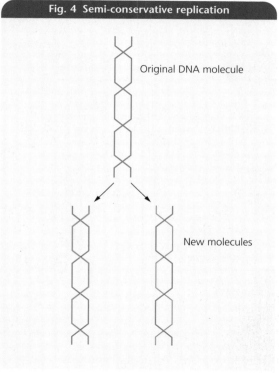

Fig. 4 Semi-conservative replication

4 Draw diagrams to show how the section of DNA in the diagram below would be replicated.

6 THE GENETIC CODE

APPLICATION Evidence for semi-conservative replication

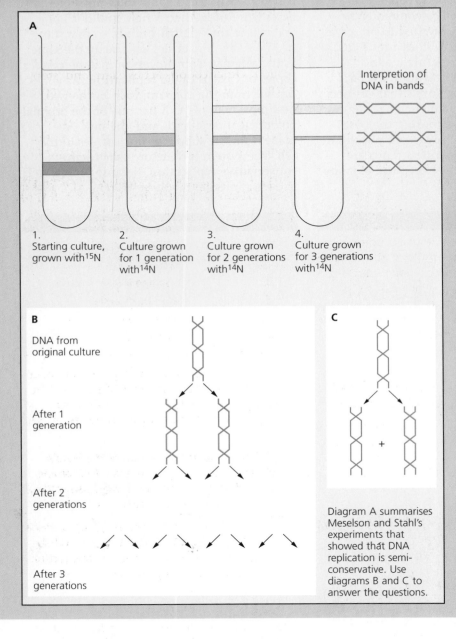

Diagram A summarises Meselson and Stahl's experiments that showed that DNA replication is semi-conservative. Use diagrams B and C to answer the questions.

Shortly after Watson and Crick published their theories and suggested semi-conservative replication, Matthew Meselson and Franklin Stahl set out to investigate whether DNA did replicate like this.

Diagram A (left) summarises the experiment. Meselson and Stahl grew bacteria in a culture medium containing 'heavy nitrogen' (the isotope ^{15}N). Some of the nitrogen in the nucleotides of the bacterial DNA was therefore ^{15}N. The researchers extracted DNA from these bacteria and centrifuged them in caesium chloride solution. The DNA molecules settled at a point in the centrifuge tube depending on the mass of the molecule. You can read about centrifuging on page 22.

After centrifuging the concentration of caesium chloride varies uniformly from the top to the bottom of the tube, with the highest concentration at the bottom. The density changes slowly from the top to the bottom and the DNA extract settles as a band at a particular level.

As the diagram shows, Meselson and Stahl then took bacteria from the ^{15}N medium and grew them on medium containing the common 'light' isotope ^{14}N.

1. Copy and complete daigram B (far left) to show Meselson and Stahl's prediction. Use different colours for the ^{15}N and the ^{14}N strands for the 2nd and 3rd generations.

2. Suppose that DNA replicated by producing a new molecule made completely of new nucleotides, as suggested in diagram C? What results would you expect to find in tube 2 after one generation?

KEY FACTS

- DNA molecules consist of two polynucleotide strands linked together.
- The sequence of bases in the nucleotides enable the DNA to store information.
- The double-stranded structure of DNA and the way in which the bases pair up enable this stored information to be copied precisely and with a high degree of accuracy.
- The large size of the DNA molecules allows a great deal of information to be held in one molecule. This makes it easier to ensure that all the information is passed from generation to generation.
- DNA replicates by a semi-conservative mechanism, which means that half of each new molecule comes from the original molecule.

6 THE GENETIC CODE

6.3 Coding for proteins

A protein is a polymer built up from units called amino acids (see Chapter 1). The order of amino acids in a protein determines its three-dimensional structure and therefore its function. The first protein to be sequenced was insulin, the hormone that regulates blood glucose concentration. In 1959, Frederick Sanger worked out the order of the 51 amino acids that make up the insulin molecule.

How does the cell put the amino acids together in the correct order? There are only four different bases in the nucleotides of DNA, but there are 20 naturally occurring amino acids. So how does the code work? After many years of research we now know that a sequence of bases, rather than a single base, codes for each different amino acid. A single-base code could code for only four amino acids. A two-base code could code for more amino acids but a sequence of three bases is necessary to provide a code for all 20 amino acids. In fact, by using three bases, there are plenty of sequences to spare.

5a How many different amino acids could a two-base sequence (eg AA, AC, AG) code for?

b What is the maximum number of amino acids that a three-base sequence can code for?

The triplet sequences of bases on the sense strand of the DNA molecule, that code for different amino acids are called **codons**. Since there are surplus codes, some amino acids have more than one codon. Some amino acids have six different codons. For others there is only one codon. Other codons act as 'start' and 'stop' signals to indicate where a gene begins and ends. Table 2 shows some examples of the codons in DNA that correspond to specific amino acids. The letters in brackets are the standard abbreviations for the amino acids.

Table 2 Codons and their amino acids	
Codon in DNA	**Corresponding amino acid**
AAA	phenylalanine (Phe)
GTC	glutamine (Gln)
ACG	cysteine (Cys)
GTG	histidine (His)
TTG	asparagine (Asn)
GAG	leucine (Leu)
CAC	valine (Val)

Fig. 5 shows a section of the gene that codes for the first four amino acids of insulin. The first triplet of bases, AAA, is a codon for phenylalanine, the next for valine and so on.

6 Fig. 5 shows the section of the insulin gene that codes for the first four amino acids of the insulin molecule. Use Table 2 to list these amino acids.

7 The nucleotid sequence in a DNA strand is: A C G T T G G T G C A C G T G. What sequence of amino acids will this section of DNA add to a protein?

Ribonucleic acid (RNA)

DNA carries the coded information to make polypeptides and proteins. But how can one small section of DNA on one chromosome in a cell nucleus make enough of a protein to supply the whole cell? Imagine trying to make enough copies of a best-selling compact disc from a single original master disc. Clearly it is much more efficient to make several copies of the DNA master template and then use these copies to produce the quantity of protein that a cell needs. And this is exactly what the cell does. Copies, or `imprints', of the DNA code are produced. These copies are **messenger**

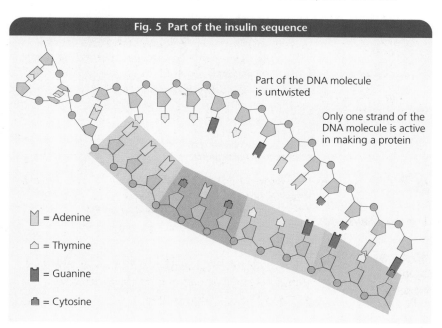

Fig. 5 Part of the insulin sequence

Part of the DNA molecule is untwisted

Only one strand of the DNA molecule is active in making a protein

= Adenine
= Thymine
= Guanine
= Cytosine

6 THE GENETIC CODE

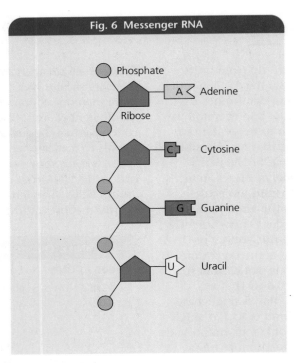

ribonucleic acid molecules (mRNA) (Fig. 6). They pass from the nucleus into the cytoplasm and are then used as guides to manufacture the protein encoded in their sequence of bases.

RNA molecules are well suited to their function. They use the same four-base system as DNA, enabling the genetic code to be copied from DNA to messenger RNA. The bases are exposed on a single strand of mRNA, and this strand can be used to assemble amino acids. The molecules of mRNA are small enough to pass through pores in the nuclear membrane. Unlike DNA, RNA molecules are quite short-lived; this enables the cell to change protein production to suit its needs.

The structure of RNA is very similar to a single strand of DNA, except that the sugar ribose replaces deoxyribose, and the base thymine is replaced by another base called **uracil**. Uracil and thymine molecules are similar in size and shape and uracil still pairs readily with adenine.

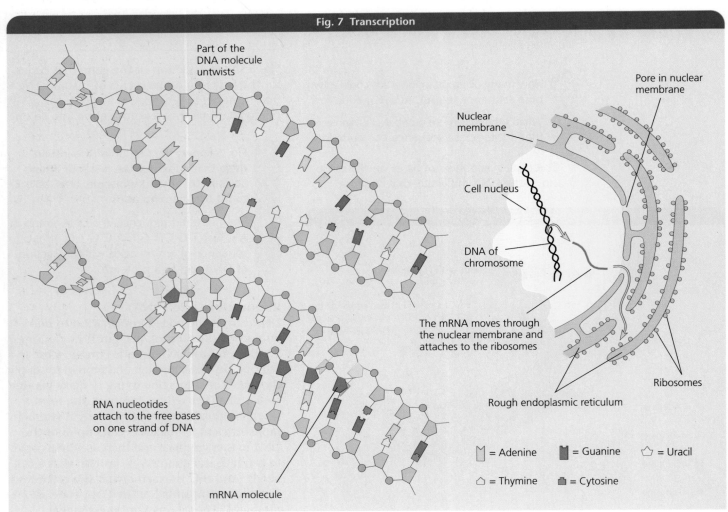

Transcription

The process of copying the code in DNA to form messenger RNA is called **transcription** (Fig.7). Transcription starts when an enzyme catalyses a reaction that makes the DNA of a gene untwist. Another enzyme, **RNA polymerase**, then assembles the RNA nucleotides along one side of the DNA molecule. This active side is called the **sense strand**. As the RNA polymerase moves along the sense strand, it produces a single-stranded molecule of messenger RNA (mRNA). mRNA carries coded information in the same way as DNA; the order of bases on a mRNA molecule is a 'mirror image' of those on the sense strand of DNA, except that uracil bases are used in place of thymine.

The sections of DNA sense strand and mRNA shown in Fig. 7 have these bases:

DNA   A A A C A C T T G
mRNA U U U G U G A A C

The three mRNA codons are therefore UUU, GUG and AAC.

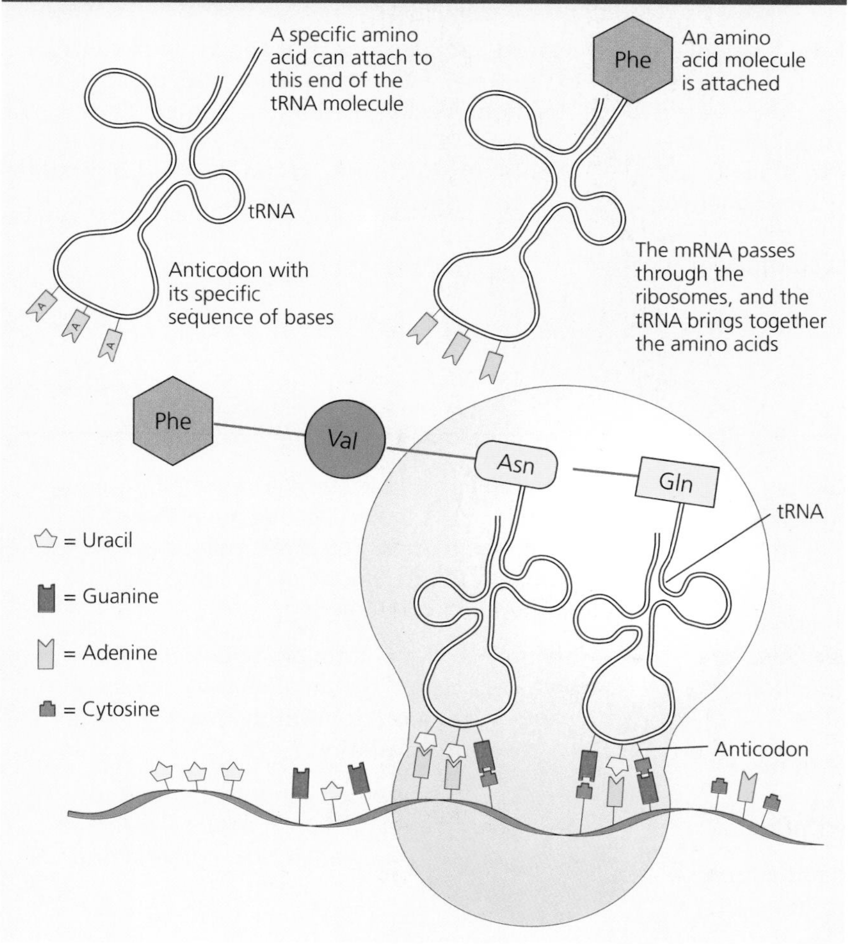

The mRNA detaches from the DNA and passes out of the nucleus through the nuclear pores and into the endoplasmic reticulum. The mRNA attaches to the **ribosomes**, which are also made of RNA. They have a specially shaped 'pocket' that is rather like the active site of enzyme molecules. The mRNA molecule fits into this pocket and the process of translation begins.

8 List four differences between molecules of DNA and mRNA.

9 What will be the order of nucleotides in the mRNA molecule produced by this section of a strand of DNA?

A C G A T T G T G C A C G A G

Translation

Once the DNA code has been transcribed and the mRNA copies have passed out of the nucleus to the ribosomes, the code on the mRNA is used to assemble the amino acids of the protein in the correct order. This process is called **translation** (Fig. 8).

Transfer RNA (tRNA) plays a key role in this process. Molecules of free tRNA are present as tRNA-amino acid complexes in the endoplasmic reticulum, near to the ribosomes. Each of the 20 amino acids has a specific tRNA molecule. The tRNA molecule is arranged in a clover leaf shape with three bases sticking out from one of the 'leaves'. The distance between these bases and the amino acid at the other end of the molecule is the same in all tRNA molecules.

When one end of the mRNA strand attaches to the ribosome, a tRNA molecule that matches the first codon on the mRNA binds to the ribosome, carrying its amino acid with it. The triplet of bases on the tRNA that binds to the codon on the mRNA is called the **anticodon**. A second tRNA molecule, also carrying its amino acid then binds to the next codon on the mRNA. So, two tRNA molecules bind to the ribosome at once.

When they are both in place the amino acids at the far end of the molecules are very close together. The amino acids are then joined together by a peptide bond. Energy from ATP is needed for this reaction to occur. The mRNA is moved across the ribosome, the first tRNA molecule, minus its amino acid, falls off and the second tRNA molecule moves across to take its place. The next codon becomes

Fig. 8 Translation

- A specific amino acid can attach to this end of the tRNA molecule
- tRNA
- Anticodon with its specific sequence of bases
- An amino acid molecule is attached (Phe)
- The mRNA passes through the ribosomes, and the tRNA brings together the amino acids

☆ = Uracil
■ = Guanine
▯ = Adenine
▮ = Cytosine

6 THE GENETIC CODE

Table 3 Insulin amino acids							
Amino acid	Phe	Val	Asn	Gln	His	Leu	Cys
DNA code in gene	AAA	CAC	TTG	GTC	GTG	GAG	ACG
Codon in mRNA	UUU	GUG					
Anticodon of tRNA	AAA	CAC					

available to bind a tRNA molecule with the next amino acid, which is then added to the growing polypeptide chain.

The order of codons on the mRNA molecule determines which tRNA molecules bind and the tRNA molecules determine which amino acids are brought together. The whole system ensures that the amino acids are assembled in the correct sequence to make the polypeptide chain encoded by the original gene on the DNA molecule.

A ribosome can translate any piece of mRNA. This means that a group of 40 ribosomes could work on 40 different mRNA molecules to produce 40 different proteins. Alternatively, the 40 ribosomes could work on 40 copies of the same piece of mRNA to produce large amounts of the revelant polypeptide very quickly.

Usually the polypeptide that is released from the ribosome after translation needs to be processed by enzymes to make it fully functional. Some proteins are made from more than one polypeptide; others need to be combined with polysaccharides or metal ions before they can work properly.

The amount of a particular protein needed by a cell varies; it does not need to make large amounts of the same protein all the time. The cell controls which genes are switched on and which are quiet. This process is incredibly complex and we are only just beginning to understand how some of the control systems work.

10 Copy and complete Table 3, showing the codes at each stage of the process in assembling the first seven amino acids of an insulin molecule.

11 A polypeptide consists of 145 amino acids. 14 different amino acids are contained in its structure.
a How many base pairs must there be in the gene that codes for this polypeptide?
b How many nucleotides are there in the mRNA that is transcribed from this gene?
c How many different types of tRNA are needed for the synthesis of this polypeptide?

KEY FACTS

- The DNA of a gene is not used to make a polypeptide in the nucleus, because this would be too slow a process. Instead, RNA copies of the gene's code are made.
- Many RNA copies of the coded information contained in a stretch of DNA can be made. This enables polypeptide products to be produced rapidly.
- One strand of the gene's DNA is used to make the copies of **messenger RNA**, which have a matching code. This process is **transcription**.
- The mRNA passes out of the nucleus and attaches to ribosomes in the endoplasmic reticulum.
- The endoplasmic reticulum has a plentiful supply of **transfer RNA** molecules that are attached to specific amino acids. The tRNA molecules have **anticodons** that recognise and bind to the corresponding mRNA codon.
- As the mRNA moves through a ribosome, the amino acids carried by the tRNA are combined in the correct sequence to form the polypeptide. This process is **translation**.
- The polypeptides formed can then be used to make a specific protein, which may be, for example, an enzyme, a membrane protein or a structural protein.

EXTENSION The genetic code

First base	G	A	C	U	Third base
G	GGG glycine	GAG glutamic acid	GCG alanine	GUG valine	G
	GGA glycine	GAA glutamic acid	GCA alanine	GUA valine	A
	GGC glycine	GAC aspartic acid	GCC alanine	GUC valine	C
	GGU glycine	GAU aspartic acid	GCU alanine	GUU valine	U
A	AGG arginine	AAG lysine	ACG threonine	AUG methionine	G
	AGA arginine	AAA lysine	ACA threonine	AUA isoleucine	A
	AGC serine	AAC asparagine	ACC threonine	AUC isoleucine	C
	AGU serine	AAU asparagine	ACU threonine	AUU isoleucine	U
C	CGG arginine	CAG glutamine	CCG proline	CUG leucine	G
	CGA arginine	CAA glutamine	CCA proline	CUA leucine	A
	CGC arginine	CAC histidine	CCC proline	CUC leucine	C
	CGU arginine	CAU histidine	CCU proline	CUU leucine	U
U	UGG tryptophan	UAG stop	UCG serine	UUG leucine	G
	UGA stop	UAA stop	UCA serine	UUA leucine	A
	UGC cysteine	UAC tyrosine	UCC serine	UUC phenylalanine	C
	UGU cysteine	UAU tyrosine	UCU serine	UUU phenylalanine	U

The table above shows which amino acids are encoded by all of the mRNA codons. You can see that several codons can code for the same amino acid. A code that has 'extra' codes like this, that are not absolutely necessary, is said to be **degenerate**. It has the advantage that every codon will result in the addition of an amino acid to the polypeptide chain. If each amino acid had one code only, a small error in the mRNA molecule would mean that the polypeptide stopped there. Every error would lead to an absence of the protein. Even if it is the 'wrong' amino acid the polypeptide may still function normally.

There are three particularly important things that you should notice about the code:

- Often it is only the first two bases of the triplet that are specific for a particular amino acid and any third base will do. This also reduces the chance that a change in the bases will alter the function of the polypeptide.
- There are three stop codes. These indicate the end of a section of mRNA, after which point translation stops.
- The codon for methionine, AUG, is also used as a start code. This means that polypeptides normally start with a methionine group when they are freshly translated. It is often removed in the processing stage that converts the polypeptide into a functional protein.

1 Look at the table.

  a A section of mRNA has the order of bases: AAG CGC UCU GCA. What will be the order of amino acids in the polypeptide it codes for?

  b What are the corresponding DNA codons on the gene that produced this mRNA?

  c Which anticodons on the tRNA molecules attach to this mRNA?

2 The first stages in deciphering the genetic code involved making synthetic mRNA. The polypeptides they produced were then analysed.

  a The researchers made mRNA in which all the bases were uracil. The polypeptide produced consisted entirely of the amino acid, phenylalanine. Explain why.

  b What amino acids would the polypeptide contain if the bases on the mRNA were all adenine?

  c The researchers then produced mRNA in which the bases uracil and cytosine alternated: UCUCUCUC. The polypeptide produced contained equal amounts of two amino acids. Which two? Explain your answer.

6 THE GENETIC CODE

6.4 Mistakes can happen

From time to time, errors occur during DNA replication. For example, one nucleotide in a strand may be replaced by another, or extra nucleotides may be added in. As a result, the sequence of bases in the DNA is changed. A change in the order of bases in a gene is called a **gene mutation**. Gene mutations can result from a change of just one base.

There are three basic types of gene mutation:

- **Addition**: an extra nucleotide is inserted, so an extra base is added to the sequence.
- **Deletion**: a nucleotide is removed.
- **Substitution**: a nucleotide is replaced by one with a different base.

These sentences illustrate the three types;

Original: THE OLD MEN SAW THE LAD
Addition: THE COLD MEN SAW THE LAD
Deletion: THE OLD MEN SAW THE AD
Substitution: THE OLD HEN SAW THE LAD

Other errors include the **inversion** of a sequence (THE OLD MEN WAS THE LAD), or **duplication** (THE OLD OLD MEN SAW THE LAD). Sometimes errors involve several nucleotides and thus a significant chunk of the 'message'. The effect of the mutation depends on how much the code is disrupted. A single substitution will only affect one codon, whereas an addition or deletion may affect all the codons beyond the error.

> **12** One strand of DNA has the following sequence of nucleotide bases:
> C A T C A T A G A T G A G A C
>
> **a** Which type of mutation could have produced each of the following mistakes during replication of the original DNA sequence?
> C A T C (G) T A G A T G A G A C
> C A T C A T A (A) G A T G A G A C +
> C A T C A (C) A G A T G A G A C
> C A T (A) T A G A T G A G A C −
>
> **b** Use the genetic code in the table on page 99 to work out the amino acid sequence that the original code and each of the mutations would code for. Don't forget that the table shows the mRNA codons, not DNA.
>
> **c** Describe the effect that each of these mutations would have on the polypeptide produced by the gene.

The consequences of mutation

The change in the code caused by a mutation may mean that a different polypeptide, and hence a different protein, is produced by a gene. This protein may not have the same properties as the original, and often does not work in the same way. A mutation can produce a different form of the gene, and so a new allele.

When the mutated DNA replicates the new form is copied, so the mutation passes on to other cells. If a mutation occurs in an ovary or testis as the gametes are being produced, the new allele may be passed on to offspring, and may spread to many individuals. Often an allele cannot spread because its effects are too damaging, for example, if the protein the original allele coded for is vital and the mutation cannot produce it, then the organism will not develop. The albino thrush in the photograph has a gene mutation which means it cannot make black pigment. It probably has a poor chance of survival because it is so conspicuous. Sometimes the absence of the correct protein may be either harmless or at least not too serious a problem. Occasionally mutations can increase survival chances, and such mutant alleles provide the genetic variation that permits natural selection and evolution.

Gene mutations occur naturally at random. As we get older, more and more cells will contain gene mutations. Mutations in body cells cannot be passed on to offspring. Mutations that occur during development may cause abnormal growth of the parts

An albino thrush.

A horse chestnut tree showing a patch of leaves that cannot make chlorophyll.

small doses have the same effect as one large dose. Radioactive substances, such as uranium and plutonium, release particles with energy levels higher than radiation, so they can have an even greater mutagenic effect. Atomic particles do not penetrate tissues in the same way as radiation, but absorbing radioactive substances into the body in food or breath is very dangerous, because they continue to decay and emit particles. Many chemicals, especially organic compounds such as those that occur in tobacco tar, cause mutations. All new drugs and pesticides must be tested to see if they are likely to be mutagenic.

formed from the cell with the mutation, as you can see in the photograph of a horse chestnut with a patch of leaves without chlorophyll.

Mutations and cancer

When mutations occur in the genes that control cell division, unchecked irregular growth takes place and a tumour develops. The frequency of mutations that lead to cancer is increased by certain **mutagenic agents**. Mutagens may cause DNA molecules to break, or change a small section of DNA chemically. Breaks in a DNA molecule in a cell are mended by an enzyme, **ligase**, which joins the broken ends together, but in this process it is possible for a nucleotide to be deleted or for some other defect to occur. High energy radiation, including X-rays, gamma rays and ultraviolet light, are mutagens, as are high energy radioactive and ionised particles. X-rays and gamma rays can penetrate deep into the body and may cause mutations in any tissue. Damage is especially serious in tissues where cell division is rapid, such as the bone marrow where blood cells are made. The effect is cumulative so many

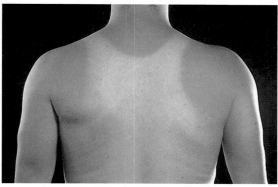

This person is experiencing the discomfort of severe sunburn. This will pass quickly but repeated exposure to strong ultraviolet light causes an increase in the mutation rate in skin cells. Skin cancer is then much more likely to develop. Thinning of the ozone layer, which normally acts as a shield, is allowing more ultraviolet rays to reach the earth's surface, especially in regions nearer the Poles. This is increasing the incidence of skin cancer in white-skinned people in several countries.

13 Explain why mutations in skin cells in a woman would not be passed on to her children.

14 Mutations can occur in mRNA molecules as well as in DNA. Explain why a mutation in an mRNA molecule is not likely to have serious consequences.

KEY FACTS

- A **gene mutation** occurs when there is a change in the sequence of bases in the DNA of a gene. Bases may be added, deleted or substituted. Segments of DNA may be inverted or duplicated

- A mutation produces a change in the DNA codons and is likely to result in a polypeptide with a different amino acid sequence.

- New alleles arise from mutations in existing alleles.

- Mutations in reproductive cells can be passed on to following generations, but mutations in body cells will only affect the tissues in which they occur.

- Mutations occur naturally at random, but the rate of mutation is increased by mutagens such as radiation and some organic chemicals.

6 THE GENETIC CODE

EXTENSION: Cancer and oncogenes

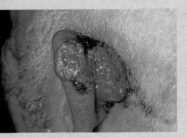
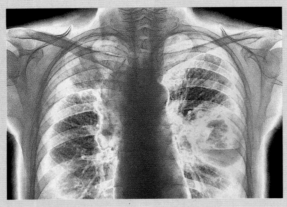

A skin tumour behind the ear of an elderly man (top) and a coloured chest X-ray (right) showing a cancerous tumour in the left lung.

Mutations occur spontaneously in cells at a rate of between 1 in 100 000 and 1 in a million at each gene locus per generation. Every sperm and every ovum is likely to have at least one new mutation. Similarly, in every tissue, at least one in a million cells will have a mutation, but the chances are much increased by exposure to mutagens. Most of these will be of little consequence, but serious effects arise when the mutations occur in the genes that control growth.

Cancer is a general term used to describe a wide range of growth disorders, and there is no single cause of cancer. Cancerous cells grow and divide much more rapidly than normal cells. The tumour that develops consists of a mass of unspecialised cells that are unable to carry out the normal functions of the tissue affected. The cells spread into neighbouring areas, block blood vessels and passageways, destroy nerves and slowly disrupt the normal function of the organ in which they appear.

Normally, cell division is tightly controlled. The growth process is controlled by genes, called **proto-oncogenes**. Only when specific proto-oncogenes are switched on by a growth factor does a cell grow and divide. For much of the time the activity of tumour-suppressor genes inhibit the proto-oncogenes. Sometimes, however, something goes wrong with this 'switching-off' mechanism A mutation of the relevant tumour-suppressor gene can also allow a proto-oncogene to keep cell division going. It continues to stimulate continuous cell division, long after it has ceased to be necessary for normal body function.

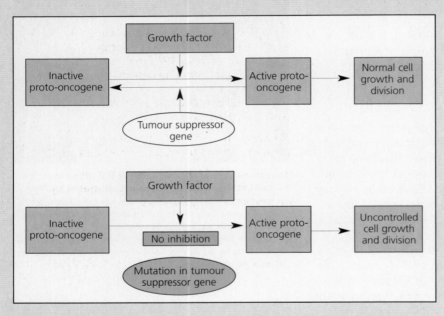

1. Cancer research scientists are discovering more genes that can affect cell growth and cancer development. For example, a gene called p53 has been found to activate a tumour suppressor gene. This gene is often mutated in cells taken from a colon cancer tumour. Draw a flow diagram to show how mutation of the p53 gene could cause colon cancer to develop.

2. In a metabolic pathway a series of reactions takes place. Each reaction is catalysed by a different enzyme. Look at the following pathway:

   Substance W $\xrightarrow{\text{Enzyme A}}$ Substance X $\xrightarrow{\text{Enzyme B}}$ Substance Y $\xrightarrow{\text{Enzyme C}}$ Substance Z

   A mutation of the gene that codes for an enzyme may result in the protein produced having a different tertiary structure so that it cannot function. Suppose that the gene for enzyme B mutates, and no enzyme B is produced.

   a Explain why production of substance Z stops.

   b Explain why substance X accumulates.

   c Explain what would happen if substance Y were then supplied.

EXAMINATION QUESTIONS

1

a Draw and label a simple diagram of an RNA nucleotide containing uracil. (3)

In 1961 biologists made synthetic mRNA. When they produced mRNA containing only uracil nucleotides, it coded for one type of amino acid, phenylalanine. When mRNA was produced with alternating uracil and guanine nucleotides, two types of amino acid were coded for, valine and cysteine. This is summarised in the table below:

Nucleotide sequence in mRNA	Amino acids coded for
UUUUUUUUUUUUUUU	phenylalanine
UGUGUGUGUGUGUGU	valine and cysteine

b For the amino acid alanine what is
   i) the corresponding DNA base sequence? (1)
   ii) the tRNA anticodon? (1)

c Explain how the information in the table above supports the idea of a triplet code? (3)

EOC Jun 98 Paper 1 Q2

2 The drawing below shows a section of a DNA molecule:

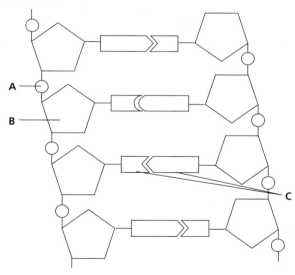

a Name the parts labelled A, B and C. (3)

b The mRNA code for the amino acid serine is UCA.
   i) Give the DNA code for serine (1)
   ii) Give the tRNA code for serine (1)

c i) What type of molecule is the end product of translation? (1)
   ii) Describe the role of tRNA in the translation process. (2)

BY02 Feb 97 Q2

3 The polymerase chain reaction is a process which can be carried out in a laboratory to make large quantities of identical DNA from very small samples. The process is summarised in the flowchart.

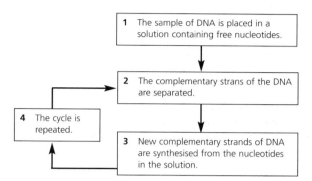

a i) At the end of one cycle, two molecules of DNA have been produced from each original molecule. How many DNA molecules will have been produced from one molecule of DNA after 5 complete cycles? (1)
   ii) Suggest **one** practical use to which this technique might be put. (1)

b Give **two** ways in which the polymerase chain reaction differs from the process of transcription (2)

c The polymerase chain reaction involves semi-conservative replication. Explain what is meant by *semi-conservative* replication. (2)

BY02 Mar 98 Q4

4 The diagram below shows a molecule of tRNA.

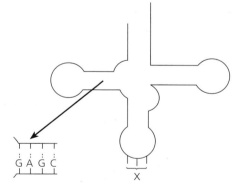

a Copy and complete the enlarged section of the diagram by inserting the correct bases. (1)

b What is the importance in protein synthesis of the part of the molecule labelled **X** on the diagram. (2)

c Give **two** ways in which the structure of a molecule of tRNA differs from the structure of a molecule of DNA. (2)

BY02 Jun 98 Q2

6 THE GENETIC CODE

EXAMINATION QUESTIONS

5 This is the sequence of bases in a short length of mRNA:

A U G G C C U C G A U A A C G G C C A C C A U G

a i) What is the maximum number of amino acids in the polypeptide for which this piece of mRNA could code? (1)
ii) How many different types of tRNA molecule would be used to produce a polypeptide from this piece of mRNA? (1)
iii) Give the DNA sequence which would be complementary to the first five bases in this piece of mRNA. (1)

b Name the process by which mRNA is formed in the nucleus. (1)

c Give two ways in whch the structure of a molecule of tRNA differs from the structure of a molecule of mRNA. (2)

BY02 Mar 99 Q5

6 In 1984 the first useful DNA sequences were extracted from the dried muscle tissue of the quagga, a zebra-like animal that became extinct in 1883. Copies of the DNA sequences were obtained by inserting the DNA into bacteria which replicated them – a process called cloning. Among the clones of the quagga DNA a sequence of 229 base pairs was obtained. This was compared with the corresponding DNA sequence in the mountain zebra and was found to differ by only 12 base pairs.

a Explain how the differences between the DNA sequences from the quagga and the mountain zebra may have arisen as a result of mutation. (3)

b Describe the molecular structure of DNA and explain how a sequence of DNA is replicated in bacteria. (9)

BY02 Mar 99 Q9

7 The DNA coding system contains the information for the production of polypeptides by a cell.

a In the DNA coding system, describe:
i) the form in which the message is transmitted from the nucleus to the cytoplasm;
ii) precisely where in the cell the message is translated.

b Explain why different alleles of the same gene produce similar, but not identical, polypeptides.

BY02 Jun 96 Q5

8 In 1958, Meselson and Stahl published the results of an experiment which provided strong evidence that cells produce new DNA by a process of semi-conservative replication.

a Why is replication of DNA described as semi-conservative?

Meselson and Stahl's experiment is outlined in the diagram below (^{15}N is a heavy isotope of nitrogen).

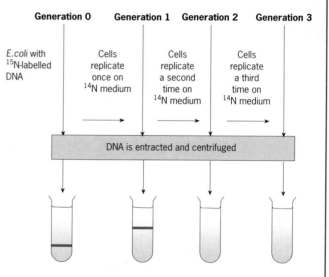

b Which component of the DNA was labelled with the 15N?

c Explain why centrifugation separates the DNA labelled with different isotopes of nitrogen.

d Copy this diagram and draw in the results you would expect for generations 2 and 3.

BY02 Feb 96 Q4

KEY SKILLS ASSIGNMENTS

Could dinosaurs live again?

The novel, Jurassic Park, by Michael Crichton, is based on the idea that DNA from dinosaurs could be used to recreate them. The following text from the book illustrates the principles being suggested.

'Regis introduced Henry Wu, a slender man in his thirties. "Dr Wu is our chief geneticist. I'll let him explain what we do here."

Henry Wu smiled. "At least I'll try," he said. "Genetics is a bit complicated. But you are probably wondering where our dinosaur DNA comes from."

"As a matter of fact," Wu said, "there are two possible sources. Using the Loy antibody extraction technique, we can sometimes get DNA directly from dinosaur bones."

"What kind of a yield?" Grant asked.

"Well, most soluble protein is leached out during fossilisation, but twenty percent of the proteins are still recoverable by grinding up the bones and using Loy's procedure. As you can imagine, a twenty percent yield is insufficient for our work. We need the entire dinosaur DNA strand in order to clone. And we get it here." He held up one of the yellow stones. "From amber – the fossilised resin of prehistoric tree sap." ...

"Tree sap," Wu explained, "often flows over insects and traps them. The insects are then perfectly preserved within the fossil. One finds all kinds of insects in amber – including biting insects that have sucked blood from larger animals."

"Sucked the blood," Grant repeated. His mouth fell open. "You mean sucked the blood of dinosaurs ..."

1 Use information from this chapter and any other useful sources to construct an article for a Sunday newspaper explaining the technology behind the newly-opened Jurassic Park. Your article should be as scientifically accurate as possible but written in a way that non-scientific readers can understand.

2 a Explain why you would need a sample of all its DNA in order to produce a dinosaur.

b Wu suggests that only 20% of protein is recovered by Loy's procedure, and that this would not yield enough DNA. Explain the mistake.

c The second method of obtaining dinosaur DNA was to extract it from the nuclei of the red blood cells found in the insects preserved in amber. This method could, in theory, work for reptiles, but not for mammals. Explain why not.

3 After extracting the DNA it would be necessary to make copies of the DNA strands. Which enzyme could be used to do this?

Still from the film Jurassic Park.

4 In Jurassic Park, the dinosaur DNA that was extracted was placed into a crocodile egg from which the nucleus had been removed.

a Explain why the egg nucleus would be removed.

b Explain what the egg would provide that could make it possible for a young dinosaur to develop.

5 The ideas in the book are not as far-fetched as they might seem. Similar techniques are being proposed as ways of re-creating more recently extinct animals, such as the dodo. Suggest the practical difficulties that would be likely to arise when trying to use these methods to re-create dinosaurs.

6 Your article led to a television appearance on a TV chat show to discuss not how dinosaurs can be recreated but if we should do it. Prepare for this discussion by working out a five minute presentation on the benefits and dangers of using modern genetic techniques to resurrect extinct animals. You will need to cover all of the points below:

- whether it might be appropriate for some organisms but not others;
- how the population of organisms created might compare with original populations;
- what might go wrong when using the techniques;
- how resurrected organisms might fit into modern ecosystems;
- how increasing the variety of organisms could be of benefit, e.g. for breeding;
- the balance between the cost of the research and the benefits;
- whether it is morally right to undertake such genetic manipulation.
- Try to present a balanced argument, rather than to describe 'doomsday' scenarios, and don't be too influenced by Jurassic Park!

7 The cell cycle

> **Talented? Intelligent? Beautiful? Successful?**
>
> *Why not make a copy of yourself? Preserve for another generation the combination of genes that makes YOU so special.*

Might we soon be seeing adverts like this? Will there be a day when the routine cloning of humans becomes possible? Cloning of plants has been carried out for many years and cloning of animals is almost certain to become common. The first ever clone of an adult mammal was Dolly, the sheep, in 1997. Dolly was grown from a single udder cell from a six-year old sheep. As you learned in Chapter 6, every cell has a full set of genes in its nucleus. Before Dolly's birth it was thought that once a cell had been specialised in an organ such as the heart or udder, it was impossible to reprogram it to go through the same full cycle of development as a fertilised egg.

At first it was thought that Dolly might be genetically damaged and would already be an 'old' ewe when born. It is still not certain whether Dolly's DNA had more than average disruption as a result of mutation, but she seems healthy so far. In 1998 she gave birth to a perfectly normal lamb.

Although widely publicised as a scientific success story, Dolly's birth was achieved only after many failures. Would it be acceptable for deformed or handicapped babies to be born in the pursuit of a perfect clone? What will be the legal status of a child born by cloning one individual, and will such children have problems accepting their own identity? Will cloning only be available to the super-rich who can afford the expense? These issues need debate and legislation and the future is uncertain. The only thing that seems sure is that there is no possibility of reversing the scientific understanding and technological advances that have made cloning a reality.

Dolly, the world's first cloned sheep with her lamb Bonnie.

7.1 Clones in nature

Human cloning is the sensationalised face of a process that is common in nature. A clone is a group of genetically identical organisms. Identical twins are, therefore, a clone, since both came from one fertilised egg that divided into two genetically identical cells that then separated. Identical twins are comparatively rare. A much more commonplace source of clones is asexual reproduction. Any organism that reproduces asexually, that is by dividing or separating off part of a parent organism, produces clones. Many plants are able to reproduce in this way. All the potato tubers from a single plant are genetically identical. Blackberry and strawberry plants spread by producing branches that put down roots and grow into identical young plants. Crop plants that are able to reproduce asexually are a enormous advantage to growers. All the new plants will be of the same variety as the parent plant with exactly the same genetic potential to produce fruit and vegetables with the same characteristics and of the same quality.

7.2 Mitosis

The key to producing genetically identical offspring is the process of cell division called **mitosis**. Every time a cell divides by mitosis the instructions contained in its DNA are copied faithfully, and each new cell receives a complete set of the genetic code from the original cell. This means it can make all the proteins that its parent cell could.

When a cell is not dividing individual chromosomes are not visible. When stained, the whole nucleus appears as a dark mass because the chromosome material, called **chromatin**, is spread out (Fig. 1). This makes it easier for transcription to occur and for mRNA molecules to move away from the DNA molecules and out of the nucleus.

Fig. 1 The stages of mitosis

These light micrographs show the stages of mitosis in cells of the hyacinth root.

Interphase
For most of the time the chromosomes in a nucleus cannot be seen. The DNA molecules are stretched out and busy synthesising proteins. At the onset of mitosis the DNA replicates. This happens during interphase, before any sign of cell division can be seen.

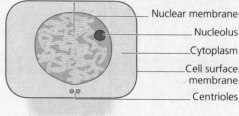

Chromatin threads; Nuclear membrane; Nucleolus; Cytoplasm; Cell surface membrane; Centrioles

Prophase
After the chromosomes have replicated, they coil up and contract. They then become visible. The replicated chromosomes appear as double strands. In fact they consist of the two new chromosomes, at this stage called chromatids, still firmly joined together at the centromere.

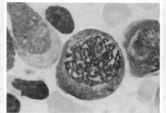
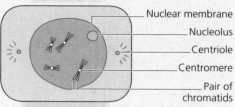

Nuclear membrane; Nucleolus; Centriole; Centromere; Pair of chromatids

Metaphase
The membrane of the nucleus breaks down and a web of protein fibres called the **spindle** forms from one end of the cell to the other. The centromeres attach to the spindle in the middle of the cell.

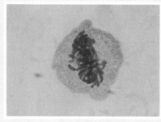
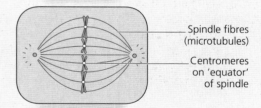

Spindle fibres (microtubules); Centromeres on 'equator' of spindle

Anaphase
The centromeres now split and the chromatids separate. The chromatids move along the spindle fibres to opposite ends of the cell.

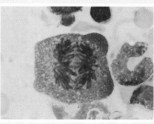
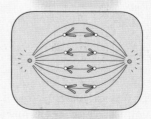

Daughter chromosomes move apart, led by their centromeres

Telophase
The separated chromatids, which are exact copies of the original chromosomes, group together at opposite ends of the cell. New nuclear membranes develop and the chromosomes uncoil. Mitosis is complete when the cytoplasm divides and new cell membranes form. Plants also form new cell walls.

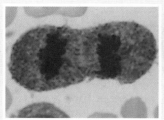
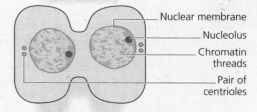

Nuclear membrane; Nucleolus; Chromatin threads; Pair of centrioles

7 THE CELL CYCLE

In a dividing cell, there is a clear sequence of events and, for much of the process, the chromosomes exist in a condensed state and are easier to distinguish. Replication of the DNA (see Chapter 6, p. 93) takes place just before the visible stages of mitosis begin. At this point, each chromosome has two molecules of DNA. The chromosomes shorten and thicken, and the DNA forms very tightly packed coils of coils called **supercoils**. This shrinkage of chromosomes into smaller packages makes it less likely that sections of DNA will break away and be lost during cell division.

After the chromosomes have contracted, they become visible when stained. It is possible to see that each chromosome consists of two separate threads, called **chromatids**. The chromatids are identical and have been produced from the original chromosome DNA by replication. The two chromosomes, remain attached to each other at the **centromere** until the nucleus divides. Each dividing cell has two sets of chromatids because it has two of each chromosome. When the cell divides by mitosis, the chromatids separate and each new cell nucleus gets a full complement of all of the chromosome pairs in the original cell.

Mitosis is a continuous process, but for convenience the cell cycle is described as having five stages. These stages are shown in Fig. 1 on page 107. Fig. 2 on this page summarises the complete cell cycle.

1a How does the amount of DNA in a nucleus differ between the start and the end of interphase?

b There are 46 chromosomes in a human body cell. How many DNA molecules are there in a cell nucleus at the start of prophase?

c Suggest why it is an advantage for the chromatids to contract independently during prophase.

d Suggest why it is important that the chromatids remain attached at the centromere until anaphase.

2 Draw diagrams of metaphase, anaphase and telophase for the cell shown below.

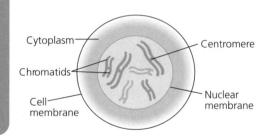

Fig. 2 The cell cycle

KEY FACTS

- **Asexual reproduction** is a form of reproduction in which a single parent organism produces offspring by simple division or by splitting off a part of itself.

- **Clones** are genetically identical organisms. The offspring of plants and other organisms that reproduce asexually are clones.

- **Mitosis** is a type of cell division. When body cells divide to increase their number, or an organism reproduces asexually, cell division occurs by mitosis.

- The cell's DNA is replicated in mitosis and each new cell produced receives an exact copy of the DNA in the parent cell.

- Replication of the DNA in the chromosomes occurs during **interphase** before the chromosomes contract and become visible in the nucleus.

- Replication produces two identical **chromatids** from each chromosome. The chromatids are separated during mitosis in a process that guarantees that each daughter nucleus has one of each pair.

7 THE CELL CYCLE

EXTENSION

Controlling the cell cycle

What makes your fingers grow to a certain size and then stop? How is that some cells in your fingers make bone while others produce skin, nerves, muscles, blood vessels and so on? As the skin on your finger tips wears away cells underneath produce new layers, following exactly the same pattern of fingerprints. How do these cells 'know' what to do? And when to stop doing it?

Cells produced by mitosis may grow and become specialised, or they may have a relatively short growth period before dividing again. Some, for example most nerve cells, may never divide again. They nevertheless remain active and their genes continue to produce the proteins necessary for their function as nerve cells. Until recently it was thought that no new brain cells are made in humans after the age of about 16, and that as cells die they are never replaced. However, even with a loss of several thousand per day, there are still plenty to spare. More recently this idea has been challenged, and there is evidence that even specialised nerve cells may be stimulated to divide again.

Cells in embryos, and in tissues that have a high cell turnover, such as the skin, gut lining and bone marrow, have quite a short interphase. During the first part of interphase the genes are actively involved in transcription and growth occurs. New organelles are formed, and some of these, such as mitochondria and chloroplasts, contain their own small sections of DNA that enable these organelles to reproduce independently. After a time the protein **cyclin** builds up in the cell. This seems to stimulate production of another protein that in turn stimulates the replication of the DNA and initiates mitosis. This second protein also breaks down cyclin.

This is only part of a much more complex process. There are many factors that start and stop division. Research suggests that some cells in a developing embryo secrete growth-promoting substances that diffuse out into the embryo, to produce a concentration gradient. The concentration in different parts of the embryo determines whether or not further cell division takes place. Perhaps the tip of a finger stops growing because it is so far from the source of a growth promoter and the concentration of the promoter is too low. Experiments in which the nuclei of cells that have stopped dividing are transferred into embryonic cells show that the chromosomes can recover their ability to replicate and divide.

APPLICATION

Cell division in an insect embryo

Stage	Mean duration /minutes
Interphase	20
Prophase	105
Metaphase	13
Anaphase	8
Telophase	54

The data shows the length of each of the stages of the cell cycle in an insect embryo.

1 a How long is the complete cell cycle?
   b Calculate the percentage of time spent at each stage of the cycle.
   c Describe the evidence that the cells in this tissue are dividing rapidly.

2 a In which of the stage is DNA replicated?
   b When do the DNA molecules uncoil?
   c When do the centromeres split? When do the centromeres attach to protein fibres?
   d When does transcription take place?

3 The graph shows how the position of centromeres change during mitosis. Line X is the distance between the centromeres and the ends of the spindle. Line Y is the distance between the centromeres of pairs of chromatids. Measurements started at the beginning of metaphase.

a How long was metaphase?
b How long was anaphase?
c What is the distance between the poles of the spindle in this cell?
d Explain the shape of lines X and Y.

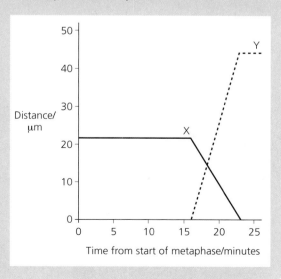

7 THE CELL CYCLE

7.3 Growing quality crops

Most of our food crops belong to a single botanical group – the flowering plants. The flowering plant group includes all plants that produce flowers and seeds during sexual reproduction. Sexual reproduction is different to asexual reproduction; it produces offspring that are not genetically identical. Sexual reproduction in a flowering plant therefore produces seeds that develop into plants that are genetically different from their parent plants and from each other. This genetic variation is vitally important to the process of evolution but is not always helpful to farmers growing commercial crops. They do not want variation – they want to preserve fruit and vegetable varieties that are of proven quality and that behave predictably when treated with agrochemicals. Farmers who use large machines to harvest crops also prefer plants that all reach maturity at the same time.

Vegetative propagation

Fortunately, as well as producing seeds, many flowering plants also reproduce asexually. Potato tubers, strawberry runners and onion bulbs are all structures of asexual reproduction. One potato plant can produce many tubers, all of which are genetically identical. Growers select varieties of potato that have particularly useful qualities, such as having flesh that remains firm when boiled, resistance to disease, and a good flavour. These varieties were originally bred by sexual reproduction but, once tubers with the right qualities were obtained, all future crops could be produced by collecting and planting those tubers. Because the plants grow from vegetative structures, this method of asexual reproduction is called **vegetative propagation**. Plants produced in this way are genetically identical to the original plant.

Vegetative propagation has several advantages for growers:

- it maintains genetic stability, so all the fruit or vegetables are of the same quality;
- mature plants usually develop more quickly because the vegetative parts that they grow from have larger food reserves than seeds;
- the plants grow at the same rate and so can all be harvested at the same time;
- it is often easier and therefore cheaper than collecting and sowing seeds;

But there are also disadvantages. The relatively large vegetative structures are more difficult to transport and distribute, and, since they are not dehydrated as seeds are, they are more prone to damage or rotting during storage. They are also more likely to carry disease – they may harbour microbes that can then be transmitted to a whole crop after planting. Potato tubers, for example, often carry serious fungal diseases, such as potato blight. The problem of disease is worse in plants that are all genetically identical because if one plant succumbs to the disease, all the rest are likely to be susceptible to it too. In a population of plants produced by sexual reproduction, genetic variation often produces individual plants with a degree of disease resistance.

Fig. 3 Asexual reproduction in plants

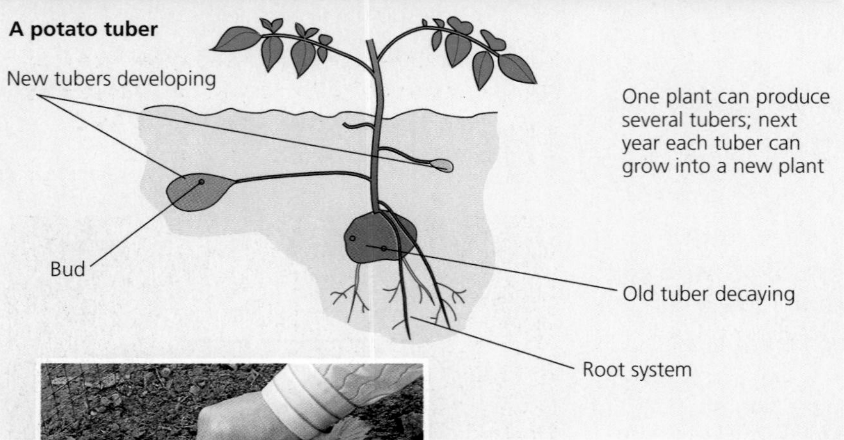

A potato tuber
- New tubers developing
- Bud
- One plant can produce several tubers; next year each tuber can grow into a new plant
- Old tuber decaying
- Root system

This strawberry plant has grown several runners with a small plant at the end. The plantlets will take root and grow into new plants. Each will have exactly the same genes as the parent.

3 Suggest the advantages to a flowering plant of reproducing both asexually and sexually.

4a Most plants that are produced asexually are genetically identical. What process could cause some to have a genetic difference?

b Even plants that are genetically identical can produce fruit or vegetables that are not exactly the same. Explain what causes these differences.

7 THE CELL CYCLE

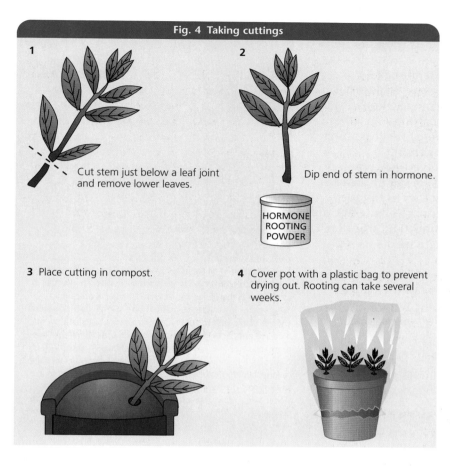

Fig. 4 Taking cuttings

1. Cut stem just below a leaf joint and remove lower leaves.
2. Dip end of stem in hormone.
3. Place cutting in compost.
4. Cover pot with a plastic bag to prevent drying out. Rooting can take several weeks.

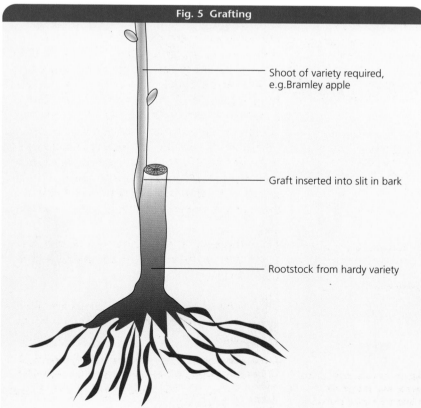

Fig. 5 Grafting

- Shoot of variety required, e.g. Bramley apple
- Graft inserted into slit in bark
- Rootstock from hardy variety

Not all crop plants reproduce naturally by vegetative propagation. For these plants, growers often use artificial propagation to maintain varieties. One of the simplest techniques is to take cuttings from a parent plant (Fig. 4). This is method is often used for flowering shrubs, such as roses.

For many plants, especially fruit trees such as apple, it is difficult to get cuttings to root. In these cases it is necessary to **graft** the cutting onto the lower part of an existing plant, called the rootstock, so that they grow as one plant (Fig. 5). This is how varieties such as the eating apple Cox's Orange Pippin and the Bramley cooking apple have been propagated since their discovery in a hedgerow and a back garden, respectively, more than a hundred years ago. Grafting has the advantage of combining the best characteristics of both the rootstock and the cutting. It is now common practice to grow apple varieties on rootstocks from dwarf plants to save space. More trees can be planted in the same area, so increasing the yield. It is possible to grow two different varieties of apples on a single rootstock giving people with small gardens the opportunities to have a wider variety of fruit. Even more surprisingly, some varieties of apples and pears can be grown on the same rootstock so it is possible for small gardens to have an apple-and-pear tree made up from parts that originally came from three separate plants.

5 Suggest why it is necessary to reduce water loss from a new cutting.

6 Sometimes stem branches grow from the rootstock of a Bramley apple tree. Apples that grow on these branches are not Bramley apples. Explain why not.

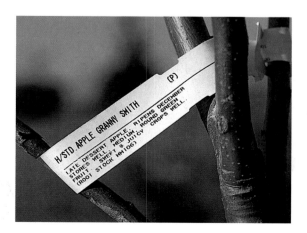

7 THE CELL CYCLE

Micropropagation

For commercial crop growing, there are some species of plant that cannot be easily produced by either a breeding programme or by vegetative propagation. To solve this problem, simple asexual vegetative propagation techniques, such as stem cuttings, have been developed as high technology tissue cultures. This is called **micropropagation**.

Micropropagation involves the growth of **plantlets** from single cells, pieces of tissue or organs, using sterile laboratory techniques (Fig. 6). Plantlets are very small plants that are produced by asexual reproduction. Plantlets grown by micropropagation are disease-free clones.

All plant cells have the ability to give rise to an identical genetic copy of the parent plant. In the case of the oil palm (right), the clones are derived from tiny pieces of leaf tissue.

7a Use the information in this section to list the advantages of using micropropagation.

b Can you think of any disadvantages?

This plantation of oil palms is 15 years old. The oil palm is an important tropical crop. The oil is has many uses, for example in foods such as margarine and in soap. Traditional breeding programmes to produce plantations take several years to general enough plants with the desired characteristics. Oil palms are now bred successfully by micropropagation and plantations developed using this technique have been grown in Malaysia since 1977. Oil palm plantlets are obtained by growing small pieces of leaf tissue, so large numbers of plantlets can be obtained rapidly and in a small space. The technique is particularly useful for obtaining clones from hybrid plants that are naturally sterile and unable to reproduce.

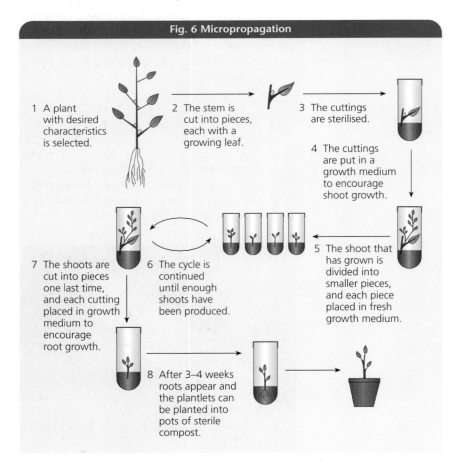

Fig. 6 Micropropagation

1. A plant with desired characteristics is selected.
2. The stem is cut into pieces, each with a growing leaf.
3. The cuttings are sterilised.
4. The cuttings are put in a growth medium to encourage shoot growth.
5. The shoot that has grown is divided into smaller pieces, and each piece placed in fresh growth medium.
6. The cycle is continued until enough shoots have been produced.
7. The shoots are cut into pieces one last time, and each cutting placed in growth medium to encourage root growth.
8. After 3–4 weeks roots appear and the plantlets can be planted into pots of sterile compost.

Constant environment growth rooms are used for raising cultured plant material such as oil palm plantlets. The conditions are kept sterile, which means the plantlets are kept free from disease. Environmental conditions, e.g. temperature, are kept at the best levels for healthy growth.

7 THE CELL CYCLE

7.4 Cloning animals

A few animals reproduce asexually but mammals do not. There is, therefore, no natural way of obtaining clones of farm animals, such as cattle or sheep. However, since every cell has a complete set of genes, it is in theory possible to grow a complete new animal from any cell or group of cells. Rapid strides are being made in research on ways to switch on the genes in individual cells so that they recommence the cycle of development. Dolly the sheep, described at the start of this chapter, was cloned.

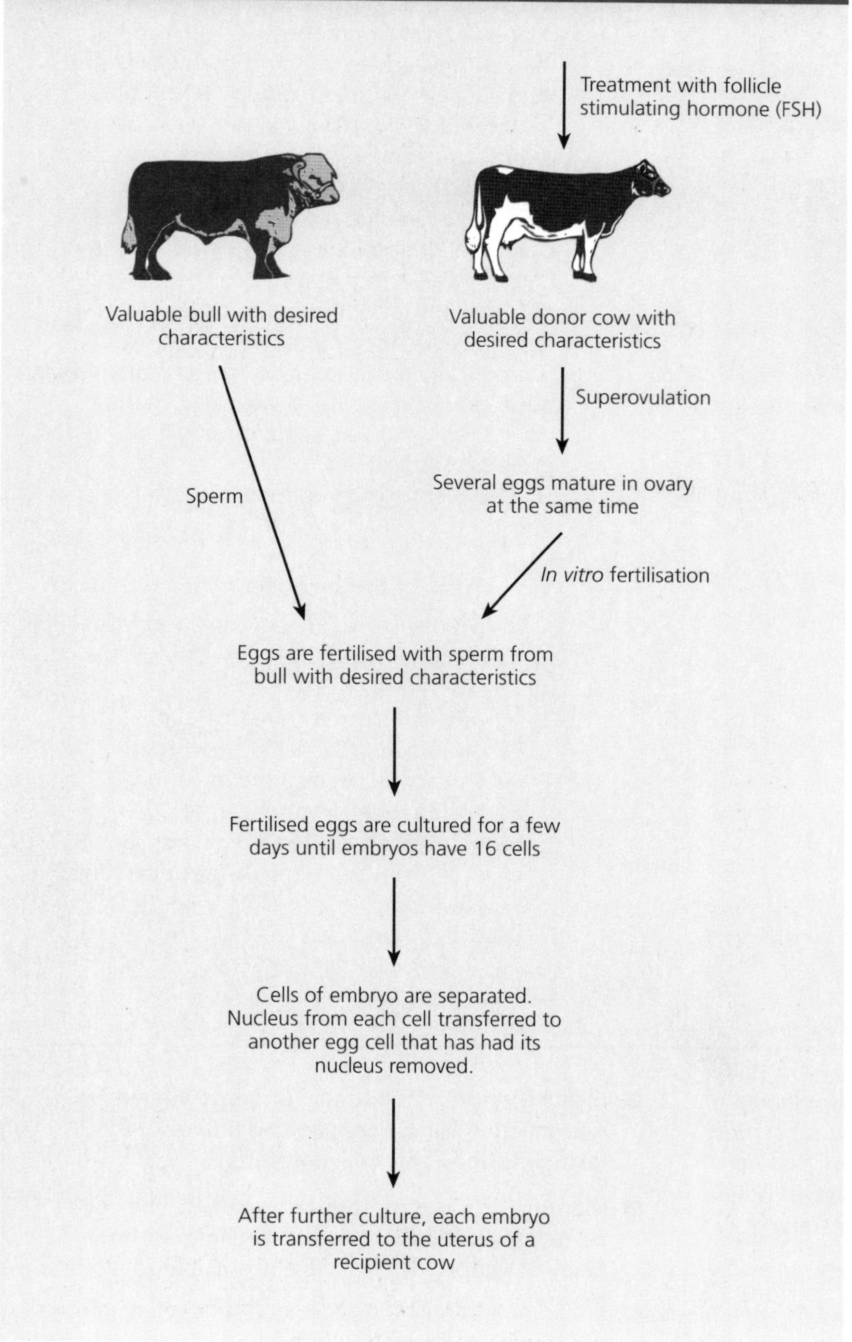

Until recently, only cells from very young embryos have shown the capacity to develop normally when separated. This happens naturally when, for some reason, the cells in a two-cell embryo split and grow independently to form identical twins. For some animals with valuable characteristics it is worthwhile prompting this process artificially. For example, clones may be obtained from calf embryos in order to increase the number of calves that can be obtained in one generation from mating particularly valuable parents. To increase further the number of offspring produced the number of eggs produced by a cow may be increased by using follicle stimulating hormone (FSH). FSH stimulates the ovaries to produce several mature eggs at the same time, instead of the normal one or two. This is called **superovulation**. To be clones the cows need to have come from the same embryo as shown in Fig. 7. The cloned cows will be as similar as identical twins. Cows produced by superovulation will effectively be brothers and sisters and will be as similar as non-identical twins.

To produce clones, the mature eggs are removed and fertilised in a laboratory. Once the fertilised eggs have grown into an embryo with 8 or 16 cells, the cell nuclei are separated and transferred to another cell. This is allowed to grow into a new embryo, which is then placed in the uterus of another cow.

Fig. 7 Embryo transplantation

Valuable bull with desired characteristics → Sperm

Valuable donor cow with desired characteristics → Treatment with follicle stimulating hormone (FSH) → Superovulation → Several eggs mature in ovary at the same time → *In vitro* fertilisation

Eggs are fertilised with sperm from bull with desired characteristics

↓

Fertilised eggs are cultured for a few days until embryos have 16 cells

↓

Cells of embryo are separated. Nucleus from each cell transferred to another egg cell that has had its nucleus removed.

↓

After further culture, each embryo is transferred to the uterus of a recipient cow

8a Explain why all the calves produced from the cells of one embryo are clones.

b Explain why the group of eggs collected from one cow and fertilised are not clones.

c A calf is born by the method shown in Fig. 7. Would it have genes from: the donor cow, the donor bull, the recipient cow? Explain your answer.

d The method that is shown in Fig. 7 would normally only be used as part of a special breeding programme. Explain why it would not be used as part of regular farming practice.

7 THE CELL CYCLE

APPLICATION: Growing Dolly

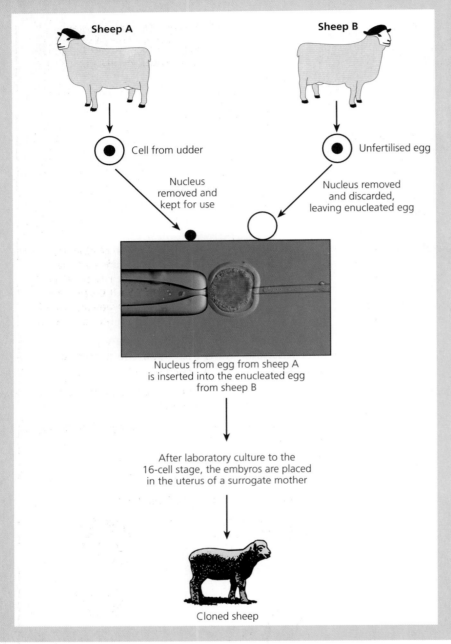

The research work that produced Dolly was done at the Roslin Institute near Edinburgh in Scotland. Fig. 9 shows the cloning process that was used. The principles illustrated may seem quite straightforward, but the practical difficulties were immense. Only one lamb was successfully produced from 277 attempts to fuse udder cells with unfertilised eggs. The major task of reprogramming the genes of the udder cell so that they re-started the process of development was overcome by freezing the udder cells and then thawing them in a very limited supply of nutrients.

After Dolly was born, other scientists were rightly sceptical, and, as is normal practice in science, challenged the Roslin Institute workers to provide further evidence of their claim. For example, could they be sure that Dolly's genes were really from the udder cells? This was confirmed by DNA analysis. Other researchers attempted to repeat the process. Since then both cattle and mice have been cloned in a similar way, and there is little doubt that the techniques will be refined over the coming years.

Use the diagram on the left to answer these questions:

1 a Which of the sheep have identical genes?
  b Why was the nucleus removed from Sheep B's egg?

2 A body cell from a sheep has 54 chromosomes (27 pairs). How many chromosomes were there in:
  a an udder cell from sheep A;
  b the unfertilised egg cell from sheep B;
  c a cell from the embryo that developed into Dolly;
  d Dolly's body cells?

KEY FACTS

- Some flowering plants, such as potato plants, reproduce asexually and produce natural clones. This is called **vegetative propagation**. Growers can maintain varieties of plants that have useful characteristics by growing them only from asexually produced structures.

- Plants that do not naturally undergo vegetative reproduction can be propagated artificially by taking cuttings and making grafts.

- Mammals do not reproduce asexually. They can be cloned artificially. Recently techniques have been developed which make it possible to clone mammals from cells in older tissues.

7 THE CELL CYCLE

EXAMINATION QUESTIONS

1 The drawing below shows how potato tubers are produced.

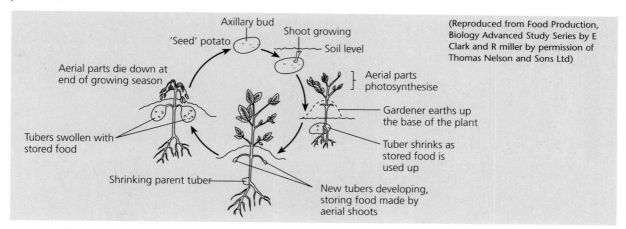

a i) Name the type of reproduction which results in the production of tubers. (1)
  ii) Explain two advantages of growing crops from tubers rather than seeds. (2)
  iii) Explain why a viral infection may destroy a whole crop of potatoes grown from tubers produced in this way. (2)

BY07 Jun 99 Q3

2 Drawings 1–5 show stages in mitosis in an animal cell.

 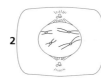

 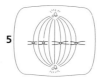

a Which of the drawings, 1–5, shows:
  i) telophase?
  ii) anaphase
  iii) metaphase? (3)

BY02 Jun 95 Q1

3
a i) Using potato tubers as an example, explain what is meant by a clone. (3)
  ii) Describe the advantages and disadvantages of growing potato crops from tubers rather than seeds. (3)

BY07 Feb 97 Q8

4 Scientists have perfected the technique of cloning sheep in recent years. The process begins with the fertilisation of eggs by sperm. Nine days later, when the zygote has divided to form about 400 cells, the cells of the embryo can be separated. When separated, each cell can grow into a new embryo that can be implanted into an adult female sheep.

a Explain why an embryo cell can divide to form an embryo, but an egg cell will not. (1)

b Explain why the sheep formed from the embryo cells are all genetically identical. (2)

c Give one advantage and one disadvantage of producing sheep in this way. (2)

BY07 Mar 99 Q4

5 The drawings show four stages in mitosis.

a Give the correct sequence of the four drawings (1)

b Explain how mitosis results in daugther cells containing copies of genetic information identical to the parent cell.

BY02 Feb 97 Q1

6 The diagram below shows an animal cell in prophase of mitosis. The letters represent alleles of a gene that codes for eye colour, and a gene that codes for body colour.

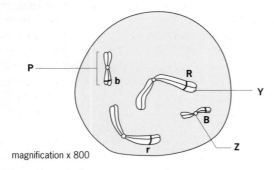

magnification x 800

a Draw the diagram, and on it,
  i) name structures Y and Z;
  ii) shade with pencil two homologous chromosomes.

b Calculate the actual length of chromosome P. Express your answer in micrometres (μm). Show your working.

UCLES March 1995 Modular Biology: Foundation Module Pape Q2

The future of cloning

An American scientist has cloned a human embryo by splitting it in a test-tube to create identical twin or triplet embryos. Dr Jerry Hall of George Washington University Medical Center, Washington DC, was attempting to devise a method of creating extra embryos for couples undergoing in vitro fertilisation (IVF) treatment. He adopted techniques common in livestock breeding to produce identical offspring, and used genetically abnormal human embryos in the experiment. None was viable and all have been discarded. However, the work raises important ethical questions if such a technique was perfected using normal embryos. Embryos can be frozen for many years and it would, in theory, be possible for genetically identical people – twins or triplets – to be born years apart to the same woman or to others implanted with the cloned embryos. American ethicists have also raised the possibility of parents keeping a 'back-up' embryo in case their child died or needed an organ donation.

Source: adapted from an article by Liz Hunt, The Independent, 1 November 1993

This article describes one of the early experiments on human cloning. Since 1993 techniques have improved rapidly. These improvements are likely to make it possible to clone cattle and other farm animals. This makes it possible to preserve special features, such as high milk or meat yield, possessed by a single individual produced by a selective breeding programme. It would also enable scientists to preserve the genetic constitution of animals that have been genetically engineered, for example to produce a useful drug (see Chapter 10). Clones of animals such as mice could be very useful in medical research, for example to find out what happens in cancer or ageing, because their genetic make-up would be known and control experiments could be carried out. There is also much interest in the possibility of cloning not whole animals but tissues and organs for transplants. Research into such possibly beneficial uses of cloning is almost certain to result in the cloning of humans from adult cells.

1 Work in a group or with a partner to develop a paper on the future of cloning. The eventual paper will need to be agreed by all members of the group and so will probably need to include a variety of views. Begin by discussing in your group the kinds of topics you will need to research and agree who is going to tackle which topic. The list of questions below provide a useful starting point. Keep a careful record of all documents used, websites visited and people consulted to produce the eventual paper. Since the group will need to agree the final paper you will also need to build in a way for everyone to comment on everyone else's work.

  a How could cloning farm animals be of benefit?
  b How could cloning benefit medical research?
  c How might human cloning be carried out?
  d Who might benefit from human cloning, and how might it be used? For example, could it be used to clone particularly talented people, or a child dying from a serious illness?
  e What would happen if the cloning technique produces defective embryos or handicapped children?
  f What would be the legal status of cloned children, and might they have difficulty in establishing their own identity?
  g Might the technique be misused by very rich people or by dictators?
  h What dangers might there be in having large numbers of people, or animals, with an identical genetic constitution?
  i Should legislation be used to ban any form of human cloning, and would it work?

Your report should be carefully organised so that your arguments are clear. You should use an appropriate style of scientific writing in which arguments are balanced and not over-stated. Include a diagram or flow chart explaining how human cloning might be carried out. You should proof-read and re-draft your report, making sure that spelling, punctuation and grammar are accurate. If possible, use IT facilities to make it simpler to redraft and correct your report.

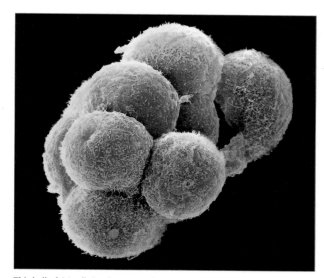

This ball of 16 cells is a human embryo 4 days old. Could these cells be used to clone people?

8 Sex and reproduction

> ## Sex tests – ancient and modern
>
> At the ancient Olympics in Greece, sex testing was a simple procedure. Athletes walked naked through the gates. No penis, no admittance.
>
> Sex tests were first introduced to the modern Olympic games in 1966. They were invasive and controversial gynaecological examinations.
>
> For the Mexico Games in 1968 a less invasive sex chromatin test was introduced. A sample of cells is taken from inside a woman's cheek. This 'buccal smear' is then stained and examined under a microscope. If a cell has two X chromosomes, one will be inactivated, and will show up as a dark blob in the cell's nucleus, a so-called Barr body. A woman fails the test if no Barr body is detected.
>
> For the Barcelona Olympics in 1992 the sex chromatin test was replaced by a more modern, and more expensive, genetic test. The new procedure tests for the so-called male-determining gene, SRY. It uses the technique of the polymerase chain reaction, PCR, to make copies of a targeted stretch of DNA from the Y chromosome, so producing enough DNA to detect the presence of the SRY gene.
>
> *Source:* adapted from articles by Alison Turnbull, *New Scientist*, 15 September 1988; Gail Vines, *New Scientist*, 4 July 1992

Is it important to determine if an athlete taking part in the Olympic Games is male or female? Are the differences between the sexes fixed biological ones, or are they, for humans at least, a result of social and cultural differences? Why do organisms need males and females at all?

If the function of reproduction were just to produce more of the same species, it would be much simpler and more efficient to use asexual reproduction. But we find it hard to imagine that a human body could reproduce by developing extensions that drop off and grow into new people, as happens in many plants. Even the idea of a single cell growing into a baby without the intervention of a second cell to fertilise it, as happens in cloning, seems strange and threatening. To most people sex is the obvious way of reproducing. After all most animals and plants do it, and there must a good reason for this. So what is it?

Trine Hattestad, Norwegian Javelin thrower.

8.1 Sex cells

Sexual reproduction always involves two cells joining together. The cells that fuse together are called **gametes**, and the process of joining is **fertilisation**. A life cycle that involves sexual reproduction must involve production of gametes to maintain the correct chromosome number in the next generation (Fig. 1). To understand why, it is necessary to learn more about gametes. Gametes are not just ordinary cells; they develop differently from normal body cells that divide by mitosis (see Chapter 6). Gametes undergo a type of cell division in which the number of chromosomes is halved. This type of cell division is called **meiosis**.

You saw in Chapter 6 that the chromosomes in body cells are in homologous pairs. One of each pair is a copy of a chromosome from the female parent and the other is a copy of the chromosome from the male parent. The two chromosomes are very similar. They carry genes for the same polypeptides in the same positions or loci, but they can have different alleles. A cell that has the full complement of paired chromosomes is said to be **diploid**.

8 SEX AND REPRODUCTION

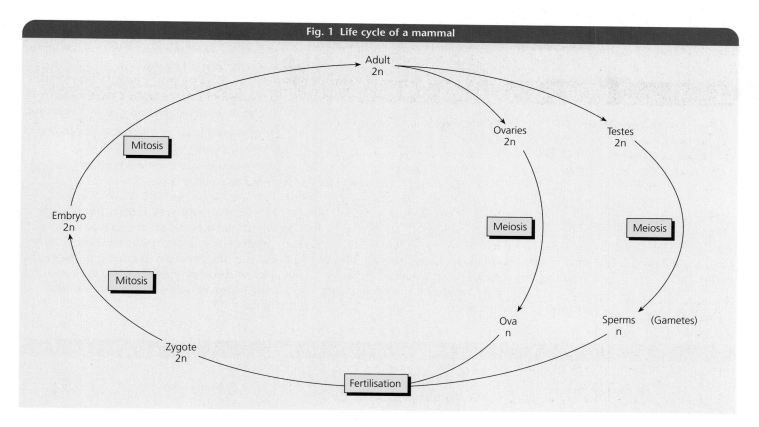

Fig. 1 Life cycle of a mammal

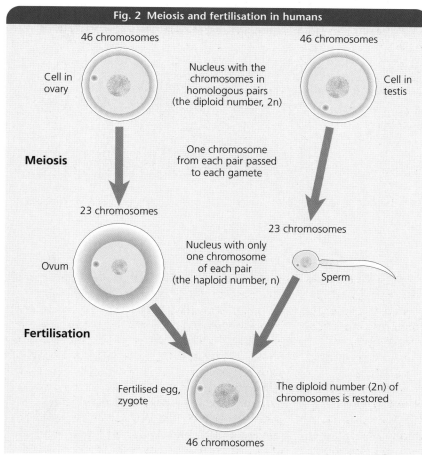

Fig. 2 Meiosis and fertilisation in humans

The diploid number of chromosomes for a species is often described as 2n. For humans the diploid number, 2n, is 46. For a cat, which has 19 chromosomes, 2n = 38.

Cell division by meiosis reduces the diploid number (2n) of homologous chromosomes to the **haploid** number (n) in gametes (Fig. 2). In meiosis, the homologous chromosomes pair up on the equator of the cell, just as the chromatids do in metaphase of mitosis. The key difference is that in meiosis, one chromosome of each pair, rather than just one chromatid, moves towards each pole. You will study the details of meiosis in the Biology A2 book.

Each gamete produced by meiosis gets one chromosome from each homologous pair. When the gametes fuse in fertilisation, homologous pairs reform to give the **zygote**. Meiosis ensures that the number of chromosomes stays constant from generation to generation. If fertilisation took place without first halving the number by meiosis, the chromosome number would double each time that fertilisation occurred. Usually all the cells of an organism are diploid, apart from the gametes. Most organisms have between ten and 50 pairs of chromosomes (Table 1). Packaging large numbers of genes

8 SEX AND REPRODUCTION

Table 1 Numbers of chromosomes

Organism	Diploid number
Human	46
Pea	14
Mouse	40
Maize	20
Fruit fly	8
Cat	38
Barley	14
Onion	16
Horse	64

1 Explain why it is essential that a gamete gets one, and only one, chromosome from each homologous pair.

2 Table 1 shows the diploid number of chromosomes in the cells of some organisms.
a What is the haploid number of chromosomes in a mouse?
b How many chromosomes would you find in a leaf cell of a pea?

into a small number of chromosomes, and reducing the size of the chromosomes before cell division, help to prevent any genes being lost as they are passed on.

Most organisms have special reproductive organs where meoisis takes place and gametes are produced. In mammals, and indeed most other animals, the reproductive organs are the **ovaries** and **testes**. Flowering plants also have ovaries, but male gametes are made in the **anthers**.

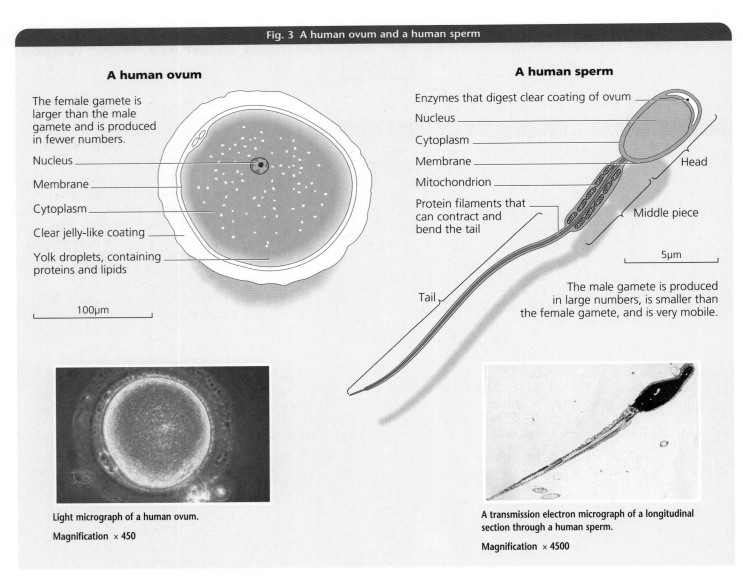

Fig. 3 A human ovum and a human sperm

A human ovum

The female gamete is larger than the male gamete and is produced in fewer numbers.
- Nucleus
- Membrane
- Cytoplasm
- Clear jelly-like coating
- Yolk droplets, containing proteins and lipids

100μm

Light micrograph of a human ovum.
Magnification × 450

A human sperm

- Enzymes that digest clear coating of ovum
- Nucleus
- Cytoplasm
- Membrane
- Mitochondrion
- Protein filaments that can contract and bend the tail
- Head
- Middle piece
- Tail

5μm

The male gamete is produced in large numbers, is smaller than the female gamete, and is very mobile.

A transmission electron micrograph of a longitudinal section through a human sperm.
Magnification × 4500

8 SEX AND REPRODUCTION

Table 2 Male and female gametes

Feature	Human ovum	Human sperms
Size	Over 100 μm diameter.	Very small.
Numbers	Only a few mature from cells already present in the ovaries at birth.	Continuous production of very large numbers.
Movement	Unable to move by itself.	Motile; its tail allows vigorous swimming movement.
Cytoplasm	Has a large amount that contains a store of nutrients (yolk droplets) that nourish the zygote.	Very thin layer around nucleus. No nutrient store.
Mitochondria	Spread throughout cytoplasm.	Concentrated in middle piece to provide energy for tail movements.

Male and female gametes differ from one another in several important respects (see Fig 3 and Table 2). In most organisms the male gametes move, or are moved, to meet the female gamete. Mammalian sperms are **motile**, which means that they can move themselves. Mammalian sperm cells swim towards an ovum by beating their tails. The pollen grains that contain the male gametes of flowering plants have no means of self-propulsion. They can, however, travel vast distances to reach the female gamete when carried by an agent such as the wind or insects to the female reproductive organs of a flower. Pollen is often described as mobile (it can move) but *not* motile.

Since the chance of a male gamete reaching an ovum is quite small, most organisms produce many more male gametes than female. Male gametes are usually very small, with the minimum of cytoplasm, since this reduces the amount of resources wasted on the large proportion that never reach the female gamete. Once the nucleus of the male gamete that reaches the ovum has combined with the nucleus of the female gamete its function is then complete. Female gametes have a store of nutrients in their cytoplasm, which provides for the initial growth of the embryo after fertilisation.

3 Explain how a human sperm is adapted for its function.

4 Many female gametes are larger than their corresponding male gametes.

a What accounts for this difference in size and why is it necessary?

b Suggest the advantages of producing small numbers of non-motile female gametes and large numbers of motile male gametes.

5 Birds' eggs are far bigger than mammals' eggs and they contain large stores of yolk. Suggest an explanation for these differences.

KEY FACTS

- Sexual reproduction involves the fusion of the nuclei of two gametes. Ova and sperms are the female and male gametes in mammals.

- Sexual reproduction combines genes from two organisms. Gametes are produced by meiosis. In this type of cell division each of the cells formed contains only one of each pair of homologous chromosomes, and therefore only one copy of each gene.

- Cells with only one chromosome from each pair are called **haploid**; cells with pairs of homologous chromosomes are **diploid**.

- In most organisms gametes are haploid while body cells are diploid.

- Male gametes are smaller than female gametes, produced in much larger numbers and are motile. In mammals, ova have cytoplasm that contains nutrient reserves.

EXTENSION: Ovaries and testes

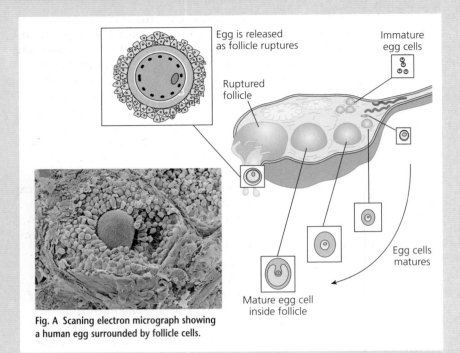

Fig. A Scaning electron micrograph showing a human egg surrounded by follicle cells.

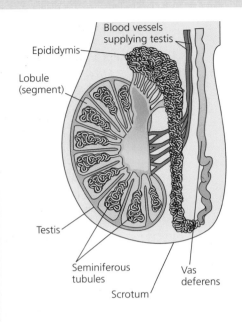

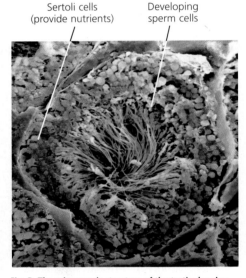

Fig. B The microscopic structure of the testis showing a section through a seminiferous tubule.

In a human female, the two ovaries are situated in the abdomen on either side of the uterus. Membranes attach each ovary to the uterus on one side and to the inner wall of the abdomen on the other. Each ovary of an adult woman contains egg cells in various stages of maturity (Fig. A). These develop from a supply of about a million potential egg cells that form in the ovary of the embryo. Many of these degenerate at an early stage, and by puberty only about 20% remain. From puberty onwards, hormones stimulate about 20 cells to mature each month, although normally only one is released from one of the ovaries. As an egg cell matures it grows much larger than other cells in the ovary and nutrients accumulate in its cytoplasm. The surrounding cells break down and a large, fluid-filled follicle forms around the developing ovum. The follicle protrudes from the wall of the ovary and then finally bursts to release the egg cell. This event is called **ovulation**.

Like the ovaries, testes also develop in the abdomen of an embryo. However, at birth or just before, the testes descend into the **scrotum**, a sac that holds them outside the abdomen. Sperms cannot develop properly at normal body temperature, and keeping them just outside the body, at 2°C below body temperature, is necessary for normal sperm production. Quite why this temperature difference is so critical is not clear. The temperature of the testes in the scrotum is controlled by muscles that draw the testes close to the body in cold conditions and relax to allow them to hang more loosely when warm.

Each testis consists of a coiled mass of narrow tubes called seminiferous tubules (Diagram B). The inside of the seminiferous tubules is lined with cells that develop into sperms, a process that takes about 8 weeks. Meiosis takes place during their production, so that, like the ova, the mature sperms are haploid. Since each testis has over 100 metres of tubules, the number of sperms that a male can produce is enormous. In the human male, sperm production is continuous from puberty until old age. Sperms can be stored for up to a month in the epididymis, but if not ejaculated they degenerate and the nutrients they contain are recycled. A single ejaculation may contain 500 million sperms.

1 Describe two differences between egg and sperm production.

2 There is evidence that wearing tight underpants may reduce sperm production. Suggest an explanation for this.

8.2 Fertilisation

Fertilisation is the process in which the male and female gametes fuse to produce a single cell. This diploid cell is called the **zygote**. During sexual intercourse several hundred million sperms are released into the female vagina. The sperms then have to swim an incredible distance for their size to reach the oviduct that contains the freshly released egg. The human oviducts, also called the Fallopian tubes, are 20 cm away from the vagina. Many sperms start off the journey but few make it. Those that reach the egg do so by a combination of their own swimming activity and the muscular contractions of the uterus walls.

Fig. 4 shows the events that lead to fertilisation. Once sperms reach an egg cell in the oviduct they release digestive enzymes from their **acrosome** (see Fig 3). These enzymes break down the coating that surrounds the egg cell and allows the head of the sperm to penetrate to the cell surface membrane. The membrane has specific protein receptors that combine with proteins in the head of the sperm, allowing the head to penetrate the membrane and to enter the cytoplasm. The cytoplasm immediately produces other enzymes that cause the membrane to thicken and so prevent any other sperm heads getting in. This is important because the egg must not be fertilised by more than one sperm. The entry of the sperm nucleus stimulates the egg cell to complete meiosis. Strictly speaking, only now that its nucleus is haploid has the egg cell developed into the female gamete. Fertilisation is complete when the sperm nucleus fuses with the nucleus of the ovum, producing a zygote with the diploid number of chromosomes.

Only one sperm actually fuses with an ovum, but vast numbers are produced. Many sperm are needed to ensure fertilisation because:

- The vagina is a hostile environment for sperms. The cells lining the vagina produce acid that is lethal to sperms;
- A high proportion of sperms are defective in some way – they have two heads or tails – and they are unable to swim properly;
- Many of the sperms do not get all the way through the uterus. Others enter the wrong oviduct (normally only one ovary ovulates each month);
- Several hundred sperms must accumulate around the egg cell to release sufficient enzymes to digest a passage through to the membrane.

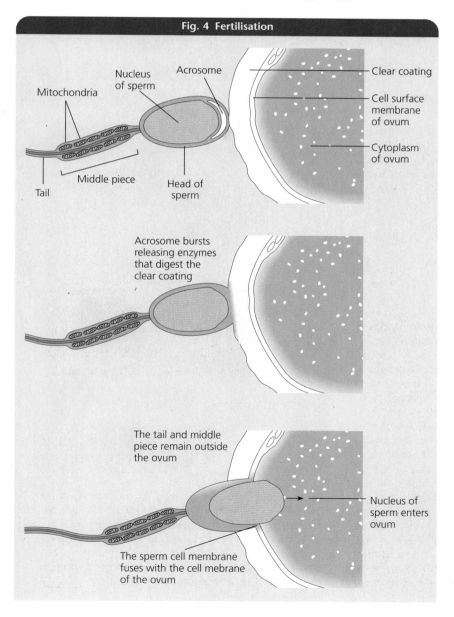

Fig. 4 Fertilisation

6 An ovum is much larger than a sperm. How do the amounts of DNA in the nucleus of each compare?

7 The vagina has a pH of about 3.5 to 4. Suggest the advantage of the vagina being acidic, even though many sperms are killed.

8 Suggest two types of digestive enzyme likely to be released from the acrosome.

123

8 SEX AND REPRODUCTION

KEY FACTS

- Formation of gametes by meiosis, followed by fertilisation, maintains a constant chromosome number from generation to generation.

- After the ovum is released from the ovary, it moves slowly along the oviduct. The sperms, which have limited energy stores because of their tiny cytoplasm, must swim up the oviduct to reach and fertilise the ovum. Many sperm fail to complete the journey.

- Mammalian sperms release digestive enzymes that break down the coating of the ovum and allow one sperm to reach and penetrate its membrane.

- Fertilisation is fusion of the nuclei of male and female gametes. It produces a diploid zygote.

APPLICATION Sperms swim for their life

A sperm swims by moving its tail from side to side. The tail contains a ring of long protein filaments that can contract. Contraction of the filaments on one side of the ring contract causing that side of the tail to shorten, forcing the whole tail to bend. When the filaments on the opposite side contract, the other side bends. This bending one way and then the other allows the tail to push against the fluid in the same way as a tadpole's tail.

The contraction of the filaments requires energy. This comes from the ATP produced by mitochondria in the middle section of the sperm (see Fig 3). Sperms have very little cytoplasm and no reserves of nutrients to act as substrates for respiration. Sugar, however, is provided in the secretions that make up semen. The seminal vesicles secrete a fructose-rich fluid into the urethra as sperms are forced from the testes during orgasm. The female secretes additional fructose into the vagina.

You may find it helpful to refer to earlier chapters in this book to answer some of these questions.

1 What type of sugar is fructose?

2 Explain how fructose fuels the contraction of the protein filaments in sperm tails.

3 As they swim along the oviduct, sperms produce large amounts of lactate. Suggest why.

4 While they are stored in the epididymis of the testis, sperms have no fructose supply. Suggest why it is an advantage not to add fructose to sperms until ejaculation occurs.

EXTENSION *In vitro* fertilisation

Some couples are unable to have a baby. In some cases, this is because the woman's oviducts are blocked, and her partner's sperms cannot reach an egg cell to fertilise it. Such cases of infertility can now be helped because it is possible to remove a mature egg cell from the woman's ovaries and to fertilise it outside the body 'in vitro'. *In vitro* literally means 'in glass', but today, the procedure is usually performed in laboratory grade plastic. After fertilisation the embryo is replaced in the woman's uterus.

The woman receives injections of hormones to stimulate the ovaries to bring several egg cells to maturity. The mature egg cells are removed from the woman's body without major surgery by inserting a thin viewing tube and suction device through the abdomen wall or vagina. Sperms from the male partner are added to the egg and fertilisation takes place under sterile conditions. The zygotes are incubated for about week, and checked for any abnormalities with a microscope. Only the healthiest embryos are chosen for implantation. Usually two embryos

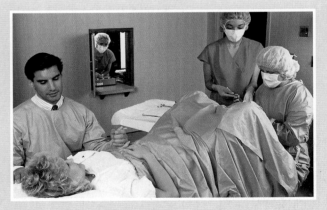

are placed in the woman's uterus, as the chances of one implanting are fairly low. Even so, failure to implant or miscarriage is common and although the technique's success rate has improved steadily during the last 20 years, it is still less than 50%. Unused embryos can be frozen and stored without apparent damage in liquid nitrogen, at -196°C, for possible use at a later date.

8.3 Why is sexual reproduction a good option?

Sexual reproduction is a much more complicated process than asexual reproduction. The fact that it occurs in such a wide range of species shows that it must have major evolutionary advantages. What are these?

Imagine a simple single-celled organism such as a bacterium that has only a single circular chromosome and therefore only one copy of each gene. When this organism reproduces asexually, the two daughter cells produced are identical copies of the parent cell. Variation does not occur unless there is a mutation in one of the genes in the single chromosome. If this mutation is of immediate benefit to the organism, the daughter cell can survive and pass on the mutant form of the gene to its offspring when it reproduces asexually. But if, as is much more likely, the mutant gene is harmful, the daughter cell is unlikely to survive and the mutant gene will not be passed on. A population that reproduces only asexually is therefore unable to maintain a variety of different alleles. Without this variation it is difficult for the population to respond to environmental change.

Now imagine that at some early stage in evolution two haploid cells joined together to produce a cell with two copies of each gene. In a diploid cell there can be one 'normal' gene and one mutated gene. Unless the mutated gene was positively damaging, the diploid cell could still survive and reproduce and the mutant form of the gene could remain in the population. This would tend to increase genetic variation. When these organisms reproduce sexually, they make haploid gametes. Each time gametes join together genes may be combined with new partners. In later generations these new combinations may produce adaptations that make the organsim better equipped to survive as conditions change. This variation enables rapid adaptation and is the great advantage that sexual reproduction has over asexual reproduction.

Although some animals can reproduce asexually, most depend on sexual reproduction to increase the numbers of individuals in their populations.

9a Which stage in Fig. 6 shows gamete nuclei, and which shows zygote nuclei?

b Which nuclei will be diploid and which will be haploid?

c Only one of the nuclei produced in stage 3 grows into a new mould. The nucleus undergoes meiosis before growth of a new hypha starts. Will the nuclei of the bread mould hyphae be haploid or diploid? Explain your answer.

Fig. 6 Bread mould

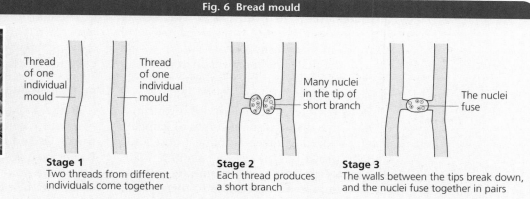

Stage 1 Two threads from different individuals come together

Stage 2 Each thread produces a short branch

Stage 3 The walls between the tips break down, and the nuclei fuse together in pairs

A scanning electron micrograph of the bread mould Rhizopus at the sexual stage.

Magnification × 200.

8 SEX AND REPRODUCTION

EXTENSION

Male exhibitionism

Males compete with each other to ensure it is their gametes that fertilise female gametes. Males of many species expend a great deal of energy trying to maintain sole access to a group of females. Some develop exaggerated features that reinforce their dominance. For example, red deer stags grow huge antlers and then cast them off after each mating season. Many birds, mammals and fish have complex mating rituals. The males that make the most impressive display are the ones that are most likely to father the most offspring. But does this mean that there is simply a rather fruitless arms race between males to produce the most effective display, or is there a benefit for the species as a whole?

Research amongst peacocks suggests that it may be more than just a competition to produce the most dazzling tail. The peacock's tail seems to be an absurdly extravagant structure. It makes flying difficult and so would appear to be a disadvantage. A researcher investigated the possibility that the splendour of the tail is somehow linked with other advantageous genes. She randomly bred males and females, using the size of the eyespots as a measure of the magnificence of the tails. After 12 weeks she weighed the chicks produced. In all cases there was a clear relationship between the size of the peacock's eyespots and the mass of the chicks. Larger-tailed fathers produced heavier, and apparently healthier, chicks. Two years later she found that more of the heavier chicks had survived. Perhaps the peacock's tail is not just a means of grabbing more than a fair share of female attention? Perhaps the female is unconsciously attracted to the fittest male that is most likely to lead to the genes of both partners surviving into the next generation.

A male peacock displaying.

Do we need different sexes?

Sexual reproduction needs two different gametes, but why have male and female? There would be just as much variation if the gametes looked the same. There are organisms in which there is no obvious difference between gametes. For example, in many fungi, such as bread mould (Fig. 6), the male and female gametes look exactly the same, and the organisms that produce them also appear identical.

However, the great majority of organisms do have two distinguishable types of gamete. The function of gametes is to meet, fuse and start the growth of the next generation. Male and female gametes have become specialised to contribute in different ways to this overall task. Male gametes tend to be smaller, motile and produced in larger numbers than the female gametes. They are specialised for the role of moving to the female gametes. The female gametes, on the other hand, are adapted for the role of starting the growth of the next generation – they tend to have the food reserves that enable growth to get going, so they are larger and less mobile than the male gametes.

Female adaptations

Not only do most species have different gametes, many also have marked differences between the males and females. Once again the reason is that each is specialised for a different role. The males produce large numbers of gametes to ensure their gametes stand a good chance of fertilising female gametes. Females are more likely to be responsible for helping the fertilised eggs to start their development. Female mammals have additional adaptations that enable the young to develop internally up to a fairly advanced stage, thus increasing the offspring's chances of survival. Male animals, including men, are biologically necessary only to provide the male gametes that generate the variation that permits adaptation to changing conditions. In human society though, males are often involved in raising their offspring. The ways in which this happens depends on the society and its cultures but a father can hunt for and gather food, he can provide financially and he can care for children directly.

10 Suggest ways in which a human female is adapted for internal development of her child.

Getting the best of both worlds

Some organisms include both asexual and sexual reproduction in their life cycles. The aphid is a good example. Aphids, often called greenflies, are insects that feed by sucking sap from the phloem tubes of plants. They are serious pests of many crop plants. Not only do they take valuable nutrients, they can also carry viral diseases which they can pass to the crop plants as they feed. Aphids survive the winter as fertilised eggs, not as adult insects. Eggs are better able to survive the cold conditions of winter when food is scarce. The eggs all hatch into females when the weather improves and food becomes more plentiful again in the spring. There are no males around, so the eggs that these females produce cannot be fertilised. Instead the eggs remain diploid because they do not undergo meiosis during their development. They grow without fertilisation into females. Throughout the summer, while there are plenty of juicy food plants, this process is repeated,

and enormous numbers of female aphids are produced. In autumn some of the eggs develop into males. These males produce sperms and this time meiosis does occur. At the same time females also produce haploid eggs. Mating and fertilisation take place to produce the fertilised eggs that are going to overwinter.

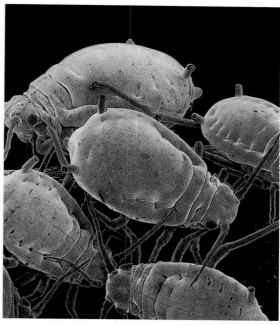

False colour scanning electron micrograph of a group of aphids feedling on a plant stem.

11
a Explain the advantages to the aphid of reproducing asexually in spring and summer.
b Explain the advantage of sexual reproduction in the aphid life cycle.
c Suggest why the spring and summer aphids develop from eggs rather than from sperms.

KEY FACTS

- The main advantage of sexual reproduction is the creation of variation in the offspring.

- Variation in a species provides a significant survival advantage. When environmental conditions change, it is more likely that there will be some individuals that are adapted to the changed conditions, and so the species will not be wiped out.

- In most organisms there is a clear difference between male and female gametes.

- It is an advantage to produce relatively small quantities of female gametes that have food reserves and are non-motile and large numbers of motile male gametes without reserves.

- Some species include both asexual and sexual reproduction in their life cycle. This has the advantage that they can reproduce and spread rapidly in the asexual stage and introduce variation in the sexual stage.

8 SEX AND REPRODUCTION

EXAMINATION QUESTIONS

1 The diagram below shows the life cycle of a moss.

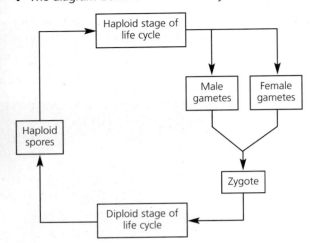

a Copy the diagram and mark it with a cross to show where meiosis occurs. (1)

b A spore of this organism contains 16 chromosomes. How many chromosomes would you expect to find in:
  i) a female gamete? (1)
  ii) a cell taken from the moss during the diploid stage of its life cycle? (1)

c Some DNA was extracted from cells during the haploid stage of the life cycle. It was found to contain 14% adenine.
  i) What percentage of thymine would you expect this DNA sample to contain? (1)
  ii) What percentage of cytosine would you expect to find in this DNA sample? (1)

d Suggest two ways in which the male gametes of this organism are likely to differ from female gametes (2)

BY02 Feb 96 Q5

2 Artificial insemination is the transfer of a semen sample to the female reproductive tract without the use of a male partner, the male acting simply as a donor of sperm. The table below shows some of the characteristics of semen from cattle.

Characteristic	Value
Ejaculate volume/cm³	5
Sperm concentration/number per cm³	1.1×10^9
Percentage of motile sperm	70

a Ten million motile sperm are required per cow for artificial insemination. Use the information in the table to calculate how many cows can be artificially inseminated per ejaculate. Show your working. (2)

b Suggest two advantages of using artificial insemination in cattle. (2)

c Explain how a constant chromosome number is maintained from one generation to the next during sexual reproduction. (3)

BY02 Jun 99 Q2

3 The drawings below show the sperm and eggs of three different animals.

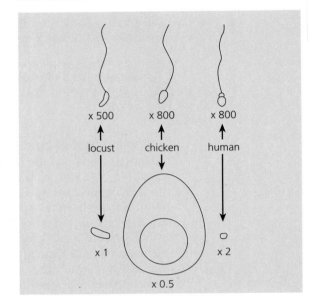

a i) How many times longer than the chicken sperm is the chicken egg? (1)
  ii) Calculate the actual length of the head of the chicken sperm in millimetres. Show your working. (2)

b What is the main advantage to the locust in producing eggs which are much larger than sperm? (2)

c Explain why human eggs are smaller than chicken eggs, even thought adult humans are much larger than adult chickens. (2)

BY02 Jun 97 Q1

KEY SKILLS ASSIGNMENTS

Sperm counts (and other sexy sums)

The composition of semen varies considerably between individuals and at different times. The average man has 78 to 100 million sperm per cm³ of fluid. Typically a man ejaculates between 2 and 5 cm³ of semen containing between 300 and 500 million (3.0×10^8 to 5.0×10^8) sperm. However, normally only about 70 per cent are motile and potentially able to reach an egg cell. If a man's sperm count is less than 20 million in an ejaculate, he may be infertile. About eight per cent of men have fertility problems.

1 A man's sperm count is 84 million sperm per cm³ of semen. He produces 3.8 cm³ of semen in each ejaculate, and on average he ejaculates 4 times per week. Calculate:
  a the total number of sperm in an each ejaculate;
  b the number of motile sperm in each ejaculate;
  c the minimum number of sperm that must be produced in his testes per hour in order to sustain 4 ejaculations per week.

2 A sample of another man's sperm is counted with a microscope. The volume of semen used is 0.004 mm³. The number of sperm counted in this volume is 24. The man produces 3 cm³ of ejaculate. Calculate the number of sperm that this man produces in an ejaculate. Would he be fertile?

3 a Use the diagram of a sperm in Fig. 3 on page 120 to work out the length of a sperm from the tip of its head to the end of its tail.
  b A sperm may swim 20 cm to reach an egg cell. How many times its own body length is this?
  c If a man who was 180 cm tall swam the same number of body lengths, how far would he swim?

4 The table below shows some information about sperm and semen production in male cattle, sheep and horses.
  a Calculate the remaining figures for the last column of the table.
  b Present the data in the table as a chart or diagram, so that the differences between the three animals can be seen clearly.

Animal	Volume of ejaculate /cm³	Mean concentration /million per cm³	Mean total number of sperm per ejaculation
Bull	6	$1\,500 \times 10^9$	9×10^9
Ram	1	2 500	
Stallion	90	200	

  c Suggest why the mean concentration of sperm in the stallion's ejaculate is lower than in other animals.

5 Semen may be collected from a prize bull 4 times per week. Each ejaculate is diluted and divided into portions which are stored in narrow plastic tubes called straws. Each straw contains enough sperm for artificial insemination of one cow. They are frozen and stored in liquid nitrogen. After thawing, assume that only 50% of the sperm are viable. To ensure successful fertilisation of the cow, a straw must contain about 10 million sperm. Calculate how many straws can be obtained from the prize bull each week.

6 At birth a human ovary contains about 1 million potential egg cells. By puberty 80% have degenerated. A woman has a regular 28 day menstrual cycle. She releases only one egg cell in ovulation in each cycle.
  a If this woman has regular cycles for 33 years, and no children, how many egg cells will she release?
  b How many of the potential egg cells present at puberty will not be released?

Infertility

One in 12 couples has trouble conceiving a baby but medical advances in the last 30 years have made infertility less of a problem for some couples. However, assisted reproduction is not without its controversies.

7 The first baby produced by in vitro fertilisation was born in 1978. The term 'test-tube baby' was coined by the Press. Explain why this term is misleading as a description of the method used.

8 *In vitro* fertilisation has raised a number of ethical problems. For each of the following issues, consider your own views and list the arguments for and against:
  a The technique is expensive and rather unreliable. Should it be available free for infertile couples?
  b Sperm can be obtained from any donor, not only the woman's partner. Should the child be told about the donor of the sperm when old enough?
  c Unused embryos that have been frozen may not be required for implantation. Should they be destroyed or could they be used for medical research?
  d Mature egg cells can be obtained from any fertile woman. Women with special attributes, such as super-models, can sell eggs for implantation into other women. Should such practices be legal?

9 Genetic engineering and microbes

About 60 000 people in the UK suffer from insulin-dependent *diabetes mellitus*. The condition is caused by the body's inability to synthesise the hormone insulin. In a healthy person, the islet cells in the pancreas produce insulin in the right quantity and at the right time to keep the concentration of glucose in the blood more or less stable. When the cells of the pancreas are damaged, they cannot produce enough insulin and so control of blood sugar breaks down. Sugar levels can soar to levels that cause damaging dehydration, high blood pressure and loss of valuable glucose into the urine. At other times sugar levels can fall so low that the cells of the body cannot obtain enough glucose for respiration. Brain cells are particularly vulnerable and extreme glucose levels, either too high or too low, can cause a collapse. Both situations can be fatal if untreated.

Today, the blood sugar levels of someone with *diabetes mellitus* can be controlled by regular, carefully measured injections of insulin. Until the early 1980s all insulin for injection came from cattle and pigs. Although it controls blood glucose perfectly well, animal insulin is not exactly the same as human insulin and some people are allergic to it. The development of genetic engineering techniques made it possible to use microbes to manufacture an exact copy of human insulin.

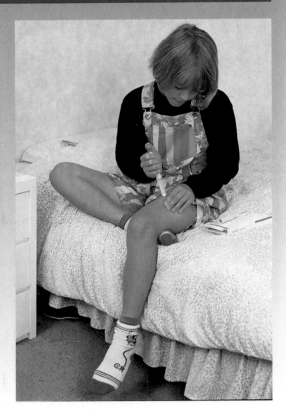

9.1 Manipulating genes

The gene for human insulin was one of the first to be inserted into microbes to manufacture drugs for human use. Other substances now made by genetic engineering include human growth hormone and Factor VIII. Growth hormone is used to treat children that fail to grow properly because of a pituitary gland disorder. Factor VIII is the blood clotting factor that is used to treat haemophilia. Using microbes to make these proteins instead of extracting them from animal or human organs or human blood reduces the possibility of contamination by, for example, viruses that cause AIDS or hepatitis. It also increases drug yield and so makes manufacturing cheaper.

Genes can also be inserted into crop plants to improve their qualities. For example, genes have been added to tomato plants to slow down the ripening process so that the fruit will stay fresh longer. Genes have been transferred to soya bean and maize crops in the USA to make them more resistant to insect pests. Other suggestions for the future include adding genes to plants to enable them to make plastics.

However, there is growing concern about the possible consequences of genetic engineering, especially in crop plants. Might genetically modified foods be a danger to health? Might genes transferred to crops or microbes spread to other organisms and create environmental havoc? Might the widespread use of pest and herbicide resistant crops devastate wildlife? Some of the arguments of opponents of genetic engineering may appear emotive and unscientific at times but the caution they demand may be wise. Indeed, many of the researchers in the field itself urge a careful assessment of the use of the technology. There are serious issues that need to be considered calmly and carefully. In this chapter we look first at how genetic engineers transfer genes from one organism to another, and then we consider some of the possible benefits and dangers of this novel technology.

9 GENETIC ENGINEERING AND MICROBES

The use of genetic engineering in food production has provoked strong feelings in many people.

Transferring genes

Fig. 1 shows the stages involved in removing a gene from an organism such as a mammal and inserting it into a microbe.

Isolating the DNA

To remove DNA from a cell, the cell membrane needs to be disrupted and the nucleus broken open. The method used depends on the type of cell. In eukaryotes, the cell surface membrane and the nuclear membrane both need to be broken open. In plants, the cell wall must also be disrupted. In prokaryotes, the cell wall needs to be broken but the absence of a nuclear membrane makes the second stage easier. One common way to disrupt a cell uses a detergent called **sodium dodecyl sulphate** (SDS). This breaks down cell membranes and cell walls. Once the DNA is free, the surrounding proteins are removed with digestive enzymes.

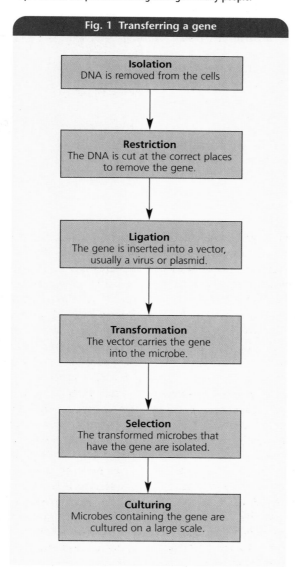

1a Suggest how the detergent breaks down the cell membranes.

b What type of digestive enzyme could be used to remove the proteins in the chromosomes of a human cell?

Cutting up the DNA

Once the DNA has been isolated from the rest of the cell, the part of the DNA molecule that contains the required gene has to be cut out and the rest of the DNA discarded. This is important because genetic engineering must be as precise as possible; only known genes should be transferred to the donor organism.

Genetic engineers isolate genes by using enzymes that cut across DNA molecules at particular positions. These are called **restriction endonuclease enzymes**. Several different

9 GENETIC ENGINEERING AND MICROBES

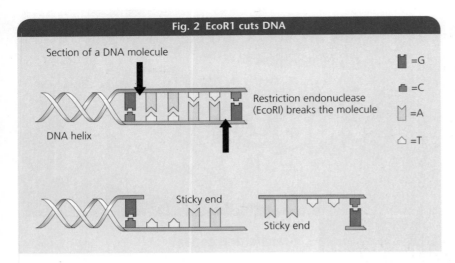

Fig. 2 EcoR1 cuts DNA

Getting the gene into bacterial DNA

The next stage is to insert the isolated gene into a **vector**. A vector is a piece of DNA that can take the gene into the chosen microbe. A common vector is a small circular molecule of DNA called a **plasmid**. Plasmids occur naturally in bacteria in addition to the larger molecule of chromosomal DNA. Genetic engineers find plasmids very useful because these loops of DNA can replicate independently from the bacterial chromosome.

The same restriction enzyme used to cut out the gene from the donor DNA is used to cut open the plasmid DNA. This creates a broken loop of DNA with sticky ends that match those on the donor gene. The donor gene can then be inserted into the plasmid loop using the enzyme **ligase**. Ligase catalyses the **ligation** reaction that joins two sections of DNA together.

restriction enzymes occur naturally in bacteria. Their function is to chop up and destroy the DNA of any viruses that infect the bacterial cell. Each enzyme cuts across the double stranded DNA molecule at a different point in a nucleotide sequence. (Look back at Chapter 6, page 91 if you need to revise the structure of DNA molecules). For example one enzyme, known as EcoR1 cuts the strands only at the sequence shown in Fig. 2. The names of restriction enzymes seem strange when you first come across them but they are actually quite logical. EcoR1 was the first restriction enzyme found in the R strain of the bacterium *Escherichia coli*.

You will see that this enzyme does not slice straight across a DNA molecule. It separates the strands over a stretch of four bases, leaving each part of the broken DNA molecule with a short single-stranded tail. These tails are called **sticky ends** because they are easy to join up with other sticky ends to make complete DNA molecules

2 Use your knowledge of enzymes to explain why EcoR1 only cuts DNA at one particular position.

3 One of the sticky ends produced by cleavage with EcoR1 in Fig. 2 has nucleotides with the bases:

A T T G
T A A C T T A A

a Which bases would attach to the sticky end to make a new DNA molecule?

b A new DNA molecule can only be made by joining this sticky end with a section of DNA with these bases on a sticky end. Use your knowledge of DNA structure to explain why.

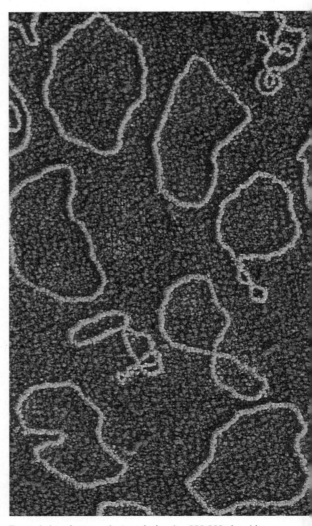

Transmission electron micrograph showing PBR 322 plasmids from Escherichia coli.

9 GENETIC ENGINEERING AND MICROBES

APPLICATION: Restriction endonucleases

The table below shows the base sequences of DNA that are cut by four different restriction endonucleases.

Restriction endonuclease	Cutting points
Bam 1	C↓C T A G G G G A T C↑C
EcoR11	C↓G G A C C G G C C T G G↑C
Hind111	T↓T C G A A A A G C T↑T
Pst1	G↓A C G T C C T G C A↑G

1 Draw diagrams to show the sticky ends produced when each of the restriction endonucleases cuts a DNA molecule.

2 A section of a DNA molecule has the following sequence of bases:

```
T C C G G A C C G A C G T C G G T T C G A A T C
A G G C C T G G C T G C A G C C A A G C T T A G
```

This DNA is treated with a mixture of all four enzymes in the table. How many DNA fragments will be produced? Draw the fragments produced and name the enzymes involved at each cut.

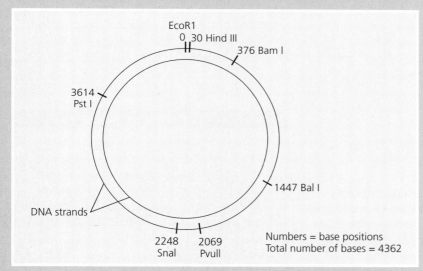

3 The diagram above shows a bacterial plasmid with 4362 nucleotide bases in each strand. The cutting sites of seven restriction endonucleases are shown in one strand. The numbers indicate the position of the base; the cutting site of EcoR1 is counted as 0.

a A genetic engineer incubates the intact plasmid with two enzymes, Bam1 and Pst1. How many bases would there be in the smaller section of DNA that is cut out?

b If the intact plasmid is incubated with all seven enzymes, how many fragments of DNA would be produced. How long would each fragment be?

c One gene in the plasmid extends from base 1876 to base 2134. Which enzymes should be used to remove this gene with as few extra bases as possible?

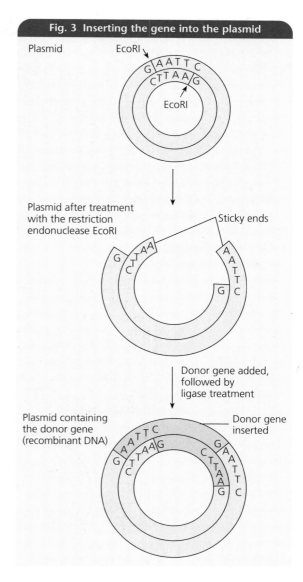

Fig. 3 Inserting the gene into the plasmid

In practice, the DNA from the donor organism and the plasmids from the bacterial recipient are incubated with the same restriction endonuclease in separate tubes for two to three hours to create identical sticky ends. The tubes are then heated to denature the restriction endonuclease. The contents of the tubes are mixed and ligase is added. The sticky ends of the donor DNA join with the corresponding sticky ends in the plasmids. Hydrogen bonds form between complementary bases and the ligase joins up the sugar-phosphate backbone. The new DNA is called **recombinant DNA**.

4 Explain why it is important to denature the restriction endonuclease before mixing the contents of tubes containing donor DNA and bacterial plasmids.

9 GENETIC ENGINEERING AND MICROBES

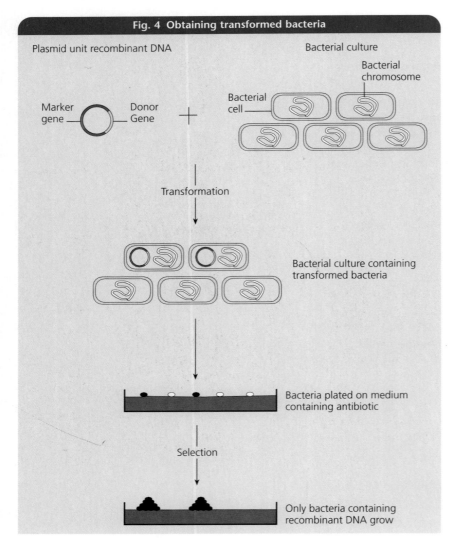

Fig. 4 Obtaining transformed bacteria

Getting the gene into the bacterium

Plasmids containing the donor gene must now be transferred into the microbe. A culture of the intended bacterial recipients is placed in cold calcium chloride solution for about 30 minutes. This changes the cell membranes of the bacteria, making them more permable. Plasmids with the recombinant DNA are added to the culture and the mixture warmed up for a short time. This shock treatment causes some of the bacteria to take up plasmids. Those bacteria that do contain plasmids with recombinant DNA are said to have undergone **transformation** (Fig. 4).

The transformation process is not very efficient and only quite a small proportion of bacteria in the culture will be transformed. The genetic engineer wants to grow only transformed cells so the next step involves identifying and isolating them. One commonly used technique is to insert a **marker gene** into the plasmids, in addition to the donor gene. A marker gene may make the bacteria resistant to a particular antibiotic. If the culture containing the transformed bacteria is grown on a medium that contains the antibiotic, bacteria with plasmids that have the antibiotic resistance marker gene (and the recombinant DNA) will survive and grow better than those that do not.

Replica plating can be then be used to produce several cultures. Fig. 5 shows how a single replica plate is made. The process is repeated several times to eliminate colonies containing non-resistant bacteria that start to grow before the antibiotic takes effect. Replica plating also increases the supply of bacteria that have the added gene.

Another problem involves getting the donor gene to start working to make its product once it has been transferred to the recipient bacteria. Not all genes in a bacterium are switched on all the time. Fortunately, it is possible to overcome this by using **promoter genes** that control the expression of the main gene. These extra genes start transcription and ensure that the mRNA is translated by the ribosomes. The genetic engineer inserts promoter genes, as well as the gene from the donor and the antibiotic resistance gene, into the donor plasmid before transformation.

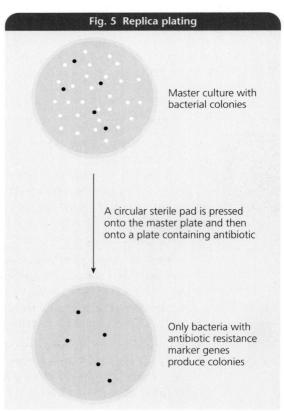

Fig. 5 Replica plating

Obtaining the gene product

The gene product is often required in large amounts so the transformed bacteria are then cultured on a large scale in an **industrial fermenter**. The medium in the fermenter contains all the nutrients that the bacteria need for rapid growth and reproduction and also supplies the oxygen they need for respiration. Fig. 6 shows one type of fermenter that may be used.

In the favourable conditions in the fermenter, transformed bacteria reproduce very rapidly, dividing asexually as often as every half an hour. A huge clone of the bacteria can be produced within days. The plasmids inside the bacteria replicate at each cell division, so the gene inserted in the plasmid is also copied many times. As the bacteria grow the gene within the plasmids synthesises its protein. This product, the object of all this genetic engineering, accumulates and is extracted from the fermenter for commercial use.

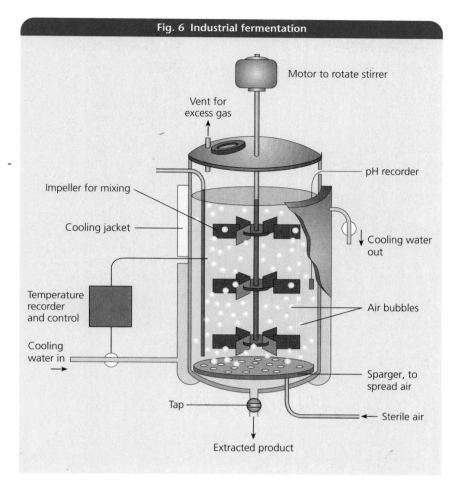

Fig. 6 Industrial fermentation

5a Explain why air is bubbled through an industrial fermenter.

b Explain why the air must be sterilised before it enters the fermenter.

c In large fermenters, the temperature of the bacterial culture rises during the fermentation process, so they have a cooling jacket.

d Explain what makes the temperature rise. What would happen if the temperature of the fermenter was not controlled?

KEY FACTS

- In genetic engineering genes are removed from one organism and inserted into another. Genes that code for useful substances, such as hormones, enzymes and antibiotics, are often transferred into microorganisms, which then produce large quantities of these substances.

- A gene is isolated from the DNA of the donor organism using a **restriction endonuclease enzyme**. This cuts out the relevant section of the organism's DNA, leaving sticky ends that will enable the gene to be inserted into a small circular piece of bacterial DNA called a **plasmid**.

- Plasmids are often used as **vectors** to incorporate the selected gene into bacterial cells. Plasmids occur naturally in cells and replicate independently of the main bacterial DNA.

- The same restriction endonuclease is used to cut the plasmid. This leaves complementary sticky ends to which the selected gene can be attached by another enzyme, **ligase**.

- The plasmids are then introduced into the bacteria, and transformed cells are selected and cloned.

- Genetic markers in the plasmids, such as genes that confer antibiotic resistance, enable genetic engineers to identify bacteria that have successfully taken up the selected gene.

- Transformed bacteria are cultured on a large scale in industrial fermenters and the useful product is then extracted.

9 GENETIC ENGINEERING AND MICROBES

EXTENSION: Extracting human genes

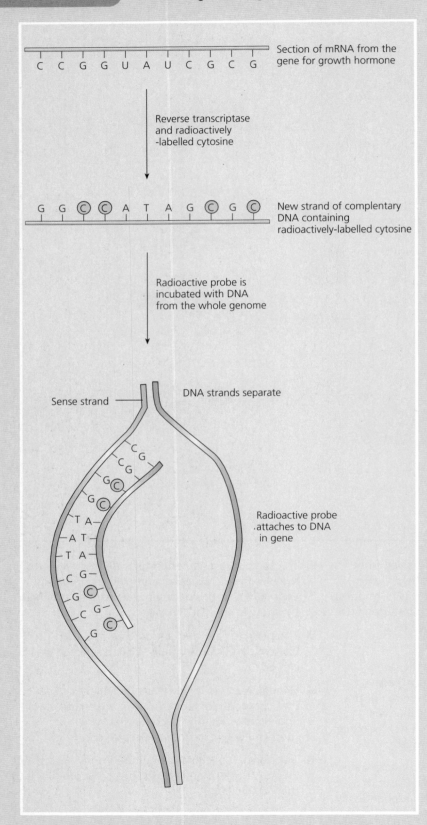

It is not easy to isolate a specific gene from human cells. The first difficulty is finding the gene. Once the human genome has been fully mapped this will be easier. In the meantime genetic engineers use **genetic probes**. A genetic probe is a marker that can reveal the position of a gene in the human genome.

The first step in making such a probe is to find cells that actively produce the gene product. Human growth hormone, for example, is synthesised in the anterior lobe of the pituitary gland. The gene for growth hormone is actively expressed in these cells and their cytoplasm contains mRNA for growth hormone. This mRNA can be extracted and used to make a complementary strand of DNA using an enzyme called **reverse transcriptase**. This enzyme reverses the usual process by which a DNA template is used to make an mRNA copy. It is obtained from viruses that have RNA instead of DNA as their genetic material. Such viruses, including the AIDS virus, HIV, use the enzyme to replicate inside their host cells. Genetic engineers use the enzyme in association with radioactive nucleotides to produce a single strand of radioactive DNA.

Radioactive DNA made from the mRNA produced originally by the growth hormone gene is cultured with DNA from the whole human genome that has been split into its individual strands. The DNA probe attaches to the matching strand of the section of DNA that carries the gene for growth hormone. When the position of the radioactivity is identified, the growth hormone gene has been located.

An alternative method is to use the extracted mRNA to synthesise an artificial gene. The order of nucleotide bases in the mRNA can be determined, and from this the order of bases in the DNA of the gene can be worked out. DNA which has the nucleotides in the correct order can then be made in the laboratory. This method has been used to synthesise the gene for human insulin, which is an unusually short protein with only 51 amino acids in the active hormone. Recombinant gene technology has then been used to incorporate this artificial gene into the bacteria that produce insulin.

1. Explain how the order of bases in DNA can be worked out from the order in the corresponding mRNA.

2. How many bases are in the section of single stranded DNA that corresponds to the 51 amino acids in the active human insulin molecule?

APPLICATION

Genetic engineering and food

Fruit sold in supermarkets is often picked well before it is ripe and transported long distances. It is then ripened artificially just before being sold. It doesn't have the same taste as freshly ripened fruit but it can be displayed and sold for longer as it does not become soft too quickly.

As tomatoes ripen they produce **polygalacturonase**, an enzyme that breaks down the pectin that normally holds the cell walls together. As the cells separate the fruit goes soft and squishy and rots. Genetically modified tomatoes been given an additional artificial gene. The bases on the sense strand of the DNA of this artificial gene are exactly opposite to those on the sense strand of the gene that codes for polygalacturonase. The two strands of mRNA are therefore attracted to each other and bind together to form a double strand. This prevents the normal mRNA joining on to the ribosomes and being translated to make polygalacturonase.

Other parts of the ripening process are not affected, so the flavour of a ripe tomato still develops. However, since the modified tomatoes do not produce polygalacturonase, this happens without the tomato going soft. This means it is possible for growers to leave the tomatoes to ripen naturally on the plant; they can be sure that the fruit will remain in good condition for several days longer than traditional tomato varieties, allowing plenty of time for transport to the shop and a few days of display on the shelves.

1a Draw the mRNA that would be produced by the sense strand of the section of normal gene shown in the diagram.

b Draw the mRNA that would be produced by the sense strand of the section of artificial gene shown in the diagram.

c Explain why the two strands bind together to form a double strand of mRNA.

d Explain how the artificial gene stops the tomatoes going soft.

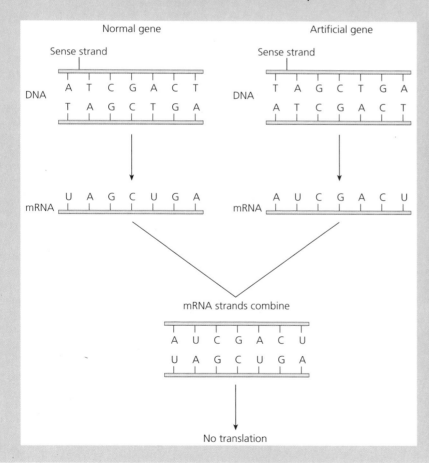

9.2 Investigating DNA

In the early days of DNA research, sequencing a nucleotide fragment or analysing a sample of DNA was a slow process. It was technically difficult and expensive to obtain a small fragment of DNA in large enough quantities to make analysis possible. It could take months to sequence a short section of DNA and extracting sufficient DNA from spots of blood at the scene of a crime was impossible. The whole field of DNA analysis has been revolutionised by the development of a technique that can make a billion copies of a strand of DNA in a few hours.

The polymerase chain reaction (PCR)

The **polymerase chain reaction** (PCR) is like a nuclear chain reaction in that it proceeds at an ever-increasing rate. PCR can amplify tiny amounts of DNA into quantities large enough for scientific analysis. It is the basis of **genetic fingerprinting**, a technique that allows forensic scientists to identify a criminal from a microscopic blood spot or single hair that they leave behind at the scene of a crime. Even dried blood or semen stains that are several years old may contain enough DNA for PCR, enabling an individual to be identified. PCR is the technique that was used in the book *Jurassic Park* by Michael Crichton to obtain the DNA from which the dinosaurs were recreated. Its more realistic uses include amplifying DNA from:

- samples of tissue from extinct animals, such as the Tasmanian wolf, to establish their closest relatives;
- buried human bodies where some soft tissue has been preserved. These studies are helping us to understand the migrations of early human populations.

PCR uses the enzyme DNA polymerase. This enzyme occurs naturally in cells; it catalyses the replication of DNA in the nucleus. The first stage in PCR is to heat DNA until the two strands of the molecule separate. DNA nucleotides are added and the DNA polymerase attaches them to each strand, just as it does in normal replication. The process can then be repeated endlessly. The two new double-stranded molecules make a total of four, then eight, sixteen and so on. PCR can now be managed by machines so that each cycle of replication takes only a matter of minutes. Fig. 7 shows the stages in a single PCR cycle.

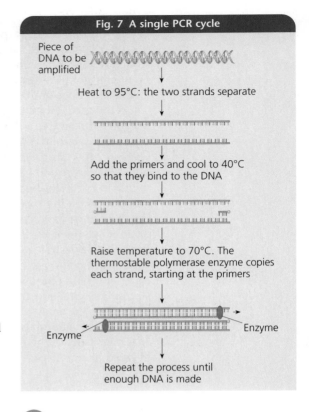

Fig. 7 A single PCR cycle

6a Look at Fig. 7. How many different types of DNA nucleotide must be added to the mixture in the final stage? Why?

b Suggest why using DNA polymerase from bacteria living in hot springs is an advantage.

c How many cycles of the process would be needed to produce a million copies of a DNA molecule from one?

Analysing samples of DNA

Once a large enough sample of DNA has been generated by PCR, it can be analysed and compared with samples from known sources. This allows us to compare the DNA from different species, or from individuals from the same species, or even from members of the same family. Forensic scientists often have to compare the DNA from samples found at a crime scene with the DNA from suspects. In every case, the first stage in the process of analysis is to cut the DNA into short lengths with restriction endonuclease enzymes (see page 131). This produces DNA fragments of varying lengths, because the sequences of bases where the enzymes cut occur at irregular intervals along the DNA molecules.

9 GENETIC ENGINEERING AND MICROBES

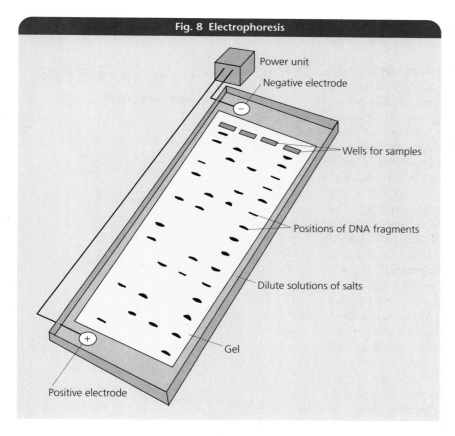

Fig. 8 Electrophoresis

The DNA fragments are then separated by electrophoresis (see page 22, Chapter 1) (Fig. 8). The mixture of DNA fragments is placed at one end of a long piece of agar gel in a trough containing a dilute solution of ionic salts. Electrodes are placed in the solution at either end and a voltage is applied. The phosphate groups in the fragments of DNA give them a negative charge, so they are attracted through the gel towards the positive electrode. The smaller fragments move more rapidly than the larger ones, so the different sized fragments are separated in much the same way as in chromatography (see page 70, Chapter 5).

The pattern of fragments is a sort of 'fingerprint' but to be able to compare one sample with another, the fingerprint must be made visible. The pattern of DNA fragments in the gel is first transferred to a nitrocellulose sheet, which binds DNA. This nylon-based sheet is placed on top of the gel and then a stack of paper towels is used to press it down onto the gel (see the diagram below). The gel and nitrocellulose sheet are left in place for several hours, usually overnight. The DNA fragments in the gel transfer to the nitrocellulose sheet to form an 'imprint' of the pattern of the fragments in the original gel.

A probe labelled with radioactivity is then used to reveal the position of the bands on the sheet that contain the DNA sequence that you are interested in. A general probe can be used to make all the bands show up or a specific probe can be used. A specific probe is a piece of single-stranded DNA that is complementary to the base sequence of the specific stretch of DNA. The nitrocellulose sheet is incubated with the probe in a sealed plastic bag containing a buffer solution. The sheet is removed and placed next to an unexposed piece of photographic film. The radioactivity in the probe causes a band to show up on the film.

7a Explain why a restriction enzyme cuts a DNA molecule at specific positions.

b PCR is used to amplify a section of DNA. When this amplified DNA is treated with a restriction enzyme, eighteen DNA fragments of different lengths are always detected by electrophoresis, no matter how many different samples are run. Explain why the samples have a fixed number of DNA fragments.

c Phosphate groups give the DNA a negative charge. Explain why DNA fragments contain phosphate groups.

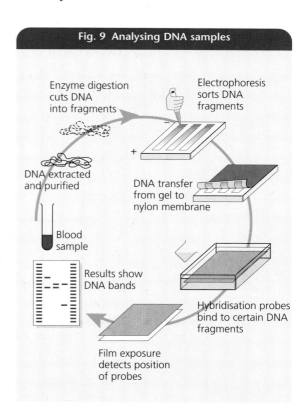

Fig. 9 Analysing DNA samples

9 GENETIC ENGINEERING AND MICROBES

APPLICATION: DNA sequencing

The order of nucleotide bases in a fragment of DNA can be found using several methods. Dideoxy sequencing (also called chain-termination or Sanger method) uses enzymes and radioactively labelled bases to synthesize DNA chains of different lengths. Four separate reactions are carried out and DNA replication is stopped in one tube at cytosine, at one tube at guanine, at one tube at adenine and at one tube at thymine. The final base in each case is radioactively labelled. The resulting fragments in each tube are run on a gel that separates them according to their size. The shorter fragments moving faster and appearing at the bottom of the gel. An autoradiograph is produced by placing a photographic film next to the resulting gel for several hours as shown. Sequence is read from bottom to top. The diagram shows the result of a sequencing gel carried out for stretch of DNA containing only 12 bases.

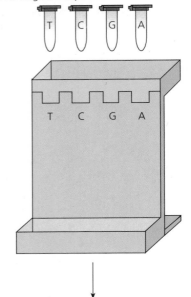

1. Sequencing reactions loaded onto polyacrylamide gel for fragment separation

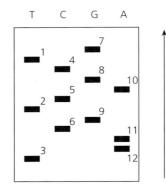

2. Sequence read (bottom to top) from gel autoradiogram

1a Look at the autoradiograph. Which of the labelled patches at positions 1 to 12 contain sections of DNA with only one nucleotide? Give a reason for your answer.

b How many cytosine bases did the original DNA strand contain?

c What was the order of the 12 bases in the original DNA strand?

d The original DNA was a section of the sense strand. What would be sequence of the mRNA strand that it would transcribe?

KEY FACTS

- DNA can be replicated artificially by the polymerase chain reaction. The enzyme DNA polymerase is used to make new double stranded DNA by synthesising a new complementary strand to a pre-existing strand, just as in natural replication.

- PCR is an amplification reaction that is self-sustaining which makes it possible to synthesise large numbers of copies of very small samples of DNA.

- DNA fragments can be separated by gel electrophoresis. A voltage is applied to the gel and the negatively charged DNA fragments move towards the positive electrode. Smaller fragments move faster than large ones.

- The bands of DNA can be seen if radioactive nucleotides are used in the PCR. The pattern of bands in the gel can be made visible by placing the gel next to a sheet of unexposed photographic film overnight. The radioactive bands cause the film to turn black.

EXAMINATION QUESTIONS

1 Tobacco plants do not grow well in salty soil. Scientists have used genetic engineering techniques to insert a gene for salt resistance from the bacterium *Escherichia coli* into the DNA of tobacco plants.

a Describe how scientists could
  i) remove the gene for salt resistance from the DNA of the bacterium *Escherichia coli*. (2)
  ii) insert this gene into the DNA of a tobacco plant. (2)

b Briefly describe how you could test whether these genetically engineered tobacco plants do better than normal tobacco plants in salty soil. (3)

BY02 Mar 99 Q3

2 Scientists have shown that kidney beans are resistant to cowpea weevils and adzuki bean weevils, two of the most serious pests of African and Asian pulses (vegetables related to peas and beans). This is because the beans produce a protein that inhibits one of the weevils' digestive enzymes. Weevils that eat the pulses soon starve to death. The researchers have identified the gene that produces the inhibitor and removed it from the kidney bean DNA. They inserted the gene that produces the inhibitor into the DNA of a bacterium called *Agrobacterium tumefaciens*. Using this bacterium, they have been able to add the inhibitor gene to peas. They hope soon to be able to add the gene to African and Asian pulses.

a Describe how scientists could:
  i) remove the gene that produces the inhibitor from kidney beans. (2)
  ii) insert this gene into the DNA of a bacterium. (2)

b The DNA in the bacterium is able to replicate to produce many copies of itself for insertion into pea cells.

Describe the structure of a DNA molecule and explain how this structure enables the molecule to replicate itself. (3)

BY02 Jun 95 Q8

3 One technique used to produce human insulin by genetic engineering involves inserting a gene for human insulin into the DNA of a bacterium.

a Name the enzyme which would be used to
  i) cut the bacterial DNA;
  ii) insert the DNA for human insulin into the cut bacterial DNA. (2)

b There are 51 amino acids in insulin, made up of 16 amino acids of the 20 that are coded for by DNA.

What is the minimum number of different types of tRNA molecule necessary for the synthesis of insulin? Explain your answer. (2)

c The base sequence below is part of the DNA sequence which codes for insulin.

CCATAGCAC

  i) Write down the corresponding mRNA sequence.
  ii) A mutation occurred which replaced guanine in this DNA sequence with a different base. Explain the possible effects of this mutation on the structure of the insulin molecule. (3)

BY02 Jun 96 Q6

Genetically modified foods

Rapid developments in genetic engineering are making it possible to alter living organisms in ways that were not dreamt of even a few years ago. It is, in theory, possible to transfer genes from almost any organism to any other. Scientists are investigating a myriad of potentially useful applications of DNA recombinant technology, not least ways of increasing crop yields and improving food supplies. However, many people have become suspicious of these developments and concerned about the possible ways in which the technology could go horribly wrong. Some people think its a very bad idea to tamper with natural food sources and they feel that new technology is being imposed without adequate safeguards by companies who just want to make money. Some people are worried that eating genetically modified (GM) foods may in itself be harmful. Others are concerned about the impact of GM crops on the environment.

The arguments for and against
Consider the following summaries of some of the arguments put forward by advocates and opponents of GM foods.

"Foods made from GM crops may be a danger to health."

Anti It is unnatural to swap genes from one species into another. The food we eat contains foreign genes, perhaps from bacteria or even humans, and we don't know what we are eating. The new combinations of genes may have unknown effects, such as making poisonous or carcinogenic substances. Also, when a foreign gene is inserted into a crop plant additional genes are normally added too. These might stimulate other genes in the crop to make harmful products. GM crops modified to produce substances intended to protect them, for example from insect pests, may have dangerous long-term effects.

Pro Genes are not harmful to eat. We eat large amounts of nucleic acids in our food. These are all digested, so we could not somehow incorporate foreign genes from food into our body. All crop plants have been genetically altered by selective breeding, so no food plants are naturally occurring wild plants. Genetic modification speeds up what has been a slow and expensive process of cross-breeding. Genetic modification is precise; only a small number of genes are transferred. Cross-breeding mixes large numbers of genes with unpredictable effects. Food products are extensively tested for safety before being sold, and this includes checking for the presence of possibly harmful products.

"Genes might be transferred from GM plants to others, perhaps creating 'superweeds'."

Anti Genes can be passed from one organism to another, and from one species to another. Bacteria regularly exchange genetic material, including plasmids. Genes in a crop that protect it from the effect of weedkillers may be passed to other plants, including weeds, and these might also become resistant to weedkillers. Pollen from GM crops may be distributed by wind or insects for long distances, and natural cross pollination may pass on the transferred genes to non-GM crops, or possibly produce harmful hybrids. Organic farmers are particularly concerned about cross-contamination.

Pro This is no different from the interbreeding that occurs anyway. Weeds containing genes that confer resistance to a particular herbicide would have no competitive advantage except in environments where the herbicide is being applied. Other weedkillers could still be used, so the weed could not become some sort of uncontrollable monster. Cross contamination of crops can be avoided by separating organic and GM crops by adequate distances.

9 GENETIC ENGINEERING AND MICROBES

KEY SKILLS ASSIGNMENTS

"Widespread planting of GM crops will seriously affect wildlife in the environment."

Anti Pesticides produced by genes in crops to protect them will kill not only crop pests but other useful insects. This will remove pollinating insects such as bees and therefore affect pollination of fruit trees as well as wild flower populations. The loss of wild plants and insect food for other species such as birds will also cut down the diversity of wildlife in the countryside.

Pro The use of resistant crops will reduce the amount of pesticide that farmers need to spray on their crops. Only insects that actually feed on a particular crop will be affected and there will be less environmental damage to neighbouring habitats. Reduced diversity of wildlife is an unavoidable consequence of modern intensive farming practices in which large areas are planted with the same crop, and the use of GM crops would make no difference.

"GM crop seeds are more expensive. This would affect the livelihood of poorer farmers, especially in developing countries."

Anti The use of advanced biotechnology in agriculture will make smaller farms less economical and thus favour large-scale agro-business and the companies that develop and patent particular GM crop varieties. The use of 'terminator genes' that prevent germination of the seeds from a crop will hit poorer farmers particularly hard, because traditionally they keep a proportion of their seeds to plant the following season. Restrictive practices by large GM companies, such as making crops resistant to a particular weedkiller that only one company sells, may lead to inflated prices for the weedkiller and excessive profits for the company.

Pro The development of GM crops with improved quality, greater yields and reduced wastage due to pests will be of benefit to all and will increase food supplies for an ever-increasing population. The technology allows crops that flourish in difficult environments, such as saline or very arid soils, to be developed. These will be of great benefit to poorer countries where population expansion is necessitating the use of more marginal land. Competition between companies will maintain costs at realistically low levels.

GM foods: you decide

1 Use information from the above summaries and from additional research of your own to prepare a presentation either in favour of or against the use of genetically modified foods. Your presentation should include at least one poster or over-head projector transparency to illustrate your argument. Try to ensure that your presentation follows a logical sequence and that the technical points are clearly understandable to a non-specialist audience.

Try out your presentation on a group. Ask them for feedback on your performance and use this to improve your presentation. You may even change your views about genetically modified foods as a result of this feedback.

10 Genes and Medicine

In 1993, Tracey was probably the most famous sheep in the world. She was the star performer in a family of sheep produced in the laboratories of Pharmaceutical Proteins in Edinburgh. When she was just an embryo, scientists had transferred human genes into her cells. When she grew to adulthood, Tracey was able to make milk containing 35 grams per litre of a substance called human alpha-1-antitrypsin (AAT). AAT is used to treat some people suffering from the lung disease, emphysema, and can also be useful for some children who have cystic fibrosis. It can be extracted from human blood, but only in very small amounts, so the 35g per litre available in Tracey's milk is a positively bountiful supply. A single flock of about 2 000 ewes like Tracey could satisfy the needs of every hospital in the world.

Apart from her ability to secrete AAT in her milk, Tracey was a completely normal ewe and she gave birth to two healthy lambs, as you see in the photo. Moreover, one of these lambs inherited her capacity to manufacture AAT, showing that the transferred gene could be passed on from one generation to the next. This is a great advantage compared with having to go through the complex procedures necessary to introduce the gene into every individual embryo in order to obtain extra sources of AAT.

Tracey, the world's first transgenic sheep, with her two lambs.

10.1 Genetic technology

In this chapter we explore how genetic engineering can be used to produce animals that are sources of medicines to treat human diseases. We also look at the area of **gene therapy**. This is the field of medicine that aims to treat people with genetic disorders by giving them a copy of a healthy gene to overcome the problems produced by their mutated gene. We also consider some of the implications of this technology for the future.

Genetically modified animals

Animals that have been given genes from another species are called **transgenic organisms**. Transgenic mammals are particularly useful because they can express human genes. It is not always possible for bacteria to express human genes. Prokaryotic bacterial cells do not have the ribosomes and other cellular machinery necessary to produce complex mammalian proteins.

The AAT gene that was given to Tracey codes for a protein that consists of 394 amino acids. It is actually a **glycoprotein**, a protein with an attached sugar group. The normal function of the AAT glycoprotein is to inhibit an enzyme called elastase that is produced by some types of white blood cell. Elastase, if not inhibited, breaks down elastic tissue in the lungs, causing emphysema. In emphysema the walls of the alveoli disintegrate, so that much less oxygen is absorbed into the blood and fluid leaks into the airspaces. The fluid disrupts the normal functioning of the lungs and can encourage infections which could make a weakened patient even worse. Emphysema can be caused by a number of different factors; having the Z allele for AAT is quite rare. The much more common causes include smoking and working with fine dust particles. Many coal miners and quarry workers suffer from emphysema as a result of their jobs.

There are many different alleles of the AAT gene. About 3% of people in the UK have the Z allele. The Z allele codes for a version of AAT that differs from the normal form by only one amino acid at one position. However, this defective AAT cannot function and patients with Z alleles frequently develop emphysema. Affected people can be treated with an aerosol spray containing AAT. When inhaled, this stops the breakdown of the alveoli and can help patients' breathing immensely. However, it does not provide a permanent cure for the disease. Patients must continue the treatment for the rest of their lives.

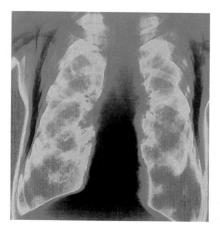

The lungs of an 83-year old man with emphysema, which shows up as red patches on this X-ray.

1a The DNA of the Z allele differs from the normal allele by just one nucleotide base. What type of mutation could have caused this difference?

b How many bases does the gene for the complete AAT protein contain?

c Suggest why the AAT protein produced by the Z allele does not inhibit elastase.

d Explain why treatment with an aerosol containing AAT does not provide a permanent cure for emphysema.

e Could treatment with AAT help someone whose emphysema is caused by smoking? Give reasons for your answer.

Making Tracey

How was Tracey persuaded to produce the AAT protein in her milk? The human gene was identified and isolated from a culture of human cells using the techniques described in Chapter 9. The gene was then combined with a promoter sequence that allows the gene to be expressed in sheep mammary glands. The protein can be collected easily and in large amounts from the animal's milk, without causing distress to the animal.

Mature egg cells were removed from the ovary of a sheep and fertilised *in vitro* (Fig. 1). The AAT gene and its promoter sequence were injected into the nucleus of the fertilised egg cells. Once the zygote had divided to form a small embryo, this was placed in the uterus of a sheep surrogate mother. Not all attempts at establishing a pregnancy were successful but Tracey demonstrated that the technique worked. Since then other transgenic animals have been produced with genes that manufacture other useful products in their milk, such as the blood-clotting protein Factor VIII, which is used to treat haemophilia.

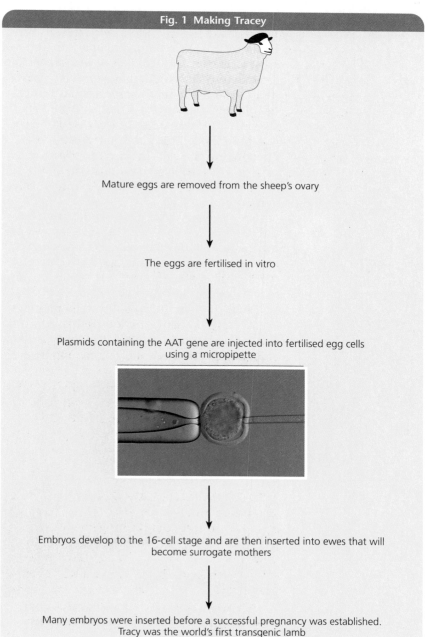

Fig. 1 Making Tracey

Mature eggs are removed from the sheep's ovary

The eggs are fertilised in vitro

Plasmids containing the AAT gene are injected into fertilised egg cells using a micropipette

Embryos develop to the 16-cell stage and are then inserted into ewes that will become surrogate mothers

Many embryos were inserted before a successful pregnancy was established. Tracy was the world's first transgenic lamb

10 GENES AND MEDICINE

> **KEY FACTS**
>
> - Animals can be genetically modified to produce substances that are useful for treating human diseases. Animals that have been given a gene from another species are called **transgenic organisms**.
> - The human gene that codes for the required protein is isolated from human cells and is then injected into a fertilised egg from the animal.
> - The tiny embryos that develop are placed in the womb of surrogate mothers, which later give birth to young that carry the human gene in their cells.
> - Transgenic sheep are used to produce alpha-1-antitrypsin (AAT), a protein that is used to treat emphysema in people who have a mutation in their AAT gene. The sheep secrete the AAT in their milk.

10.2 Gene therapy and cystic fibrosis

Using genetically modified animals to produce substances that can treat people who are unable to make the substances themselves is a great step forward. But another approach is to insert the relevant gene into the human organ where the substance is needed. In the case of AAT, this would mean putting healthy AAT genes directly into the cells within lung tissue of a patient so that they can produce the substance for themselves. This type of procedure is called **gene therapy**.

Gene therapy is being developed to treat the genetic disease **cystic fibrosis**. Most people have a gene that produces a protein, called the cystic fibrosis transmembrane regulator (CFTR). This has a complex molecule consisting of 1480 amino acids, and it is one of the essential channel proteins in cell membranes (see Chapter 2). The function of this particular channel protein is to transport chloride ions through the cell membrane.

Sally, the girl in the photograph below, suffers from cystic fibrosis, because she has a mutation in both alleles of her CFTR gene. The effect of this mutation is that one of the 1480 amino acids in her CFTR protein is missing. This affects the three-dimensional shape of the channel protein and it cannot transport chloride ions through a membrane. The mutation affects all the cells in Sally's body, but it is a particular problem in her lungs. The epithelial cells lining the airways in the lungs normally pass chloride ions out and absorb sodium ions (Fig. 2). However cells with CFTR produced by the mutant genes retain chloride ions, and the increased concentration of ions in the cells causes them to retain water that normally leaks out of the cells into the thin layer of mucus that lines the airways.

With less water the mucus layer becomes more sticky and thicker and it is not easily moved out of the lungs by the beating cilia that line the airways. For this reason Sally has to be helped to expel the excess mucus from her lungs. If her airways become blocked with mucus, she will get breathless because her lungs cannot absorb enough oxygen. The sticky mucus also traps bacteria and cystic fibrosis patients are prone to chest infections. These can lead to pneumonia and bronchitis. Although these can be treated by antibiotics, repeated infections can make a child weak and can cause scarring of the delicate lung tissue. The accumulation of mucus also affects the digestive system as the duct linking the pancreas to the duodenum tends to become blocked, preventing pancreatic juices reaching food in the gut (see Chapter 5).

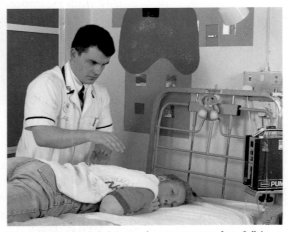

The physiotherapist is helping to clear excess mucus from Sally's lungs, a process that has to be carried out two or three times every day.

10 GENES AND MEDICINE

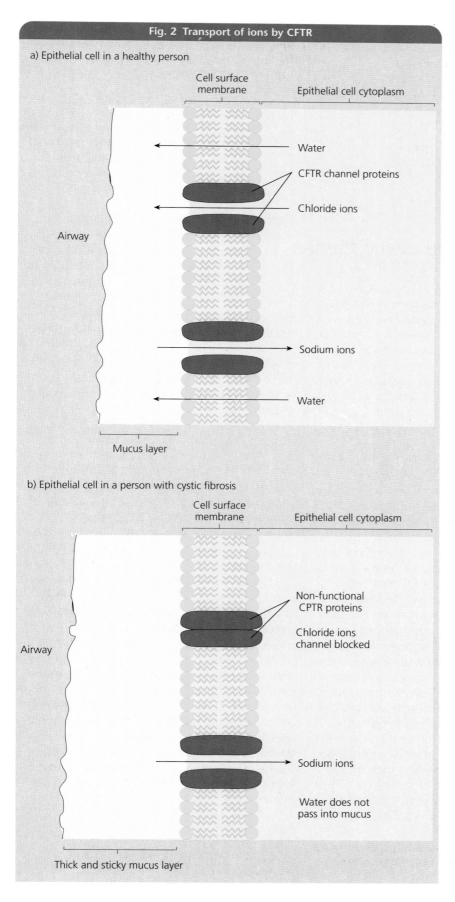

Fig. 2 Transport of ions by CFTR
a) Epithelial cell in a healthy person
b) Epithelial cell in a person with cystic fibrosis

2a Suggest how an increased concentration of ions in the cells reduces the amount of water that they lose.

b Explain how blockage of the pancreatic duct might affect digestion.

Getting healthy CFTR genes into cells

Gene therapy for cystic fibrosis involves inserting functional CFTR genes into epithelial cells of the lungs to replace the defective genes. Two techniques are being tested.

- wrapping CFTR genes in lipids that can be absorbed through the cell surface membranes
- inserting CFTR genes into harmless viruses that are then allowed to 'infect' the cells

The first method uses tiny lipid droplets called **liposomes**. Liposomes can fuse with the phospholipid membranes that make up a cell surface membrane and so carry genes into target cells. CFTR genes are isolated from healthy human tissues, cloned and inserted into plasmids using recombinant DNA technology as described in Chapter 9. The genes are then inserted into liposomes. The next hurdle is to get the liposomes into the area of the body where the genes are required. Aerosol sprays are used to get the liposomes into the lungs, in much the same way as inhalers are used by asthma sufferers. In the lungs, the liposomes fuse with the surface membrane of the epithelial cells, and the genes are transported across the membrane into the cells. The technique is not without its difficulties. One problem is developing an aerosol that can product a fine enough spray to get the liposomes through the very narrow bronchioles of the lungs and into the alveoli. Another is that only a small proportion of the genes that are absorbed into cells are actually expressed. These problems must be overcome before gene therapy for cystic fibrosis can become widely available.

3 In one clinical trial, doctors used a gene probe to test for the presence of CFTR mRNA to find out if CFTR genes absorbed into cells were being expressed. Explain how this would confirm that the genes were working.

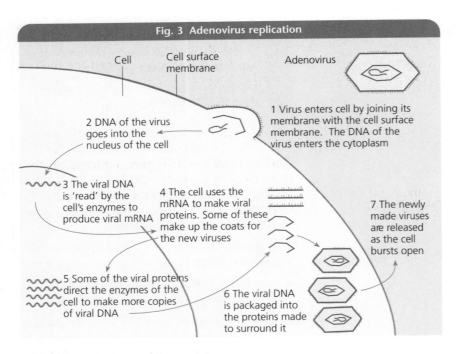

Fig. 3 Adenovirus replication

Adenovirus vectors
In the second method, viruses called **adenoviruses** are used. Adenoviruses reproduce themselves by injecting their DNA into host cells. The viral DNA uses the cell's enzymes and ribosomes to replicate and produce copies, which then reconstruct the rest of the virus before being released from the cell.

Adenoviruses infect the cells of the airways in the lungs, and they are adapted for replication in the epithelial cells. They normally cause colds and other respiratory diseases. The adenoviruses used for gene therapy have been modified so that they infect cells but do not cause disease. The modification involves disabling the genes that allow the virus to replicate. Modified adenoviruses are cultured in epithelial cells grown in the laboratory and are exposed to plasmids that have the normal CFTR gene. The CFTR gene is incorporated into the viral DNA. The adenoviruses containing the CFTR gene are then extracted from the epithelial cells, purified and sprayed into the lungs. Here they infect the epithelial cells, taking the CFTR gene into the cells. The CFTR channel protein is synthesised as normal, but as the viruses can not replicate, they do not damage the cells.

Because adenoviruses are well adapted for entry into lung cells they are a highly efficient vehicle for getting the CFTR gene into target cells. However, there are some concerns about possible dangers of infection by adenoviruses. Infection could occur if the genes that allow the virus to replicate have not been completely inactivated. Also patients treated repeatedly may develop antibodies that make them immune to the viruses and so resistant to the treatment.

 4 Use the information above to compare the advantages and disadvantages of each of the two techniques for getting healthy CFTR genes into lung epithelial cells.

Perfecting gene therapy techniques
Neither of the gene therapy techniques described provides a permanent cure for cystic fibrosis, but both have the potential to alleviate its most distressing symptoms and improve patients' quality of life. Before using either method on people, research workers had to carry out careful testing to make sure that gene therapy is safe. The first tests, carried out in laboratory animals such as mice, had the following objectives:

- to find the best ways of getting the gene into cells;
- to check that the gene was expressed in the target cells;
- to detect any ill effects of the therapy itself.

Following encouraging results in animals, the next step was to start testing the gene delivery systems in people. Although mammals are similar in many respects, each species is different and scientists needed to check that the techniques that worked with mice, also worked in humans. This involved human volunteers, who, for example, had small doses of liposomes containing the recombinant gene applied to the lining of the nose. The nose was chosen because the epithelial cells here are much easier to study than cells deep in the lungs. The next stages, which are still on-going, involve clinical trials with cystic fibrosis patients. Many trials need to be analysed before gene therapy for cystic fibrosis can become an officially approved treatment, available to all CF patients. All of these very necessary precautions take time, but progress in the basic science underlying gene therapy continues at a rapid pace.

5 Explain why gene therapy does not provide a permanent cure for cystic fibrosis.

KEY FACTS

- Cystic fibrosis is a genetic disorder caused by a mutant allele that produces a defective form of the channel protein, called CFTR. This protein normally transports chloride ions out of cells.

- The defective CFTR protein causes chloride ions to build up in cells. This causes those cells to retain water. In the lungs and intestines this is a particular problem. Water fails to pass into the mucus that lines the airways and gut, causing the mucus to become thick and sticky. In the lungs, this leads to breathing difficulties and the risk of infection; in the gut, the mucus blocks ducts that carry digestive enzymes.

- Some severe genetic disorders can be treated by gene therapy. Healthy genes are cloned and then transferred to target cells in the body to take over the function of defective genes that cause the disorder.

- Two forms of gene therapy are being developed to treat cystic fibrosis. In the first, healthy CFTR genes are inserted into liposomes, which fuse with cell membranes and take the genes into the cells. In the other, harmless viruses are used to insert the CFTR genes into the cells.

APPLICATION

Genetic testing

Genetic testing can be used to find out whether a person's cells carry normal CFTR genes or defective, mutant CFTR genes. This is pointless for people who suffer from cystic fibrosis – they must have the mutant genes. Genetic testing can be useful for a couple who are planning to have a child; if the test reveals that they are both carriers of one mutant CFTR allele, there is a one in four chance that they will have a child that has cystic fibrosis. If they decide to go ahead with a pregnancy, the fetus can also be tested. Fetal cells are obtained by amniocentesis or chorionic villus sampling, two ante-natal screening tests that are performed routinely. Amniocentesis is performed at between weeks 15 and 22 of the woman's pregnancy. A small amount of the fluid that surrounds the fetus is removed. This contains cells that have come from the fetus. Chorionic villus sampling is done earlier, at about weeks 10 to 12 of the pregnancy. This method involves removing a small piece of placenta, the organ which links the fetus with the mother and which has the same genetic makeup as the fetus.

Once the parents know the results, they can decide whether to go ahead with the pregnancy. This is never an easy decision but it does provide the family with some choice about their future and the future of their child.

1 About 1 in 2000 babies in the UK are born with cystic fibrosis. Do you think genetic screening of fetuses could be used to eliminate cystic fibrosis? Explain your answer.

2 In theory genetic screening of a fetus could be used for each of the following:
- to detect Tay-sachs disease; a genetic disease for which there is no treatment, and which causes serious brain damage, paralysis and death in early childhood.
- to detect a genetic condition which results in deafness.
- to detect the possession of a gene which increases the chances of developing breast cancer in later life.
- to choose the hair or eye colour of a child.
- to choose the gender of a child.

a In which of these situations, if any, do you think it is acceptable to screen the genetic make-up of the fetus?

b How would you justify where to draw the line?

10 GENES AND MEDICINE

EXTENSION

Gene therapy for SCID

One of the first examples of gene therapy being used in humans was in the treatment of a rare inherited condition called **SCID** (**S**evere **C**ombined **I**mmuno**d**eficiency). Children who have this condition have no effective immune system, so they are very susceptible to infections that can develop into life-threatening illnesses. The only treatment for such children has been to be protected in an isolation 'bubble' to prevent exposure to germs.

SCID results from the child inheriting a defective gene and as a result being unable to make an enzyme called adenosine deaminase (ADA). Without this enzyme, toxins build up and destroy the white blood cells, called T-lymphocytes, that normally recognise infectious microbes that enter the body. The aim of gene therapy for SCID is therefore to replace the child's T-lymphocytes with cells containing the normal ADA gene. The method used is shown the diagram below.

This child is playing inside a biological isolation garment. It is a type of space suit that has been customised for patients who are highly vulnerable to inection. It is often used by children suffering from SCID.

1 a What is the advantage of inserting the ADA gene into developing lymphocytes from the bone marrow, instead of mature lymphocytes in the blood?

b Suggest why the treatment has to be repeated at regular intervals.

c Suppose a child that is given healthy ADA genes in this way eventually has his or her own children. Could the healthy ADA genes be passed on to the children? Explain your answer.

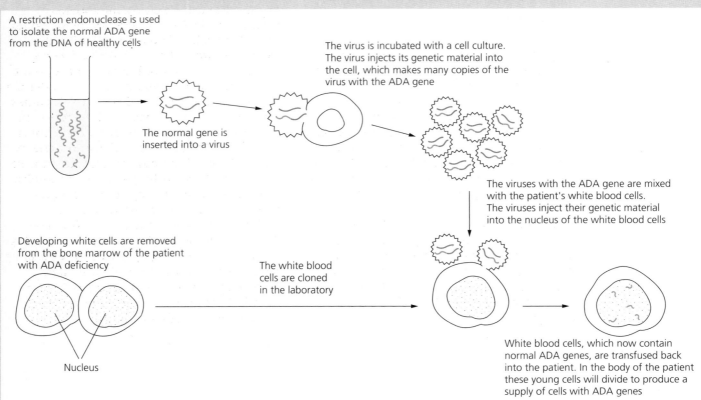

A restriction endonuclease is used to isolate the normal ADA gene from the DNA of healthy cells

The normal gene is inserted into a virus

The virus is incubated with a cell culture. The virus injects its genetic material into the cell, which makes many copies of the virus with the ADA gene

Developing white cells are removed from the bone marrow of the patient with ADA deficiency

Nucleus

The white blood cells are cloned in the laboratory

The viruses with the ADA gene are mixed with the patient's white blood cells. The viruses inject their genetic material into the nucleus of the white blood cells

White blood cells, which now contain normal ADA genes, are transfused back into the patient. In the body of the patient these young cells will divide to produce a supply of cells with ADA genes

The future of gene therapy

Progress in genetic research is proceeding at a great rate. It may, for example, soon be possible to:

- incorporate a healthy gene into one of the chromosomes before putting it into a cell. The chromosome would be replicated as the cell divides, thus providing a longer-lasting treatment;
- treat rapidly dividing cancer cells with genes that kill them;
- provide protection from viral infection of cells by inserting genes that interfere with the replication of the virus;

More effective ways of getting genes into cells will make the treatment of genetic diseases such as cystic fibrosis more effective. Liposomes or adenoviruses get the CFTR gene into target cells where it is expressed, but treatment has to be repeated at regular intervals because copies of the gene are not passed from cell to cell during mitosis. Epithelial cells in the lining of the respiratory tract divide often, so this is a serious limitation. However, if the healthy gene was inserted into a chromosome, it would be replicated each time the cell divided. One significant problem is that it might insert at a point in the DNA that could damage other genes.

Explain how inserting a gene into one of the cell's chromosome might cause damage to other genes.

Gene therapy might also become an important weapon in the war against cancer and disease. There are genes that cause cell death. If cancer cells could be targeted reliably so that only actively dividing cells took up or expressed these genes, cancerous tumours could be destroyed.

Reproduction of viruses in cells could also be prevented by inserting genes that produce mRNA that is complementary to the viral RNA. This would inactivate the viruses in much the same way as the ripening enzyme is inactivated in genetically modified tomatoes (see Chapter 9).

Use the information about inactivating the ripening enzyme in tomatoes in Chapter 9, and the information from Fig. 3 in this chapter (Viral replication), to explain how cells could be protected from viral infection by gene therapy.

Difficult choices ahead

The most effective treatment for genetic disorders would be to replace the defective genes completely. This could be done using *in vitro* fertilisation methods similar to those already used to produce transgenic animals (see Chapter 9). However, the danger of damaging the developing embryo has meant that this technique is not acceptable in humans. Such a genetic change would affect the **germline**. This means that the genes of the sex cells would be altered and therefore this genetic change would be passed on to future generations.

One proposed method of improving the technique is to create an artificial chromosome. This chromosome would contain only healthy extra genes together with the promoter genes needed for them to be expressed. Once added to a nucleus this chromosome should be replicated and take part in mitosis in the normal way. There is less danger that it will cause damage to other chromosomes and genes.

Geneticists also need to take into account that most human characteristics are under the control of several genes. Once the full human genome has been decoded and the functions of many more genes are understood, there is the possibility of adding packages of genes to such artificial chromosomes. This might raise the possibility of dealing with more complex genetic disorders than just the single gene defects that can be treated at present.

However, some of the possibilities raise difficult moral and ethical questions. Balancing the benefits of this emerging technology against the potential for its abuse will take many years of consultation and consideration by the whole of society, not just the scientists and patients directly involved.

10 GENES AND MEDICINE

EXAMINATION QUESTIONS

1

a Suggest why it is important that there are many different types of restriction endonuclease enzymes available to genetic engineers. (2)

People suffering from haemophilia need treatment with the blood-clotting protein, Factor VIII, because they are genetically unable to produce their own.

Factor VIII used to be extracted from human blood but a genetically engineered kind has just been made available. An artificial version of the human gene has been produced which codes for the same sequence of 2338 amino acids that is present in natural factor VIII. This is inserted into hamster kidney cells which are grown in large fermenters. The cells secrete large amounts of factor VIII into the surrounding fluid.

b i) Why was it necessary to know the amino acid sequence of factor VIII before an artificial version of the human gene that codes for it could be produced? (1)
ii) What is the minimum number of nucleotides which must be present in the gene for factor VIII?

c Suggest one advantage of the use of genetically engineered factor VII over that extracted from blood. (1)

BY02 Feb 95 Q4

2 Read the following passage on gene technology (genetic engineering), then replace the dashed lines with the most appropriate word or words to complete the text.

The isolation of specific genes during a genetic engineering process involves forming eukaryotic DNA fragments. These fragments are formed using enzymes which make staggered cuts in the DNA within specific base sequences. This leaves single-stranded 'sticky ends' at each end. The same enzyme is used to open up a circular loop of bacterial DNA which acts as a ----------- for the eukaryotic DNA. The complementary sticky ends of the bacterial DNA are joined to the DNA fragment using another enzyme called -----------. DNA fragments can also be made from a template. Reverse transcriptase is used to produce a single strand of DNA and the enzyme ----------- catalyses the formation of a double helix. Finally new DNA is introduced into host ----------- cells. These can then be cloned on an industrial scale and large amounts of protein harvested. An example of a protein currently manufactured using this technique is -----------.

ULEAC 1996 Biology/Human Biology, Specimen Module Paper Q4

3 The Colorado beetle is a pest of potato crops. A soil bacterium, Bacillus thuringiensis, produces a substance called Bt which kills Colorado beetles but is harmless to humans. Scientist have isolated the gene for Bt production from bacteria and inserted it into potato plants so that the plant produces Bt in its leaf tissues.

a i) What is a gene? (2)
ii) Suggest how the gene for Bt production could be isolated from the bacteria and inserted into cells of the potato plant. (4)

b Bt can also be used as a spray. Colorado beetles may be killed if they ingest potato leaves which have been sprayed with Bt.

Suggest and explain one reason why using Bt-producing potato plants might increase the rate of evolution of Bt-resistance in the beetles compared with using Bt as a spray. (2)

NEAB Feb 97 Modular Biology: BY2 Q3

KEY SKILLS ASSIGNMENTS

Investigating gene therapy

1. A genetic disease such as cystic fibrosis could be cured by inserting a healthy gene into a fertilised egg cell.

   Prepare a presentation for an audience of people who have not studied biology, explaining how the genetic disease could be cured in this way and the possible dangers in this method.

2. Developments in gene therapy are taking place very rapidly. Use the Internet and other sources of information to find information about up-to-date techniques being used or proposed to treat one condition in which you are interested, such as cystic fibrosis.

   To prepare for your internet search make a list of words that may be useful to submit to search engines. Try these words and note down which ones gave the most useful results. As you search you will probably find unexpected results which may be useful. Note these down to keep a track of your progress. When you have finished your research prepare a simple search strategy guide for a new researcher to use. Your strategy guide should not exceed two sides of A4 (or better yet two screens on a computer) and must include:

   - some useful starting points for searches (including the search engines that gave the best results)
   - a review of some of the commonest sites (not all of them need to be good)
   - a list of key words that gave good results
   - a list of the five most useful sites (some of these may be difficult to find so may not appear in the common site list).

   Evaluate the sources of information for the likely quality and reliability of the information you find. What moral or ethical issues do the proposed treatments raise?

3. The suggestion that extra chromosomes might be inserted into embryonic cells could make it possible for parents to select a range of features that they would like their children to have. Prepare for a group discussion of the potential benefits and dangers of being able to 'design' babies in this way. You might consider some of the following aspects:

   - What features might parents select for?
   - What might the social consequences be, for example if rich parents were able to favour their children in this way?
   - How might the relationship between parents and their children be affected?
   - How might parents react if children failed to match up to expectations, for example if enhanced intelligence or sporting prowess was not achieved?
   - Are there possible benefits if such techniques became widely available? Could human society be improved in the long-term?
   - How might the system be abused, for example by totalitarian regimes?

Early diagnosis and regular treatment of symptoms has increased the life expectancy of someone with cystic fibrosis considerably during the last 15 years. At the start of the 21st century, this three-year old cystic fibrosis patient could expect to live into her 30s. The further refinement of gene therapy techniques could allow her to live a normal lifespan.

11 Transport systems

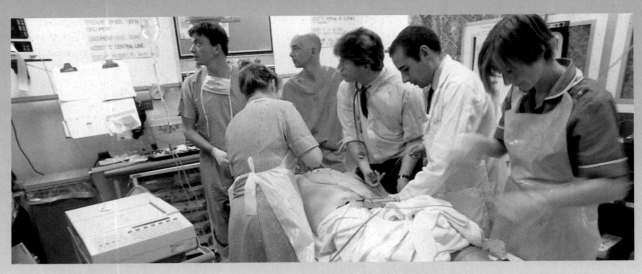

Most hospitals have a team of doctors and nurses who can respond rapidly to a patient with a heart attack. They are known as the crash team.

Doctors in hospital accident and emergency departments cope skillfully with emergencies such as heart attacks. Nevertheless, because a heart attack can cause a great deal of damage to the heart and other parts of the body, roughly 40% of patients who have a heart attack still die as a result. So the best thing to do is to prevent heart attacks from happening in the first place.

Like all muscles, heart muscle needs oxygen. Oxygenated blood is carried to heart muscle by the coronary arteries. In coronary heart disease, layers of fatty material build up in the coronary arteries, causing them to narrow. This fatty material is known as atheroma. If it narrows the arteries enough, the oxygen supply to the heart muscle can be cut off. If a muscle is deprived of oxygen we experience pain in that muscle; that is why one symptom of a reduced blood supply to the heart is chest pain. If the blood supply becomes very poor, sections of heart muscle can die; when this happens, the person has a heart attack.

But if only the heart muscle dies, why does a heart attack cause damage to other parts of the body? The heart is essentially a muscular pump that pumps blood around the body. This blood carries the oxygen and other substances that organs and tissues need, and takes away waste. When the heart is damaged, the efficiency of the pump is reduced and other organs can be damaged because their supply of oxygen fails. The brain is the most vulnerable organ; it is badly damaged if deprived of oxygen for as little as four minutes.

11.1 Mass transport systems

Every organism moves substances around its body, including dissolved gases such as oxygen and carbon dioxide, nutrients such as glucose, waste products and hormones. Diffusion is efficient only over very short distances so organisms larger than an earthworm use **mass transport**. This is the movement of relatively large amounts of material at relatively high speed. In most animals, mass transport forces liquid to flow along a system of tubes.

You may need to revise Chapters 3 and 5 to answer this question:

1a Which substance enters blood in the alveoli in the lungs?

b Which substances enter blood in the small intestine?

c Where do all these substances leave the blood again?

11.2 The human blood system

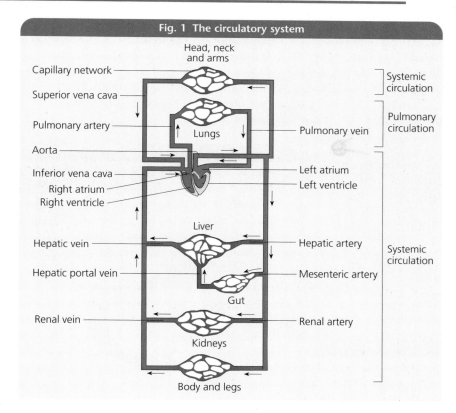

Fig. 1 The circulatory system

In the human mass transport system, materials are carried mainly by blood through a network of vessels called the **circulatory system** (Fig. 1). The heart, a muscular pump, forces blood through a series of vessels called **arteries**, **capillaries** and **veins**. As blood flows through the body, substances are exchanged between the blood and the tissues. In this chapter we look at how dissolved gases such as oxygen are moved round the body. We start by finding out more about how the heart works and then go on to investigate how blood acts as a mass transport system. Finally, we cover the lymphatic system and see how it behaves as a secondary mass transport system.

Heart structure and function

The heart contains two muscular pumps (Fig. 2). Each pump has two chambers, an upper **atrium** and a lower **ventricle**, and two valves, an **atrioventricular** valve and a **semilunar** valve. The valves prevent back-flow of blood into the atria and ventricles. The right side of the heart pumps deoxygenated blood from the body to the lungs along the **pulmonary artery**.

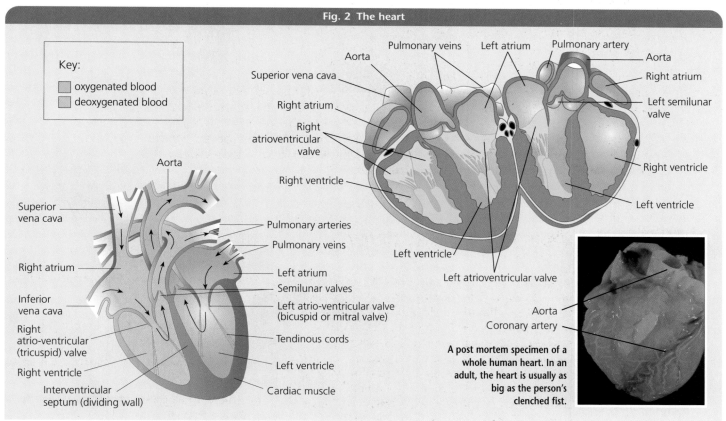

Fig. 2 The heart

A post mortem specimen of a whole human heart. In an adult, the heart is usually as big as the person's clenched fist.

155

Deoxygenated blood has a low concentration of oxygen and a high concentration of carbon dioxide. The left side of the heart pumps oxygenated blood from the lungs to the rest of the body along the **aorta**. Oxygenated blood has a high concentration of oxygen and a low concentration of carbon dioxide.

The cardiac cycle

Blood enters the atria from the veins. Contraction of the muscles of the atrial wall forces blood into the ventricles. Blood is pumped out of the heart by contraction of the ventricle muscles. Contraction of atrial muscle is called atrial **systole**, and contraction of ventricle muscle is known as ventricular systole. Relaxation of heart muscle is called **diastole** (Fig. 3).

2 Use Fig. 3 to calculate how many complete cardiac cycles (heartbeats) there are per minute.

As pressure builds up in the ventricles the atrioventricular valves are forced shut; this prevents backflow of blood from ventricles to atria. The semilunar valves prevent blood flowing back into the heart from the arteries. Fig. 4 shows the way in which valves work.

3 Suggest why tendinous cords ('heart strings') are attached to the edges of the valve flaps.

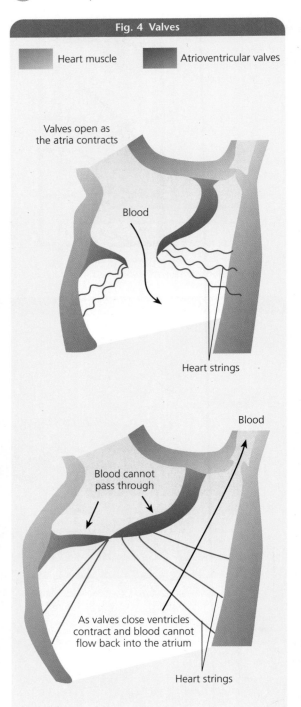

Fig. 3 The cardiac cycle

The cardiac cycle:
- Pacemaker cells initiate systole (contraction).
- The squeezing of the muscle walls reduces the volume and so increases the pressure in the chamber(s), forcing blood in a particular direction.
- The direction of blood flow causes the valves to open or close.

APPLICATION

The active heart

A fetus does not breathe air. It obtains all its oxygen from its mother's blood via the placenta. The heart of the fetus and the main blood vessels are modified as shown in the diagram below. Before birth there is a 'hole', the foramen ovale, between the right atrium and the left atrium, and a 'shunt', the ductus arteriosus between the left pulmonary artery and the aorta. Usually, when the baby takes its first breath the 'hole' and the 'shunt' close but sometimes this does not happen. Such babies are known as 'hole in the heart babies'. These defects can now be detected shortly after birth and corrected successfully by surgery.

1. **a** Suggest the advantage to the fetus of the 'hole' and the 'shunt'.

   b Suggest how the circulation of blood is affected if the hole and the shunt do not close after birth.

2. Elderly people sometimes develop problems with heart valves. Suggest how each of the following problems can affect the circulation of the blood:

   a the left atrioventricular valve does not fully close;

   b the left semilunar valve does not fully close.

3. The healthy heart is a remarkably efficient pump. Suggest why the wall of the left ventricle is thicker than the wall of the right ventricle.

4. **a** At rest the heart pumps out about 75 cm^3 of blood at each beat. Measure your own resting heart beat by taking your pulse whilst sitting down. Count how many pulse beats you can feel during a minute. Do three readings and average them. Calculate how much blood your heart would pump out if you remained sitting or lying in bed for twenty four hours.

   b Your heart pumps out far more blood than this during a day. The more active you are, the faster the heartbeat and the greater the volume of blood pumped out at each beat. What is the advantage to the body of an increased heart rate and beat volume?

5. Suggest what causes:

   a the atrioventricular valves to close;

   b the semilunar valves to open;

   c the semilunar valves to close;

   d atrioventricular valves to open.

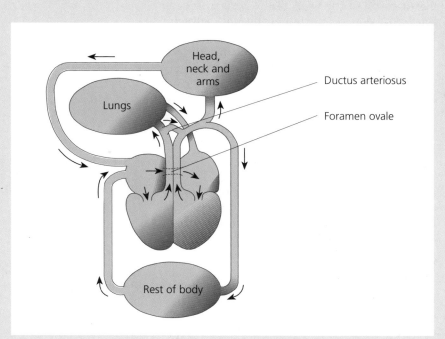

Before birth, the foramen ovale and the ductus arteriosus work together to greatly reduce the blood supply to the lungs. After birth the amount of blood that passes through the lungs is as large as the amount which travels around the rest of the body.

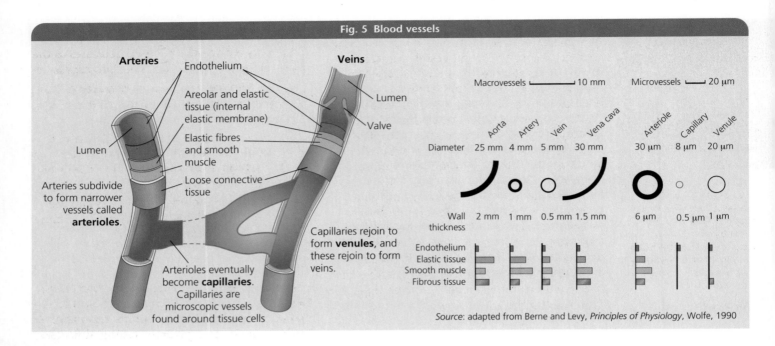

Blood vessels

Fig. 5 shows the structure of the major blood vessels. Study the difference between arteries, veins and capillaries and refer to the diagrams frequently as you read the next section.

Arteries

The wall of the aorta and other large arteries contains a thick layer of elastic tissue. This is stretched by the pressure produced by the left ventricle during systole. During diastole the elastic recoil of the walls of the aorta and large arteries squeeze on the blood and this maintains blood pressure at the right level. The expansion and recoil of the arteries helps to smooth the surges in blood flow that are caused by contraction of the left ventricle. Some signs of these surges can still be felt as the pulse. This is particularly obvious at points where arteries pass over a bone near to the skin, such as at the wrist and temple.

Arterioles

In the arterioles, the thickest layer of the wall consists of muscle fibres. This muscle enables the arterioles to act as control points for the blood system. By contracting, the muscle can restrict the flow of blood through a particular blood vessel.

As the arteries subdivide to form narrower vessels, the pressure of blood in the vessels decreases (Fig. 6). This is because the *total* cross-sectional area of all the smaller vessels is *greater* than that of all of the larger vessels.

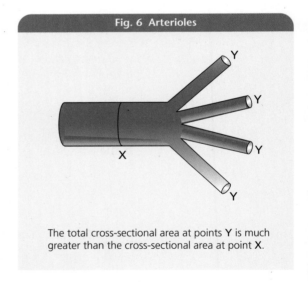

The total cross-sectional area at points **Y** is much greater than the cross-sectional area at point **X**.

4 Suggest the effects of narrowing of the arterioles that supply the capillaries in the surface layer of the skin and that supply the capillaries in the villi of the small intestine.

Capillaries

Capillary walls are made up from a single layer of **endothelial cells**. Capillaries have a diameter of 8 μm and a wall thickness of 2 μm. Blood flow and velocity are slowest through the capillaries. The velocity of the blood describes how fast it is moving, while the flow describes

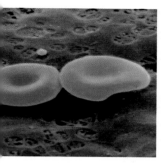

Above: A false colour scanning electron micrograph of red blood cells travelling through a capillary in the liver. The fenestrated endothlium of the capillary is characterised by a number of small holes through which nutrients can reach all the liver cells.

Magnification × 1520.

how much blood moves past a particular point. The flow depends on *both* the velocity *and* the size of the blood vessel. You can see how a large, slow-moving river (low velocity) can shift much more water than a small, fast moving stream (high velocity) in the same time. Blood flow follows the same rules (Fig. 7). So, although blood flows through the capillaries in an organ at a low velocity, a large volume of blood flows through the many capillaries that serve that organ.

Exchange of materials such as oxygen, carbon dioxide and soluble food molecules occurs between the blood and the tissues at the capillaries. Some capillaries have small gaps in between adjacent endothelial cells. These gaps, called **fenestrations**, allow even faster rates of diffusion between the capillaries and the tissues.

5a Use information from Fig. 7 to explain why the velocity of blood decreases as the blood flows from arterioles into capillaries.

b Explain why the velocity increases as the blood flows from the smaller veins into the larger veins.

c Suggest two organs other than the liver where the fenestrations shown in the photograph on the left might occur. Give reasons for your choices.

d How does the size of capillaries and the rate of blood flow through the capillaries suit them for their function?

Veins

Generally, veins have thinner walls than arteries and contain both elastic and muscle tissue. The pressure of blood in the veins is lower than in either arteries or capillaries, falling almost to zero in the largest vein, the **vena cava**, which carries deoxygenated blood back from the body into the heart. Unlike arteries, veins have **valves** to prevent backflow of blood.

Three main forces keep blood moving in the veins:

- The small residual pressure of blood coming from the capillaries;
- The action of the leg muscles and valves in the veins;
- The reduced pressure in the atria at atrial diastole.

Leg muscles act as a 'secondary heart' (Fig. 8) when they contract to move the legs. The contractions squeeze the veins in the leg and this pushes blood upwards rather like squeezing toothpaste along a tube. Valves in the veins prevent the blood moving downwards. This explains why people can faint if they stand still for a long time – particularly on a hot day.

At atrial diastole, the muscles in the atrial walls relax and the atria increase in volume. This reduces the pressure and creates a suction force in the vena cava and other main veins that draws blood towards the heart.

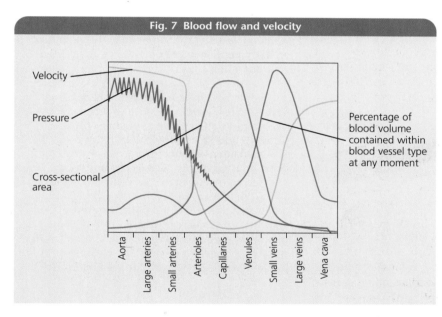

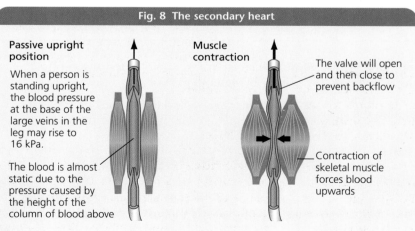

6a Use the information from Fig. 5 to calculate: the cross sectional area of the space inside an artery and a capillary.

b If the pressure inside the capillaries was exactly the same as the pressure in the artery supplying them, calculate the number of capillaries that would be supplied by the artery.

11 TRANSPORT SYSTEMS

EXTENSION

Keep those arteries open

Coronary arteries are the blood vessels that supply blood to the muscles in the walls of the heart. Like other vessels in the body, these arteries may become partly blocked by a build up of fatty tissue, called **atheroma**. This restricts the blood supply to the heart muscle. Any muscle in the body gives rise to pain if it is short of oxygen and the heart muscle is no exception. In more severe cases the pain may be experienced even at rest. This is a condition known as **angina**.

A blood clot, called a thrombosis, may develop on the surface of the plaques of atheroma. When this happens, the blood vessel can become completely blocked and oxygenated blood is no longer able to reach the heart muscle. Without oxygen, the heart muscle dies and the result is a **myocardial infarction**, or heart attack. If the area of muscle affected is small, the person may recover but heart attacks that are caused by the death of large parts of the cardiac muscle are usually fatal.

Two surgical methods are used to keep diseased coronary arteries open and so prevent heart attacks:

- **Balloon angioplasty**. Cardiologists insert a balloon into the blood vessel where blood flow is obstructed. When inflated, the balloon expands the inside walls of the vessel, compressing the fatty material blocking the vessel, and clearing some of it away. When the balloon is removed, the vessel has a much wider diameter, However, the technique involves some slight damage to the inside of the vessel and scarring can occur. This encourages more fatty material to be laid down and the narrowing recurs, sometimes within weeks. This condition known as **restenosis**.

- **Stenting**. This is a modification of the angioplasty technique in which surgeons use the balloon but they insert a coil into the blood vessel at the same time. This rigid tube is left permanently inside the vessel to keep it open.

People who have had stents in their blood vessels have a lower incidence of subsequent heart attacks and require fewer treatments to restore blood flow to the heart.

One Dutch research project studied 227 heart attack patients. 112 were randomly selected to receive stents while the remaining 115 were treated with conventional balloon angioplasty. One stented patient suffered another heart attack compared to 8 of the people who had had angioplasty and only four stented patients needed further treatment for blocked arteries, compared to 19 angioplasty patients.

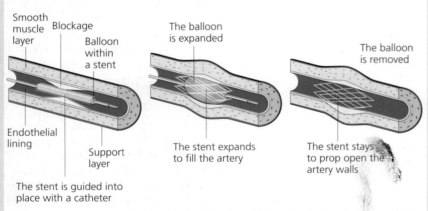

Source: adapted from an article in *New Scientist*, 25 June 1994

KEY FACTS

- The heart contains two muscular pumps: one pump sends deoxygenated blood arriving from the body's tissues to the lungs to be oxygenated; the other pumps oxygenated blood around the body.

- Blood is kept moving in the right direction by **valves** in the heart and veins.

- Elastic tissue in the walls of **arteries** smoothes surges of blood pressure.

- Muscle layers in the walls of **arterioles** control the supply of blood to each organ.

- **Capillaries** have walls one-cell thick to enable the exchange of materials between the blood and the tissues.

- Valves in veins prevent backflow of blood, particularly in the limbs.

- Contraction of the skeletal muscles that surround veins helps to return blood to the heart.

11.3 Exchanging materials

The main function of the blood system is to transport materials such as oxygen, food molecules and mineral ions to the cells of the body and to take wastes such as carbon dioxide and urea away. In this section we look at the role that haemoglobin plays in transporting dissolved gases in the blood and then go on to examine how other materials are exchanged between the blood and the tissues.

Transporting oxygen

Haemoglobin transports oxygen around the body. It is this pigment molecule that gives red blood cells their colour (see Chapter 4). Haemoglobin is such a good oxygen carrier that it can be given on its own, instead of whole red blood cells, as a transfusion for patients who have lost a lot of blood.

The **haem** part of a haemoglobin molecule can combine temporarily with four oxygen molecules. In oxygen-rich situations such as in the capillaries of the lungs, haemoglobin and oxygen combine to form **oxyhaemoglobin**. In oxygen-poor situations, such as in the capillaries of exercising muscles, the oxyhaemoglobin **dissociates**. It splits up and releases the oxygen. A far greater mass of oxygen can be carried in the form of oxyhaemoglobin than can be carried in solution. The reaction between oxygen and haemoglobin is summarised by the equation:

$$\text{oxygen} + \text{haemoglobin} \rightleftharpoons \text{oxyhaemoglobin}$$

To understand why haemoglobin is such an efficient molecule for transporting oxygen we have to consider its **oxygen dissociation curve**. If you look at Fig. 9, you can see that the percentage of haemoglobin molecules that combine with oxygen to form oxyhaemoglobin varies with the external partial pressure of oxygen.

At the oxygen levels that occur in the lungs haemoglobin readily absorbs oxygen to form oxyhaemoglobin. Part X on the graph shows this. Even if the oxygen level starts to fall haemoglobin can still absorb oxygen – this means the blood leaving the lungs is almost always completely saturated with oxygen. At point Y the graph starts to fall steeply and the oxyghaemoglobin gives up its oxygen. This occurs in the tissues. The steepness of the graph at part Z means that a slight fall in oxygen concentration in the tissues produces a large increase in the rate of oxyhaemoglobin breakdown. This delivers oxygen very effectively to the tissues that need it.

Fig. 9 The oxygen dissociation curve

Haemoglobin in lungs becomes fully saturated with oxygen forming oxyhaemoglobin

Oxyhaemoglobin dissociates in tissues releasing most of its oxygen

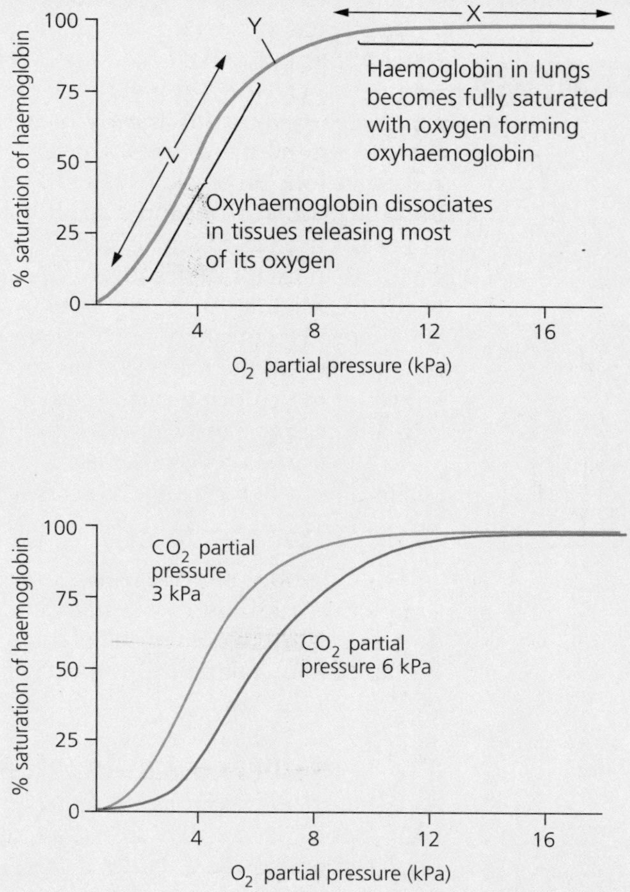

7 Look at Fig. 9 and answer the following:

a What percentage of haemoglobin is saturated at partial pressures of oxygen of 8 kPa; 12 kPa; 16 kPa?

b What is the lowest partial pressure of oxygen at which all the haemoglobin molecules combine with oxygen?

c Blood moves from a partial pressure of oxygen of 12 kPa in the lungs to a ppO_2 of 4 kPa in the muscles. What percentage of the oxygen is released?

Because haemoglobin has an 'S'-shaped oxygen dissociation curve it becomes fully saturated with oxygen in the lungs but readily gives up this oxygen to respiring muscles. The tissue fluid in an active muscle has a high partial pressure of carbon dioxide due to the high rate of respiration in the muscle cells. The effect of this is to 'shift' the dissociation curve to the right. This shift is known as the **Bohr effect**. This means the oxyhaemoglobin gives up its oxygen even more readily – exactly what is needed in actively respiring muscle.

Organisms that live in anaerobic conditions often have haemoglobin with a slightly different dissociation curve to that of humans. *Tubifex* worms live in mud at the bottom of lakes and rivers. The oxygen dissociation curve of *Tubifex* haemoglobin is shifted to the left compared with human haemoglobin (Fig. 10).

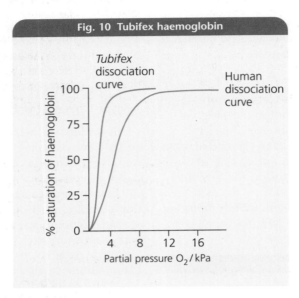

8 What is the advantage to *Tubifex* in having haemoglobin with a dissociation curve of the shape shown in Fig. 10?

Transporting carbon dioxide

As well as carrying oxygen to the tissues, blood also carries carbon dioxide from the tissues to the lungs. Carbon dioxide is much more soluble than oxygen in water. This means that about 5% of the carbon dioxide produced in the tissues can be transported in simple solution in the plasma. Some of the carbon dioxide dissolved in the plasma is converted into **carbonic acid**, which then dissociates to form hydrogen ions and hydrogencarbonate ions:

$$CO_2 + H_2O \rightarrow H_2CO_3 \rightleftharpoons H^+ + HCO_3^-$$

This reaction between carbon dioxide and water to form carbonic acid occurs very slowly in the plasma. However, red cells contain the enzyme **carbonic anhydrase**, which increases the rate ten thousand times. The hydrogencarbonate ions produced by the red cells diffuse into the plasma. This should result in an excess of positively charged ions inside the red blood cells and an excess of negatively charged ions in the plasma but this is prevented by a simultaneous movement of chloride ions from the plasma into the red cells. This mechanism, known as the **chloride shift**, keeps both red cells and plasma electrically neutral (Fig. 11).

In the lungs a reverse series of reactions occurs and carbon dioxide is breathed out.

$$H^+ + HCO_3^- \rightleftharpoons H_2CO_3 \rightarrow CO_2 + H_2O$$

Haemoglobin has one other important function in addition to the transport of gases: it helps to keep the pH of the plasma constant. It does this by 'taking up' any excess H^+ ions, preventing the blood becoming too acidic.

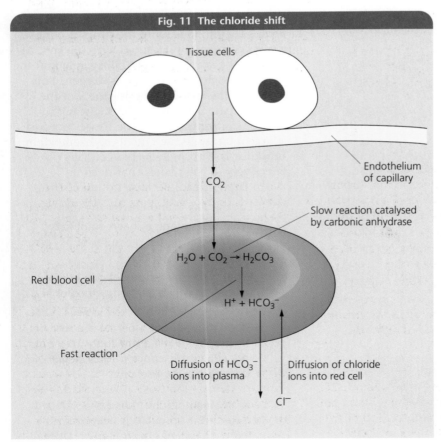

9 Blood with a reduced red cell count cannot transport oxygen as efficiently as healthy blood. However, its ability to transport carbon dioxide is hardly affected. Explain this difference.

EXTENSION

Buffers

Any decrease in the pH of the blood that results from an increase in hydrogen ion concentration is called **acidosis**. Any increase in the pH of the blood is due to a decrease in hydrogen ion concentration is called **alkalosis**. Both of these conditions can be life-threatening if they remain untreated.

In a healthy person, the pH of the blood is maintained at 7.35 – 7.45, thanks to the body's **buffer** systems. A buffer system resists changes in pH when an acid or an alkali is added. The haemoglobin molecule acts as one of these buffer systems.

Oxyhaemoglobin (HbO_2) in the red cells leaving the lungs is associated with potassium ions. This state can be represented by $KHbO_2$. In the tissues the $KHbO_2$ dissociates:

$$KHbO_2 \rightarrow KHb + O_2$$

The KHb is basic and so can take up the hydrogen ions that are formed when carbonic acid dissociates into H^+ and HCO_3^- ions.

$$KHb + H^+ + HCO_3^- \rightarrow HHb + KHCO_3$$

HHb is called **haemoglobinic acid**. By taking up hydrogen ions, even to form an acid, it enables the blood to transport large amounts of carbon dioxide without becoming too acidic.

KEY FACTS

- Red cells contain **haemoglobin**. The principal function of haemoglobin is to carry oxygen.
- The reaction between haemoglobin and oxygen that forms **oxyhaemoglobin** is reversible.
- Haemoglobin has an oxygen dissociation curve that allows it to become fully saturated as it passes through the lungs and to release oxygen as it passes through the tissues.
- The oxygen dissociation curve can be 'shifted' to the right in muscle to allow more oxygen to be released.
- The oxygen dissociation curve can be shifted to the left in animals which live in situations where there is reduced oxygen availability, resulting in maximum uptake of oxygen in these conditions.
- Carbon dioxide is carried mainly as hydrogencarbonate ions in the plasma. The hydrogencarbonate ions are produced in the red cells when carbonic acid dissociates; they then diffuse into the plasma. The enzyme **carbonic anhydrase** catalyses the formation of carbonic acid from carbon dioxide and water.
- Carbon dioxide is released in the lung capillaries by the reverse of the reaction that picks up carbon dioxide in the tissues.
- Electrical neutrality between the red cells and the plasma is maintained by the movement of chloride ions, a mechanism called the **chloride shift**.

Exchange of other substances

Exchange of other materials occurs through the endothelial cells of the capillaries between the blood and the tissues. Different materials are exchanged using different mechanisms:

- Water is literally forced out of capillaries in the tissues by the pressure of the blood and it later re-enters the capillaries by *osmosis*;
- Sugars, mineral ions and waste products such as urea are exchanged between the capillaries and the tissues by *facilitated diffusion*.

Formation of tissue fluid

When blood travels through the tissues, the exchanges that occur between the blood and the cells determines the composition of the fluid that bathes the cells. This medium is called **tissue fluid**. Because not all of the contents of the blood pass into the tissue, tissue fluid and blood are not the same. Red blood cells, platelets and plasma proteins remain inside the blood capillaries. The resulting tissue fluid is essentially blood plasma without the plasma proteins. Like blood plasma, its composition varies depending on its position in the body. Tissue fluid in the small intestine, for example, contains a high concentration of sugars in the couple of hours following a meal.

Tissue fluid also contains some types of white cells that squeeze out of the capillaries through the fenestrations. These white cells combat infection by ingesting bacteria or producing chemicals such as antibodies to combat the effects of bacteria and viruses.

How tissue fluid is formed

To understand how tissue fluid is formed, we first need to consider the composition of blood. This is shown in Fig. 12.

When blood flows into the capillaries, water and other materials are exchanged between the blood and the tissue fluid (Fig. 13). Two forces affect the exchange of water between the capillaries and the cells (Fig. 14):

- **Hydrostatic pressure.** This is the pressure that is generated by the pumping force of the heart. It tends to push water out of the capillaries;
- **Water potential.** This is due mainly to the presence of large protein molecules in the plasma. These protein molecules attract large numbers of water molecules so making the water potential inside the capillaries much more negative. This draws water into the vessels.

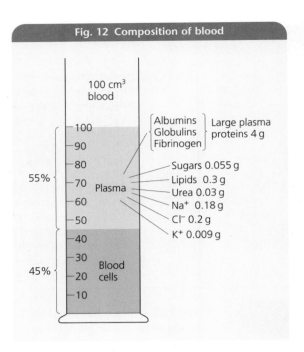

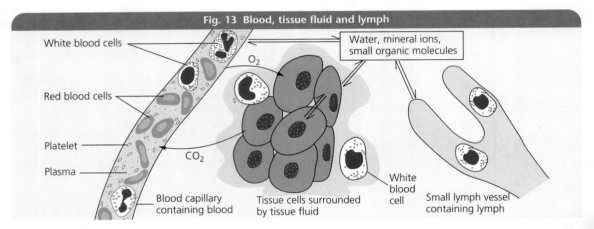

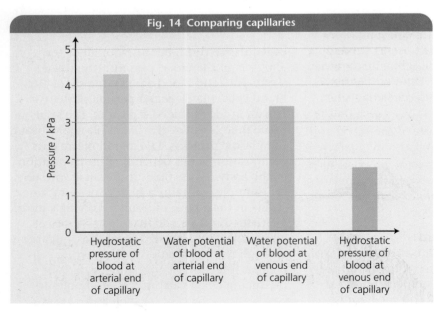

At the arterial end of a capillary, the hydrostatic pressure forcing water out is greater than the force due to water potential moving water back in, so there is a net movement of water into the tissues. The protein molecules in the plasma are generally too large to leave the capillaries so the water potential of the fluid that remains in the capillaries after water has moved out becomes more negative. This increases the force drawing water into the capillaries. The loss of water from the blood also reduces the hydrostatic pressure – again reducing the effect of the force pushing water out of the capillaries. So, by the time blood reaches the venous end of the capillary, the effect of the water potential drawing water in exceeds hydrostatic pressure pushing water out. This means water re-enters the capillaries.

APPLICATION

Serum

Serum is a blood product used widely in medicine. It is very useful in replacing the blood plasma lost by patients with extensive burns. In these patients, the loss of the outer waterproof outer layer of skin means that tissue fluid, but not blood cells, is lost from their wounds very rapidly. To produce serum, whole blood is first centrifuged to remove blood cells, then treated to remove a blood protein called fibrinogen The fibrinogen is an essential component in the clotting process and its removal prevents plasma clotting. The liquid produced is called serum. Serum is useful for burns patients but patients who have lost a lot of whole blood need a whole blood transfusion.

1. Construct a table to compare the composition of blood, plasma, tissue fluid, lymph and serum.

2. Oxygen, carbon dioxide, glucose, amino acids, urea and proteins are all carried in the blood. Which substances pass from capillaries into tissues and from tissues into capillaries at the following sites; alveoli; intestinal villi; brain; leg muscles; liver; kidney.

3. Children suffering from protein deficiency often appear to be fat, but their abdomens are swollen because of the build-up of fluids in their tissues. Suggest why this fluid accumulation occurs.

11.4 The lymph system

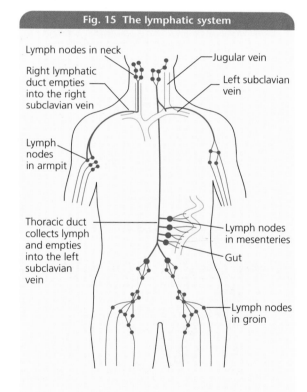

Fig. 15 The lymphatic system

Some tissue fluid passes into small capillary-like vessels called lymph capillaries, which drain into lymph vessels. The fluid inside a lymph capillary is **lymph**. It has almost the same composition as tissue fluid but it contains more lipids. Lymph vessels form a secondary drainage system that empties lymph back into the blood system. Larger lymph vessels have valves to prevent fluid flowing backwards.

Lymph vessels form a network called the **lymphatic system** (Fig. 15), which also contains **lymph nodes**. Since much of the tissue fluid drains back through the lymphatic system, these lymph nodes are situated at sites where they can intercept bacteria and viruses that may come from infected tissues, preventing these microbes from entering the blood system. The nodes consist of tissue that acts as both a filter and a white blood cell production site. White blood cells called lymphocytes mature in the lymph nodes and become ready to act against micro-organisms.

KEY FACTS

- The main function of blood plasma is to transport materials around the body, in solution.

- Plasma is forced out of the capillaries by hydrostatic pressure to form **tissue fluid**. Tissue fluid does not contain plasma proteins, but may contain white blood cells.

- Tissue fluid returns to capillaries by osmosis.

- Much tissue fluid drains into **lymph capillaries**.

- **Fibrinogen**, a plasma protein, can be removed from plasma so that the liquid will not clot. Blood plasma with the fibrinogen removed is called **serum**.

11 TRANSPORT SYSTEMS

EXAMINATION QUESTIONS

1 The bar chart shows the relative thickness of parts of the walls of two blood vessels, A and B. One of these blood vessels was an artery, the other a vein.

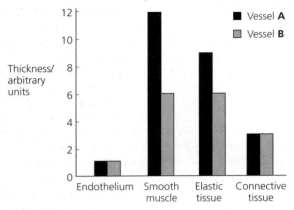

a Explain why the thickness of the endothelium is the same for both vessels. (1)

b Which blood vessel is the artery? Explain the reasons for your answer. (2)

c Explain how the structure of veins ensures the flow of blood in one direction only. (2)

BY03 June 1996 Q7

2 Read the extract then answer the questions

> **Fish breathe easy by putting the squeeze on blood cells**
>
> In the gills of a fish blood passes through structures called lamellae, two thin membranes held apart by cells which look like the pillars supporting a roof. Video recordings of this flow show that red blood cells become deformed as they pass between the pillar cells. Normally the red blood cells of a trout are oval and measure 13.5 by 8.4 micrometres. As the cells flow through the lamellae, however, they stretch to more than 18 micrometres in length, and take the shape of a letter C or S. Some of the red blood cells get jammed between the pillar cells, blocking the progress of other blood cells. This means that red blood cells passing through the gill lamellae travel about 50 per cent further than the shortest path. This helps to explain why fish gills are so good at picking up oxygen from water.
>
> (Reproduced by permission of New Scientist)

a The function of red blood cells is to transport oxygen. How is oxygen transported by red blood cells? (1)

b i) Explain the advantage to the trout of the change in shape of its red blood cells as they pass through the gills. (2)

ii) Explain the advantage to the trout of some red cells getting jammed between the pillar cells. (2)

BY03 June 1998 Q7

3 The diagram represents part of the human blood circulation.

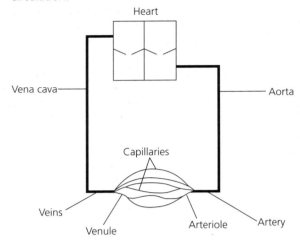

Table 1 shows the minimum and maximum blood pressures in each of the types of blood vessel labelled on the diagram. These blood vessels have been represented by letters A-F in the table.

Table 2

Blood vessel	Blood pressure/kPa minimum	maximum
A	12	18
B	4	7
C	8	11
D	1	3
E	10	16
F	9	14
vena cava	–2	0.5

a Which of the blood vessels A to F is
i) the aorta;   **ii)** a capillary? (2)

b Table 2 shows the mean diameter of some blood vessels and the mean velocity of blood travelling in those vessels. Describe and explain the relationship between the diameter of vessels and the velocity of blood. (2)

Table 2

Blood vessel	Mean diameter of vessel /mm	Mean velocity of blood/cm per second
aorta	10	40
arteries	3	40 – 10
arterioles	0.02	10 – 0.1
capillaries	0.008	less than 0.1
venules	0.03	less than 0.3
veins	6	0.3 – 5
vena cava	12.5	5 – 20

c Describe how tissue fluid is absorbed into blood capillaries. (3)

BY03 March 1999 Q3

KEY SKILLS ASSIGNMENTS

Artificial blood

Every year roughly 100 million units of donated blood trickle into patients. Recently a small but growing number of pioneers have allowed something other than human red blood cells to fill the bags hanging above their hospital gurneys. Some patients have accepted into their veins protein solutions extracted from cow's blood or fermented from genetically engineered bacteria. In others, a Teflon-like solution has displaced, for a few hours, up to 40 percent of the blood from their vessels.

This year at least six companies in the U.S. are testing so-called blood substitutes in human surgeries. "Substitutes" is perhaps too ambitious a label for these solutions, because none can replace the clotting and infection-fighting abilities of whole blood. But all six liquids can, like red blood cells, ferry oxygen from the lungs to the rest of the body and carry carbon dioxide back. Two of the products are on track to enter final, phase III clinical trials in hundreds of patients next year.

The rush to produce alternatives to blood may seem oddly timed. Tighter screening prompted by the emergence of HIV has made the blood supply safer than it has ever been. Yet donation levels have never recovered from the initial AIDS scare, and blood banks face periodic regional shortages.

The main benefit of these products will be to reduce the amount of donated blood a patient receives. That can minimise the risk of infection [because the chemicals can be sterilised more rigorously than blood] and will preserve blood for cases where it is really needed.

Synthetic substitutes should have other advantages as well. All will stay fresh for six months or more; red blood cells deteriorate within six weeks. And the artificial compounds bear none of the proteins and sugars that coat blood cells and separate them into eight distinct types. Theoretically, substitutes could be pumped into anyone, without fear of provoking a serious allergic reaction.

Of course, doctors had the same hope back in 1868, when they first extracted haemoglobin, the oxygen-bearing protein in red blood cells. Haemoglobin failed as a blood replacement because it works only when intact and when assisted by a cofactor found in red blood cells. Stripped from its protective cell and its cofactor haemoglobin is quickly snipped in two by enzymes, and the fragments can poison the kidneys.

Baxter Healthcare in Deerfield, Ill., has completed five phase II trials of HemAssist, which it makes by extracting haemoglobin from outdated human blood and chemically binding its pieces together with a derivative of aspirin. In June, Baxter became the first company to win approval in the U.S. for a phase III trial of its blood substitute. The firm started a similar trial last year in Europe and has already begun building a factory to produce the drug in Switzerland.

If the thought of having genetically engineered goo injected into your arteries makes your skin crawl, fret not: the substitutes will simply be options available-at premium prices-for those who cannot use their own previously stockpiled blood and do not trust others'. Unfortunately, prospects are slim that substitutes cheaper than blood will be able to address perhaps the greatest need for them: saving lives on battlefields and in hospitals in the more remote corners of the world where blood shortages are chronic.

This article, by W. Wayt Gibbs in San Francisco, is reproduced with permission from Scientific American.

1 What are the advantages of using artificial blood rather than natural blood in the transfusion service?

2 What advantages has natural blood over artificial blood in the transfusion service.

3 Explain why extracted haemoglobin is not used in blood transfusions.

4 Describe how the haemoglobin gene could be inserted into the DNA of a bacterium.

5 Discuss the following issues in a group:

  a Should people be paid to donate blood?

  b Should blood donated freely be sold, and if so where should the money go?

  c Write up a report of your discussion. The report should present the views of everyone involved in the discussion fairly. Submit your first draft to the group so that everyone can agree the account is accurate. If possible, agree a view the whole group can support and present this as a final statement of belief. If it is not possible to agree a single viewpoint give two, indicating which one is the majority.

12 Energy and exercise

Runners get underway at the 1998 London Marathon.

The first London Marathon was held on March 29, 1981. Of the 20 000 people who wanted to run, 7 747 were accepted and 6 255 finished. They were led home by the American, Dick Beardsley, and Norwegian, Inge Simonsen, who staged a spectacular and fitting dead-heat. Joyce Smith broke the British record to win the women's race.

The event was enjoyed by the thousands of spectators who lined the course and the millions of viewers who followed the race on BBC TV. Its success was so great that, the following year, more than 90 000 hopeful runners from all over the world applied to run. Since then, the event has continued to grow in size, stature and popularity. In its first 18 years, 382 672 people completed the race, 29 924 of these broke the marathon's record when they all finished the 1998 Flora London Marathon. The amount raised by the runners for charity has grown significantly each year and the London Marathon is now one of Britain's most successful annual charity fundraising events.

12.1 The power to exercise

Runners have high energy requirements; they can use as much as 30 kJ of energy per minute when exercising. This energy is needed mainly for muscle contraction, and muscles need a constant supply to work properly. Glucose provides the energy for exercising; the chemical energy in glucose is released during respiration inside the muscle cells and stored briefly in a molecule called ATP (see Chapter 1). It is ATP that transfers energy directly to the muscle contraction mechanism.

All the glucose supplied to the muscles comes originally from the carbohydrates, lipids and proteins in food. Excess food can be stored as carbohydrates and lipids in various parts of the body, but only a limited amount can be stored in the muscles themselves. When an athlete prepares for a marathon, he or she must build up energy stores and improve the circulation to the muscles so that glucose and oxygen can be supplied quickly from other parts of the body. In this chapter we look at how the body stores food and mobilises those stores to fuel exercise. We also see how the circulation and breathing systems respond to exercise.

Energy sources

There are three main sources of energy for exercise:

- glycogen stores in the muscle cells;
- glucose from the blood;
- fatty acids from the blood.

An average healthy adult male has about 450 g of carbohydrate and 10.5 kg of fat in his body. The total amount of carbohydrate could provide enough energy for a 20 mile run. Of the 450 g, about 325 g is stored in the muscles as glycogen, 125 g is stored in the liver, also as glycogen, and 20 g is circulating in the blood as glucose. The amount of glycogen stored in the body can be modified by diet. If someone eats a low carbohydrate diet, their glycogen store diminishes rapidly. Eating a high carbohydrate diet for a few days can almost double the amount of glycogen in the body. Glycogen is a polymer (see pages 18-19 in Chapter 1). It is formed from glucose by condensation, and can be hydrolysed back to glucose.

$$\text{glucose} \underset{\text{hydrolysis}}{\overset{\text{condensation}}{\rightleftarrows}} \text{glycogen} + \text{water}$$

When blood glucose increases, as it does after a meal, this rise is detected by the cells in the pancreas that secrete the hormone **insulin**. Insulin increases the rate at which cells take up glucose from the blood. The liver and the muscles then convert some of this glucose into glycogen.

When blood glucose falls, as it does during exercise, the drop is detected by other cells in the pancreas that secrete the hormone **glucagon**. Glucagon stimulates the enzyme **phosphatase** in the liver to hydrolyse glycogen into glucose. The liver releases most of this glucose into the blood. The actions of insulin and glucagon keep the concentration of the blood within fairly tight limits, between 4.5 and 5.5 millimoles per litre.

  Suggest why the glucose concentration of the blood is kept constant at a relatively low level.

Most of the fat in the body is stored in adipose tissue immediately under the skin. The cells of this tissue are literally crammed full with triglycerides. The 10.5 kg of fat they contain would provide enough energy for a walk of several thousand miles!

Excess fat stores start to build up if someone eats a high-energy diet for more than a few days. The liver converts excess triglycerides into a form that can be transferred to the adipose tissue. During exercise some of the triglycerides in the adipose tissue are broken down into fatty acids and glycerol. The fatty acids enter the blood stream and travel to the muscles where they can be used by aerobic respiration to release energy. They are not broken down via glycolysis, but enter the mitochondria and are used in the later stages of aerobic respiration. Fig. 1 summarises the role of the liver in energy storage.

Triglycerides release approximately twice as much energy as carbohydrates when oxidised. 1 g of triglyceride releases approximately 37 kJ of energy whereas 1g of glucose releases only 20 kJ. Triglycerides are a water-free energy store whereas 2.7 g of water are stored with each gram of glycogen. Storing energy as fat rather than carbohydrate therefore takes up less space, and there is less mass to carry around.

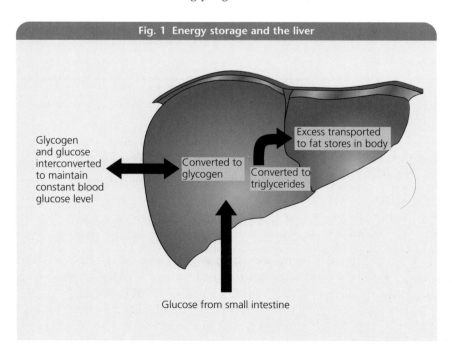

Fig. 1 Energy storage and the liver

12 ENERGY AND EXERCISE

Energy stores and exercise

In the first few minutes of exercise, muscles obtain almost all of the energy they need from blood glucose and the glycogen stored in the muscles themselves. For the next twenty minutes or so, about 50% of the energy needed is supplied by muscle glycogen and by blood glucose that has been produced by the breakdown of liver glycogen. Fatty acids derived from hydrolysis of triglycerides in adipose tissue provide the rest of the energy during this time. The triglycerides are transported to the muscles in the blood. If exercise continues beyond three or four hours, this depletes most of the liver glycogen stores. After this, blood fatty acids derived from stored triglycerides supply 80% of the muscles' energy needs.

2 Storing energy as fat rather than carbohydrate has several advantages. Give two ways in which the body benefits.

APPLICATION Putting fat to good use

The graph shows how the percentage contribution of muscle glycogen, liver glycogen and triglycerides to muscle respiration changed over a period of 4 hours.

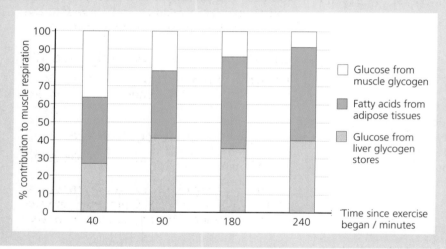

1 What percentage of the muscle's energy needs were supplied by triglycerides at 90 minutes?

2 Describe and explain the trends for the energy supplied by:
  a muscle glycogen;
  b liver glycogen.

3 An average young adult stores approximately 400 000 kJ of energy as triglycerides and 9000 kJ as glycogen. If an average young male expends about 450 kJ per mile when walking, how far could this male walk:
  a using only glycogen stores?
  b using only triglyceride stores?

KEY FACTS

- The polysaccharide **glycogen** is stored in large amounts in the muscles and the liver.
- When blood sugar levels rise following a meal, the pancreas secretes **insulin**. This hormone increases the rate at which cells take up glucose from the blood.
- In the muscles and the liver, excess glucose is converted into glycogen by condensation reactions.
- When blood sugar levels fall the pancreas secretes **glucagon**. This hormone stimulates the enzyme **phosphatase** in the liver to hydrolyse glycogen into glucose. This glucose is carried by the blood to the muscles.
- Triglycerides are stored as fat globules in adipose tissue, mainly under the skin.
- During prolonged exercise, triglycerides are hydrolysed into fatty acids and glycerol.
- Fatty acids are transported to the muscles in the blood.
- When oxidised in respiration, fatty acids release much more energy per unit mass than glucose.
- During exercise, energy is obtained first from ATP already stored in the muscle. ATP supplies are replenished first from muscle glycogen, then blood glucose, then liver glycogen and finally triglycerides.

12.2 Respiration in muscle cells

As we saw in Chapter 1, the energy released from the respiration of glucose is used to add inorganic phosphate (Pi) to adenosine diphosphate (ADP) to produce adenosine triphosphate (ATP) (Fig. 2). ATP is a temporary energy store, storing a small 'package' of chemical energy. This chemical energy can be transferred to kinetic energy by muscles. Respiration produces the ATP that muscles need to contract. However, the ATP supply in muscles is very limited. Only enough ATP is stored for a 1 minute brisk walk, a 25 second slow run, or a six second sprint. There is not enough stored ATP to run a 100 metre race. To provide the energy to run fast for this long, ATP must be synthesised continually from ADP and phosphate.

For the first few seconds after the original ATP energy store is used up, energy to re-synthesise ATP comes mainly from the breakdown of glycogen stored in the muscle. This breakdown releases glucose and then pyruvate. The Greek word 'lysis' means 'to break down' so the breakdown of glucose to pyruvate is called **glycolysis** and the breakdown of glycogen to lactic acid is called **glycogenolysis**.

What happens next depends on how much oxygen is available in the muscle cell:

- if a plentiful supply of oxygen is available, **aerobic respiration** occurs; most of the pyruvate is oxidised and this releases sufficient energy to re-synthesise large amounts of ATP.

- if insufficient oxygen is available, **anaerobic respiration** occurs; much of the pyruvate is converted to lactic acid.

$$C_6H_{12}O_6 + 6O_2 \rightarrow 6CO_2 + 6H_2O + 36\text{ ATP}$$
glucose + oxygen → carbon dioxide + water + 36 moles ATP

$$C_6H_{12}O_6 \rightarrow 2\ C_3H_6O_3 + 2\text{ ATP}$$
glucose → lactic acid + 2 moles ATP

There is a big difference in the amount of energy released by the two processes and the end products are different, as Table 1 shows.

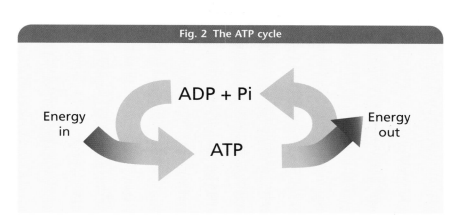

Fig. 2 The ATP cycle

3 Use Fig. 10 in Chapter 1 to find out where in the cell:
a glycolysis occurs;
b aerobic respiration occurs.

Problems with lactic acid

Lactic acid dissociates into hydrogen ions and lactate ions. The hydrogen ions lower the pH in muscle cells. This increase in acidity in the muscles has two effects:

- the pH is no longer the optimum for the enzymes involved in glycolysis, so glycolysis slows down;
- the hydrogen ions interfere with the actual mechanism of muscle contraction, making the muscles feel 'stiff'.

Table 1 Aerobic and anaerobic respiration compared

Type of respiration	Energy released from one glucose molecule	End products
Aerobic respiration	Enough to synthesise 36 molecules of ATP	Carbon dioxide and water
Anaerobic respiration	Enough to synthesise 2 molecules of ATP	Lactic acid ($C_3H_6O_3$). No carbon dioxide is released from anaerobic respiration in muscle.

Proteins in muscle cells bring about muscle contraction. There are two types of fibrous proteins molecule involved. These protein molecules are filamentous in shape and have either 'pegs' or 'sockets'. The pegs on one type of filament fit into sockets on the other type. When the pegs on one filament move, the other filament is moved. This mechanism depends on very precise fit of the pegs and sockets. You saw in Chapter 4 how changes in pH affect the shape of protein molecules (see page 62). Lowering the pH inside muscle cells alters the shape of the proteins slightly. This does not stop the muscle working altogether, but is enough to affect the contraction mechanism and to cause pain.

One aim of athletic training is to reduce the possibility that lactic acid will build up in the muscles and so avoid pH being lowered inside the muscle cells. Training improves the circulation to the muscles, allowing lactate ions and hydrogen ions to be removed more quickly. Training also increases the rate at which oxygen is extracted from the air in the lungs and increases its rate of circulation around the body. This helps the muscles because the more oxygen that reaches them, the greater the rate of aerobic respiration and the lower the rate of lactic acid production.

The lactate removed from muscle can be used in aerobic respiration in other parts of the body. These parts do not then need as much blood glucose, leaving more available for the muscles. The lactate is converted back to pyruvate, which can then be oxidised, releasing energy to produce ATP. Two organs in particular use lactate:

- heart muscle is particularly efficient at using lactate in aerobic respiration;
- the liver recycles lactate to form glycogen, so re-stocking the liver's glycogen stores.

4 When the oxygen supply to a muscle is not optimal, the rate of glycolysis drops.

a How does a reduction in the rate of glycolysis affect a muscle?

b How does lactate get from the muscles to other organs?

APPLICATION

Lactate and exercise

The graph below shows the changes in concentration of lactate in the blood during and after exercise. The yellow shaded area indicates the period of exercise.

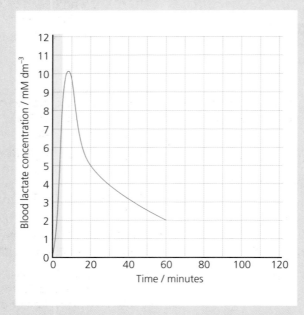

1 How long was the period of exercise?

2 For how long after the exercise finished did blood lactate concentration continue to rise?

3 By how much did blood lactate concentration rise after the exercise had finished? Explain why the rise continued after exercise.

4 Estimate how long it would take the blood lactate concentration to fall to its original value, once exercise had finished.

5 Suggest what caused the fall in blood lactate concentration.

Oxygen deficit

The graph in Fig. 3 shows typical changes in rate of oxygen uptake during and after light exercise. The yellow shaded area shows the amount of oxygen taken in during the whole period of exercise. It levels off when aerobic respiration is providing all the energy needed by the muscles. Whilst the graph is rising, much of the energy comes from stored ATP and from muscle glycogen via anaerobic respiration. This anaerobic respiration produces lactate. The blue shaded area shows the amount of oxygen that is actually needed to replenish stored ATP and to convert lactate back to glycogen. This is known as the **oxygen deficit**. After exercise, the rate of oxygen uptake remains higher than normal for quite some time. This is shown by the green shaded area in Fig. 3. This extra oxygen, the **recovery oxygen**, is used to replenish the ATP stores in the muscles and glycogen stores in both muscles and the liver. Lactate is first oxidised to pyruvate, which can then be converted back to glycogen. Some of the energy needed to convert pyruvate back to lactate comes from aerobic respiration of lactate.

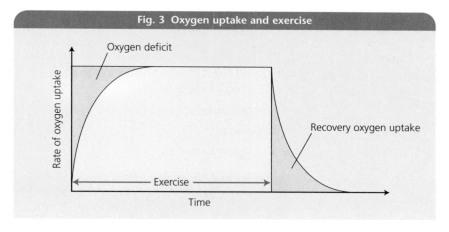

Using aerobic and anaerobic respiration

During a 100 metre sprint, the athlete hardly has time to breathe, and about 90% of the energy needed by the muscles comes from anaerobic respiration. Only 10 % comes from aerobic respiration. In an event lasting about 2 minutes, such as the 800-metre race, the proportion changes to 50:50. Fig. 4 shows the relative contributions of aerobic and anaerobic respiration for maximum rate of exercise for increasing lengths of time.

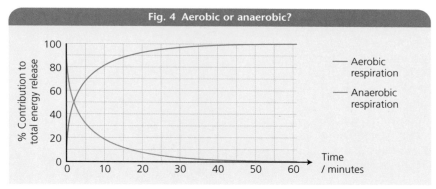

5 What proportion of energy comes from anaerobic respiration in a:

a 400 metre race lasting 50 seconds;

b 1500 metre race lasting 3 minutes 50 seconds?

c Explain the disadvantages of prolonged anaerobic respiration during a long distance race.

KEY FACTS

- Respiration releases energy that is used to produce the short-term energy storage compound ATP from ADP and phosphate.
- ATP provides the energy for muscle contraction.
- The first stage in respiration is the breakdown of glucose into pyruvate.
- In aerobic respiration, oxygen is used to oxidise pyruvate to carbon dioxide and water. This releases enough energy to produce 36 moles of ATP
- Anaerobic respiration does not use oxygen, but converts pyruvate to lactic acid releasing enough energy to produce 2 moles of ATP.
- Lactic acid dissociates into hydrogen ions and lactate. If hydrogen ions accumulate in muscles, this causes pain and the muscles do not work as efficiently.
- Lactate can be used as a fuel if it is first converted back to pyruvate.
- The muscles and the liver can convert lactate back into glycogen.
- During the early stages of exercise an oxygen deficit may build up due to anaerobic respiration.
- After exercise the rate of oxygen uptake remains high for some time, providing the recovery oxygen needed to convert lactate back into glycogen.

12.3 Control of breathing

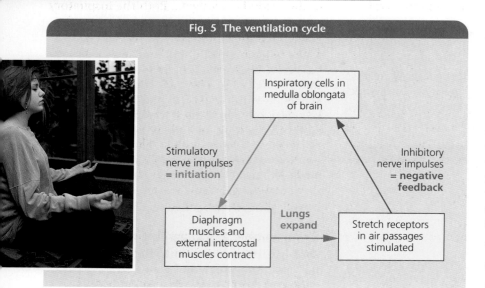

Fig. 5 The ventilation cycle

Yoga exercises like this can help you change your rate of ventilation. But you do not usually have to think about this. You breathe automatically for 24 hours a day. The more active you are the higher your ventilation rate and the deeper your breathing.

You do not usually have to think about breathing. You breathe automatically 24 hours a day due to the actions of the medulla of the brain. Control of the ventilation cycle begins with inspiratory cells (Fig. 5). These are a group of cells at the top of the medulla oblongata that pass impulses along efferent nerve cells to the diaphragm muscles and the external intercostal muscles, causing them to contract. This increases the volume of the thorax and draws air into the lungs.

The walls of the bronchi and bronchioles contain **stretch receptor cells**. These are so named because they are sensitive to stretch. As your lungs inflate, the walls of the airways distend and this stimulates the stretch receptors to send impulses along afferent nerve cells to a group of **expiratory cells** in the medulla oblongata. These expiratory cells send impulses that inhibit the production of impulses by the inspiratory cells. When this happens, the diaphragm and intercostal muscles start to relax and you breathe out. The volume of the thorax then decreases again.

6 Explain why a decrease in the volume of the thorax brings about another ventilation cycle.

Breathing and exercise

The rate of oxygen uptake into the body rises and then stabilises at a high level during exercise. This, together with increased blood flow to the muscles, increases the oxygen supply to the muscles, making them work more efficiently. The control mechanism that changes ventilation rate depends on receptor cells called **chemoreceptors** that are sensitive to chemical changes in the blood (Fig. 6). The main chemoreceptors involved are sensitive to carbon dioxide concentration and the pH of the blood. These receptors are found in three parts of the body:

- the **medulla oblongata** of the brain;
- the aorta;
- the carotid arteries.

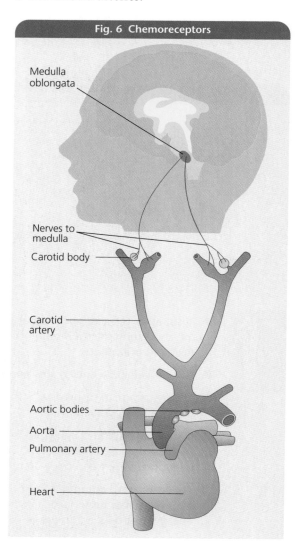

Fig. 6 Chemoreceptors

The chemoreceptors in the medulla oblongata are called the **central receptors**. Those in the arteries are called **peripheral receptors**.

During exercise the chemical composition of blood changes due to increased respiration in the muscles. There is an increase in the concentration of both carbon dioxide and hydrogen ions. The membranes surrounding the medulla are permeable to carbon dioxide but not to hydrogen ions. Once carbon dioxide has diffused into the medulla, the following reactions occur, and hydrogen ions form inside the medulla.

$$CO_2 + H_2O \rightleftharpoons H_2CO_3 \rightleftharpoons H^+ + HCO_3^-$$

The central chemoreceptors (Fig. 7) detect this rise in the concentration of hydrogen ions and send impulses to the **inspiratory cells** and another group of cells called the **ventral group**, in the medulla oblongata. Both the inspiratory cells and the ventral group send nerve impulses to the diaphragm and intercostal muscles. These impulses increase the rate and the strength of the contractions of the intercostal muscles and so increase the rate and depth of ventilation. This increases the diffusion gradient for carbon dioxide in the alveoli by lowering the level of carbon dioxide in the lungs. When the blood plasma carbon dioxide level returns to normal, the hydrogen ion concentration in the medulla falls and the central chemoreceptors stop sending impulses. The ventilation rate then returns to normal.

 What is the difference between rate of breathing and depth of breathing?

Some of the peripheral receptors send impulses to the medulla if they detect very low oxygen concentrations in the blood plasma (Fig. 8). Others are sensitive to decreases in the pH of the plasma. This change triggers the pH-sensitive peripheral receptors to send impulses to the medulla. When impulses from the peripheral receptors reach the inspiratory and the ventral cells in the medulla, they have the same effect as impulses from the central receptors; the rate and depth of ventilation increases until the carbon dioxide level returns to normal.

The effect of training on the rate of air intake into the lungs during exercise is covered in Chapter 3, on page 47.

 Draw a flow diagram to show what happens as a result of a decrease in blood pH stimulating peripheral receptors.

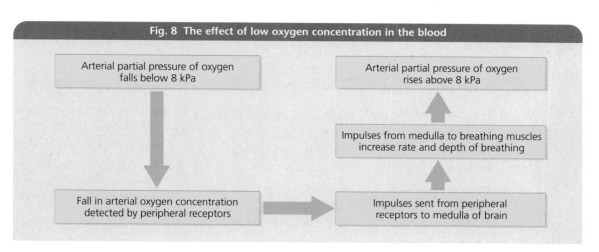

12 ENERGY AND EXERCISE

APPLICATION: Chemoreceptors and breathing

Low concentrations of oxygen in the blood plasma can affect breathing dramatically. At high altitudes, the atmospheric pressure falls and less oxygen is available to diffuse into the blood. When someone is short of oxygen, they are said to be suffering from **hypoxia**. The chemoreceptors respond to the falling levels of oxygen in the blood and the body increases its rate of ventilation so that more carbon dioxide can be breathed out. However, the concentration of carbon dioxide can fall so much that the blood pH rises. This increase in pH inhibits the activity of the chemoreceptors. This makes the problem worse by reducing ventilation rate, so that the person absorbs even less oxygen. Climbers going to high altitudes carry a supply of oxygen as a safety precaution; this is given to any climbers who start showing symptoms of hypoxia before their oxygen levels fall dangerously low.

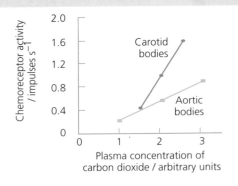

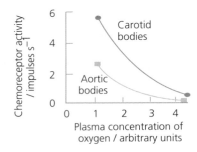

Source: adapted from Berne and Levy, Principles of Physiology, Wolfe, 1990

1. Look at the two graphs top right. Which of the aortic and carotid bodies is more sensitive to:
   a increasing levels of carbon dioxide in plasma;
   b falling oxygen concentrations in plasma?

2. Explain how the inhibition of pH-sensitive chemoreceptors affects the rate of ventilation.

3. Look at the graph bottom right. Which person, A or B, could be suffering from hypoxia?

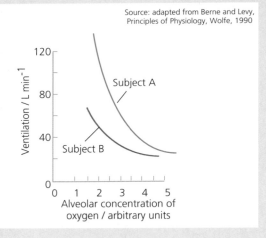

Source: adapted from Berne and Levy, Principles of Physiology, Wolfe, 1990

KEY FACTS

- The medulla oblongata controls the rate of breathing 24 hours a day.
- The ventilation cycle involves **inspiratory cells**, **expiratory cells** and **stretch receptor cells**.
- **Central receptors** in the **medulla** of the brain are sensitive to changes in pH, and therefore to changes in carbon dioxide concentration in the blood.
- **Peripheral receptors** in the aortic bodies and the carotid bodies are sensitive to carbon dioxide and oxygen concentrations in the blood and blood pH.
- An increase in carbon dioxide levels or a decrease in oxygen levels in the blood brings about increases in the rate and depth of ventilation.

12.4 Controlling heartbeat

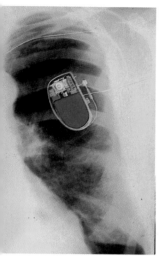

A coloured X-ray of the chest of a patient who has had a heart pacemaker fitted. The electronic battery-run device can be seen above the ribcage, with a yellow lead connecting it to the heart on the lower right of the picture. Most pacemakers are internal, implanted into the chest like this one, but they can also be external and worn on a belt fastened around the body. Pacemakers can supply a fixed rate of impulses or can discharge only when a heart beat is missed.

Like breathing, the beating of the heart is automatic; it happens without us thinking about it. However, unlike the muscles used in breathing movements, heart muscle does not need nerve impulses from the brain to keep up steady contractions. The heart has its own internal pacemaker, the **sino-atrial node** (SAN) (Fig. 9). The SAN is a group of cells in the wall of the right atrium that produces electrical impulses at regular intervals. These spread out across the muscles of the heart causing the atria to contract first, followed by the ventricles.

The electrical impulses are carried by specialised muscle cells that behave like nerve cells. As impulses travel along these muscle cells, the surrounding atrial muscles cells contract. This has a sort of 'domino' effect, causing neighbouring muscle cells to contract too. When the atria contract, the heart is said to be in **atrial systole**.

The impulses are prevented from spreading directly to the ventricle muscles by the layer of connective tissue that separates the atria and the ventricles. A group of receptor cells called the **atrioventricular node** (AVN) is found near to this junction. When impulses from the contracting atrial muscle cells reach the AVN, this starts impulses in a group of specialised conductive muscle fibres called the **Purkyne fibres**. These fibres group together to form the **bundle of His**, which passes down the wall between the two ventricles.

At the base of the ventricles the bundle of His divides into two branches, one passing to each ventricle. Fibres fan out from each of these and as impulses reach them from the bundle of His, the electrical activity causes the surrounding ventricle muscles to contract. Because the bundle of His fibres do not affect the muscle in the wall between the two ventricles, muscle contraction starts at the base of the ventricles and spreads upwards. Contraction of the ventricles is called **ventricular systole**.

The heart's natural pacemaker can develop problems - one of the most common is heart block. Damage to the electrically conductive tissue of the heart blocks the conduction of impulses from the atria to the ventricles. Hundreds of thousands of people suffer from heart block each year in Britain but they are often saved by implantation of an artificial pacemaker. This device overcomes the heart block by sensing the level of electrical activity in the atrium and delivering an electrical impulse at the correct rate to the ventricle. The pacemaker mimics the natural activity of the sino-atrial node, making the heart rate speed up or slow down, to suit the activity of the body.

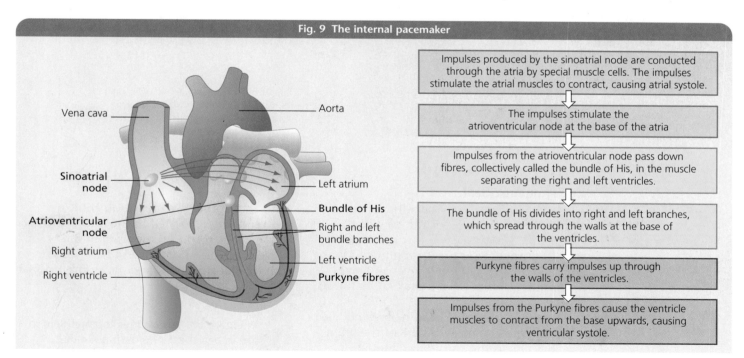

Fig. 9 The internal pacemaker

12 ENERGY AND EXERCISE

> **KEY FACTS**
> - The **sino-atrial node** (SAN) is the 'natural' pacemaker of the heart.
> - The SAN does not require impulses from the nervous system to initiate electric impulses.
> - Impulses are conducted from the SAN, first to the atria and then to the ventricles.
> - These impulses cause the atria to contract, followed by the ventricles.

EXTENSION The electrocardiogram

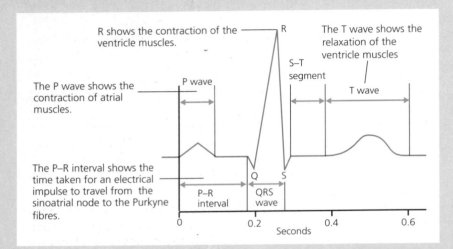

A machine called an electrocardiograph (ECG) can record electrical impulses as they pass from the SAN through the heart. The diagram shows how an ECG is related to pressure changes in the cardiac cycle.

If a thrombosis cuts off the blood supply to Purkinje fibres, they die and prevent electrical impulses from passing to healthy muscle cells. This is the main cause of deaths from heart attacks. Damage to any part of the heart's muscle can be recognised by changes to the ECG of the patient. If an ECG shows a large P–R interval, it probably means that there has been damage to the conducting fibres in the region of the atrioventricular node. On seeing such a trace a doctor would consider fitting the patient with a pacemaker.

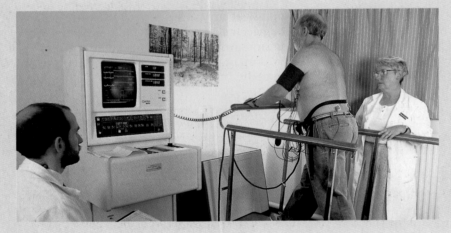

This patient is exercising whilst an ECG records the electrical activity of his heart. The various electrodes taped over his chest and back are connected to the ECG machine.

Modifying heartbeat

Although the heart can beat without nerve impulses, nerves are needed to change the rate of the heartbeat. The heart of a healthy adult beats roughly 70 times per minute. During exercise the rate may rise to over 140 beats per minute. During sleep it may fall as low as 50 beats per minute.

The rate at which the heart beats is modified by nerve signals from two areas of the brain (Fig. 10). One is the **cardioaccelatory centre**, the other is the **cardioinhibitory centre**. Both are located in the cardiovascular centre in the medulla oblongata of the brain. Nerve fibres pass from each of these centres to both the sinoatrial node and the atrioventricular node.

 What are the advantages to an athlete in a rise in heartbeat rate during a race?

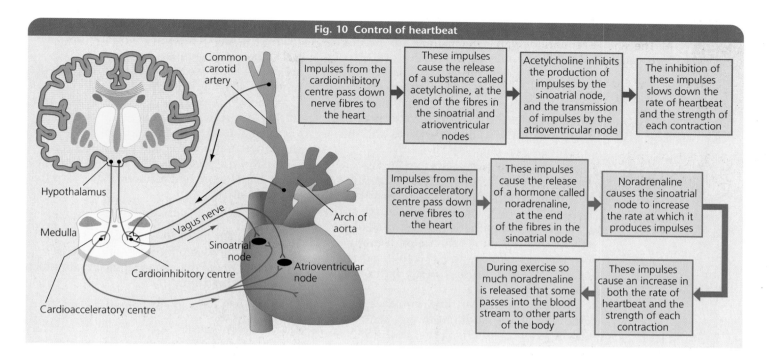

Fig. 10 Control of heartbeat

Chemoreceptors and pressure receptors

We do not know what causes the cardio-acceleratory centre to produce impulses that speed up heart rate. However, we do know it is usually linked to an increased ventilation rate. If the oxygen concentration of the blood is low, or if the carbon dioxide concentration is high, ventilation rate increases. This somehow also causes the rate of heartbeat to increase.

> **EXTENSION**
>
> ### Carbon dioxide and scuba diving
>
> As scuba divers become more adventurous and seek greater challenges, technologists produce more sophisticated apparatus for them. One of the latest machines off the production line is the rebreather. Many scuba divers use rebreathers, but few understand how they work.
>
> The principle behind them is the fact that we only extract about 20% of the oxygen we breathe in; the rest is just breathed out. In ordinary scuba apparatus this means that 80% of the compressed air in the diving tanks is wasted. So what is the problem with recycling the air breathed out? The problem is that this exhaled air contains up to 6% of carbon dioxide. If this air is rebreathed there is a very much reduced diffusion gradient from blood to air in the lungs, so the rate of diffusion of carbon dioxide out of the blood falls, resulting in an increase in blood carbon dioxide levels. You have learned in this chapter what effect this would have on breathing rate and subsequently on heart rate. There is an additional effect though – **hypercapnia**. This is too much carbon dioxide in the blood and symptoms of hypercapnia include headache, dizziness, confusion, and unconsciousness. It is obvious, therefore, that in order to recycle our exhaled breath, we need to remove the carbon dioxide from the expired gas. But how could this be done with equipment that could also be used for scuba diving?
>
> The answer came from the space programme; space engineers use carbon dioxide scrubbers to remove carbon dioxide from the air in space suits, rocket capsules, and the space shuttle. The technology is tried and tested, and the materials used are widely available. But the introduction of rebreather technology to recreational diving has been hindered by a combination of technical and logistical factors. Firstly, there was the problem of cost. NASA could easily afford space suit rebreathers that cost $250,000 but for the average scuba diving enthusiast, this was impossible. The US Navy adapted the technology for their use and brought the cost down to $50,000 for each underwater system. More recently, the cheaper cost of computer technology has made space-suit technology available to scuba divers. Their rebreathers now recycle their air supply, removing the carbon dioxide over several cycles and the oxygen tanks last much longer.

Chemoreceptors sensitive to oxygen and carbon dioxide concentrations are present in the walls of the aorta and the carotid arteries. However, these chemoreceptors do not control the heart rate directly. The cardioaccelerary and cardioinhibitory centres that affect heart rate are under the direct control of **pressure receptors** in the walls of the aorta and the carotid arteries (Fig. 11).

The heart rate increases when impulses produced by the cardioaccelerary centre pass down sympathetic nerve fibres to the SAN and the AVN. Impulses that arrive at the ends of the sympathetic fibres stimulate the release of a hormone called **noradrenaline**. Noradrenaline causes the SAN to increase heart rate and increases the strength of each systole.

To slow down heart rate, impulses produced by the cardioinhibitory centre pass down parasympathetic nerves to the SAN and the AVN. When the impulses arrive at the ends of the sympathetic fibres, this releases **acetylcholine**. Acetylcholine causes the SAN to reduce the rate of heartbeat and inhibits the transmission of impulses by the AVN. The actions of the *sympathetic* and *parasympathetic* fibres are therefore **antagonistic**.

Although these changes seem to be linked to changes in the breathing rate, no-one has yet been able to show a direct link between the concentration of respiratory gases in the blood and the production of impulses by the cardioaccelerary and cardioinhibitory centres of the medulla.

10
a What effect would an increase in blood pressure have on the rate of heartbeat? How would this change in heart rate affect blood pressure?

b How do sympathetic impulses affect the heart?

c How do parasympathetic impulses affect the arterioles?

Fig. 11 Pressure receptors

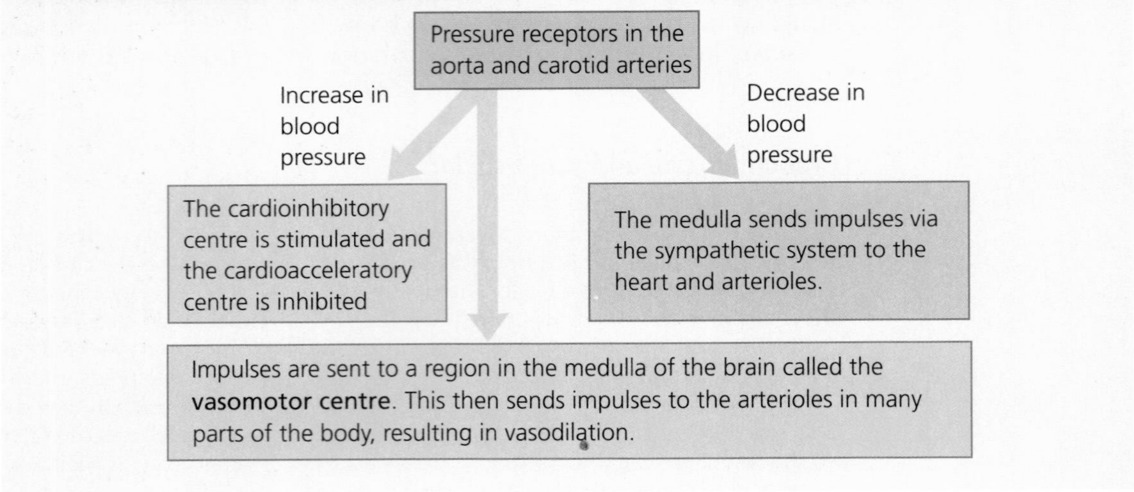

Pressure receptors in the aorta and carotid arteries

Increase in blood pressure → The cardioinhibitory centre is stimulated and the cardioaccelerary centre is inhibited

Decrease in blood pressure → The medulla sends impulses via the sympathetic system to the heart and arterioles.

Impulses are sent to a region in the medulla of the brain called the **vasomotor centre**. This then sends impulses to the arterioles in many parts of the body, resulting in vasodilation.

KEY FACTS

- The rate at which the sino-atrial node sends out impulses can be modified by the nervous system and by hormones.
- Modification of heartbeat is controlled by the medulla of the brain.
- Impulses via sympathetic nerve fibres speed up the rate at which the sino-atrial node sends out impulses.
- Impulses via parasympathetic nerve fibres slow down the rate at which the sino-atrial node sends out impulses.
- The medulla of the brain is influenced by impulses from pressure receptors and chemoreceptors in the walls of the aorta and carotid sinuses.
- There is no direct relationship between the oxygen and carbon dioxide concentration of the blood and the rate of heartbeat.

EXAMINATION QUESTIONS

1 The bar chart shows the sources of energy for 1 hour of cycling.

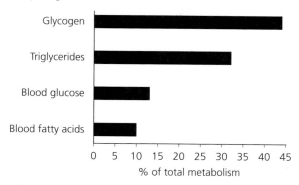

a What percentage of the energy came from carbohydrates? (1)

b Where in the body is each of the following stored:
  i) glycogen;
  ii) triglycerides. (2)

c What is the source during exercise of:
  i) blood glucose;
  ii) blood fatty acids.

d i) In which organelle is most energy released?
  ii) Name the substance, produced during respiration, that acts as an intermediate energy source for muscle contraction (2)

BY03 Jun 98 Q6

2 The graph shows changes in blood lactate concentration during and after exercise.

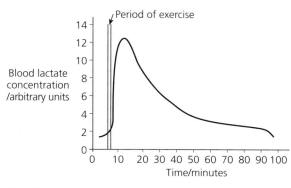

a How long after the period of exercise did it take for the blood lactate level to return to normal? (1)

b i) Where in the muscle cells is lactic acid produced? (1)
  ii) Name the process by which lactic acid is produced. (1)
  iii) Explain the advantage to an athlete of lactic acid production by muscles during exercise. (3)

c Explain how the production of lactic acid may lead to muscle fatigue. (3)

d Explain why the concentration of blood lactate:
  i) continues to rise after exercise; (2)
  ii) eventually returns to normal. (3)

Sample question written for this book

3 a Describe how the human respiratory system is ventilated. (4)

b Describe the role of the nervous system in:
  i) maintenance of regular breathing whilst the body is at rest;
  ii) increasing the rate of ventilation during exercise. (8)

BY03 Jun 96 Q9

4 The drawing shows the pathways for the conduction of electrical impulses during the cardiac cycle.

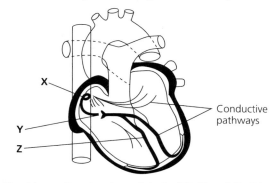

a i) Name the structures labelled X, Y and Z. (3)
  ii) In which chamber of the heart is structure X located? (1)

b Describe the role of structure X in the control of the cardiac cycle. (2)

c The table shows the pressures in the left atrium, left ventricle and aorta during a single cardiac cycle.

	Pressure / kPa		
Stage	Left atrium	Left ventricle	Aorta
1	0.5	0.4	10.6
2	1.2	0.7	10.6
3	0.3	6.7	10.6
4	0.4	17.3	16.0
5	0.8	8.0	12.0

Give the number of one stage when:
  i) blood flows into the aorta;
  ii) the atrioventricular valves are open;
  iii) the atrioventricular valves and the semilunar valves are closed. (3)

BY03 Mar 98 Q5

5 The diagram shows a vertical section through a human heart. The arrows represent the direction of the movement of the electrical activity which starts muscle contraction.

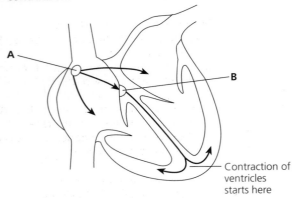

a Name structure A (1)

b Explain why each of the following is important in the pumping of blood through the heart.
   i) There is a slight delay in the passage of electrical activity that takes place at point B. (1)
   ii) The contraction of the ventricles starts at the base (1)

c Describe how stimulation of the cardiovascular centre in the medulla may result in an increase in heart rate. (2)

BY03 Jun 97 Q1

6

a i) Name one location of chemoreceptors which, when stimulated, may lead to an increase in heart rate. (1)
   ii) Name one stimulus that may trigger these receptors. (1)

b In some people narrowing of the coronary arteries leads to a condition known as angina, where sudden exertion is accompanied by chest pains. Angina sufferers can be treated with drugs called beta-blockers. The diagram illustrates how such a drug works.

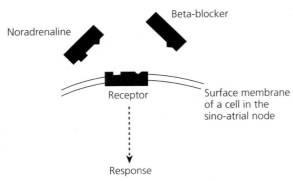

   i) Exertion leads to an increase in noradrenaline. Suggest what happens when the sino-atrial node is stimulated by noradrenaline. (1)
   ii) Suggest how beta-blockers may help angina sufferers. (3)

BY03 Jun 96 Q2

12 ENERGY AND EXERCISE

KEY SKILLS ASSIGNMENTS

Coaches corner

Here are some extracts from the official 'Coaches Corner' for the marathon.

> Food is the fuel that will help to ensure your body makes it through the miles of training in the months ahead.
>
> You should aim to get around 60 percent of all your daily calories from foods which are high in carbohydrates. Unlike protein and fat, carbohydrates are stored in your muscles so that it is readily available when you are running.
>
> A few days before the race there is still plenty to do. You should concentrate on eating a high carbohydrate diet so that your muscles are constantly refuelled.
>
> Ideally you should eat your last big meal at lunch time on the day before the race, so head down to the pasta party. Have a light evening meal and a bedtime snack. Don't go to bed feeling hungry.
>
> On race morning the ideal meal should be high in carbohydrate with a little low-fat protein to make it more digestible. Something like lightly scrambled egg on toast is ideal.
>
> After the race you should aim to eat 0.5 g of carbohydrate for every 0.5 kg of body weight 2-3 hours after you finish to top up your depleted glycogen stores.
>
> Celebrate being a hero. Well done.

1 Use a suitable diet program on a computer to design a diet which would provide about 525 g of carbohydrate for one of the days just before the marathon. This carbohydrate should provide about 65% of the body's energy requirements for the day.

Endurance athletes like this marathon runner tend to be of slight build, with small muscles and very little body fat. Anyone can train to run a marathon, providing they don't have serious helath problems but the training is long and difficult.

2 Where is the main store for the athlete's;
 a carbohydrate? **b** fat?

3 Why are the athlete's advised
 a to go to the pasta party on the day before the marathon?
 b to eat scrambled egg on toast for breakfast on the morning of the marathon?

4 a How much carbohydrate should a 70 kg runner consume 3 hours after the race?
 b Describe what will happen to this carbohydrate after it has been digested.

5 Use the internet and other information sources (printed material, CD ROMs, databases and personal interviews) to create a program for the following athletes:

> A fairly fit amateur runner preparing for the London marathon

> An unfit, middle-aged male preparing for a fun run in 3 months time

> A fit, 17-year-old female swimmer who is preparing for a large meeting in the summer but also having to revise for her GCE examinations.

Each competitor has six months to prepare for their chosen events. For each competitor you will need to research the fitness needs of their chosen events and prepare a fitness programme that matches their current physical condition. You can assume that none of them suffer from heart or respiratory illnesses.

While you are conducting your searches make sure you keep a record of all of your information sources. You will need to list these when you present your final report. The report should be presented on no more than 2 sides of A4 paper for each competitor and must include:

- a month-by-month programme for each competitor;
- four pieces of good advice to help them stick to the programme you have devised;
- a list of all your information sources indicating which were the most useful.

13 Transport systems in plants

When rains do fall, desert plants burst into flower, making the most of the water that is now available.

Much of the land that is desert today used to be fertile. The change to desert began about 5 000 years ago, when a shift in the Earth's climate caused the rainfall in some regions to decrease. The land became drier and plants withered. The soil blew away, leaving large stretches of drifting sand. In some areas, humans made the problem worse by using poor farming techniques.

Once an area has become desert, the air above the land becomes very dry and no clouds form. In the burning heat of the day, temperatures can rise as high as 50 °C. However, as there are no clouds to provide any insulation, heat radiates out into space very rapidly at night and deserts can be freezing cold. Water from occasional rain collects underground and wherever this happens, plants grow. These plants need to conserve water to survive long dry spells. Plants that have adapted to arid conditions often look spectacular.

Reclaiming areas of desert is difficult; crops that are introduced to desert environments often fail. Scientists are increasingly coming to recognise that the solution to feeding people in arid lands is to grow plants that have become adapted to dry conditions over millions of years. Research into the survival methods of plants from arid areas could help turn desert into productive land. While science cannot solve the political and economic causes of famine, it can help to increase food production. In arid areas, this can mean the difference between life and death.

13.1 Transport in plants

If you don't water a pot plant regularly the soil in the pot will soon dry out and the leaves will droop. Plants are constantly absorbing water from the soil, but, at the same time, they are also losing water vapour to the air. This creates a continual stream of water through the plant. Plants do not have complex circulatory systems with a heart to pump the fluid around the organism; movement of water and the substances dissolved in it is brought about by simple physical processes. In this chapter we will look at the way in which evaporation, osmosis and active transport can move materials round plants as small as a daisy and up trees taller than a cathedral. We shall also investigate some of the ingenious adaptations that plants have evolved for retaining water.

13.2 Transpiration

Photosynthesis is the primary function of leaves and for this they need carbon dioxide, water and sunlight. Leaves obtain carbon dioxide from the air through tiny pores. Each pore is called a **stoma**, and the plural term is **stomata**. Two cells called **guard cells**, surround each stoma and we find out more about their role in stomatal opening and closing later.

When the stomata are open, the hole that allows carbon dioxide to pass into the leaf also lets water molecules out. This is a disadvantage to the plant because it loses precious water; if the loss of water from the leaves is greater than the uptake of water by the roots, the plant wilts. For many plants, stomata are therefore a 'necessary evil'.

The loss of water vapour by evaporation from a plant to the air surrounding it, through the open stomata, is called **transpiration**. This loss of water vapour via the stomata is due to the water potential (ψ) gradient between cells inside a leaf and the air outside (Fig. 1).

Factors that affect transpiration

Water molecules diffuse along a water potential gradient towards areas with a more negative water potential value. The greater the gradient, the faster the rate of movement. The air inside the leaf is always saturated with water vapour so changes outside the leaf can alter the water potential gradient. Three factors increase this gradient and therefore the rate of transpiration:

- an increase in temperature;
- a decrease in humidity;
- an increase in wind speed.

An increase in temperature causes a higher rate of evaporation from the leaf surface by transferring more energy to the water molecules. The more energy a water molecule has, the faster it moves and the more likely it is to escape into the atmosphere outside the leaf and move away from the plant. Humidity is a measure of the number of water molecules in the air. Any *decrease* in humidity in the atmosphere around the leaf decreases the number of water molecules in the air and so *increases* the water potential gradient from the plant to the air. Warm air can hold more water vapour molecules than cold air; that is why the air in tropical forest feels so uncomfortable, and why dew forms as water condenses from cooling air. On a hot day the air in a tropical jungle may become fully saturated with water vapour, reducing the transpiration to nil as there is no longer a water potential gradient between the plants and the air. When wind moves the air and water vapour away from around a leaf, this increases the water potential gradient from the plant to the air.

> **1** Explain how an increase in temperature and wind speed and a decrease in humidity make the water potential of the air just outside the leaf more negative. Use the idea of the number of water molecules per unit volume of air in your answer.

Air always has a value of water potential more negative than that of the leaf cells, except when it is fully saturated with water vapour. So water molecules will always diffuse out through the open stomata. The only way to stop the loss of water is to close the stomata. Most plants close their stomata during the night because they do not need to take in carbon dioxide for photosynthesis when it is dark. If a plant has lost too much water during the day it gets the chance to replenish its supplies during the night.

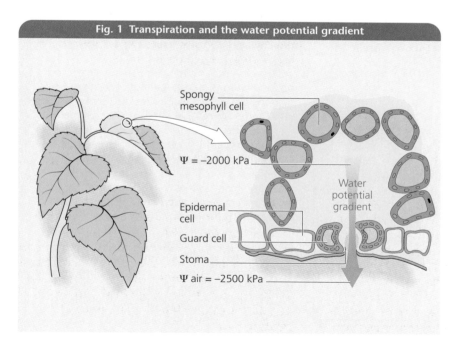

Fig. 1 Transpiration and the water potential gradient

Spongy mesophyll cell
$\psi = -2000$ kPa
Epidermal cell
Guard cell
Stoma
ψ air $= -2500$ kPa
Water potential gradient

13 TRANSPORT SYSTEMS IN PLANTS

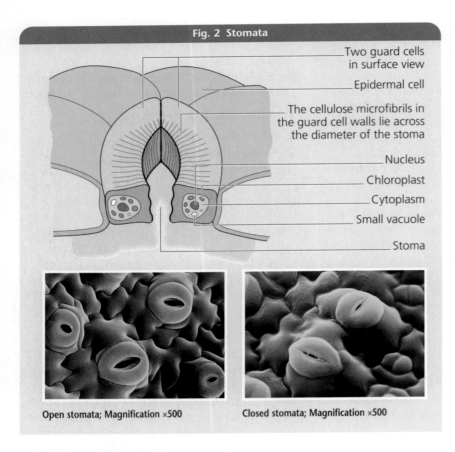

Fig. 2 Stomata

Open stomata; Magnification ×500

Closed stomata; Magnification ×500

How do stomata open and close?

The guard cells surrounding the stoma have two special features that allow the pore to open and close (Fig. 2):

- the guard cells are not connected along the whole of their length;
- there are additional cellulose microfibrils in the inner walls of the guard cells that line the stoma.

When the guard cells take in water, they become turgid but they cannot expand in diameter because the cellulose microfibrils will not stretch. They therefore increase in length particularly along the thinner, outer walls. The outer walls stretch more than the inner walls, causing the cells to change shape and to form a gap between them. This gap is the stoma.

EXTENSION — Stomatal opening and closing

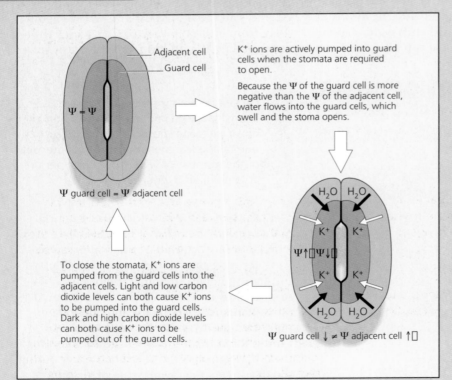

Ψ guard cell = Ψ adjacent cell

K^+ ions are actively pumped into guard cells when the stomata are required to open.

Because the Ψ of the guard cell is more negative than the Ψ of the adjacent cell, water flows into the guard cells, which swell and the stoma opens.

To close the stomata, K^+ ions are pumped from the guard cells into the adjacent cells. Light and low carbon dioxide levels can both cause K^+ ions to be pumped into the guard cells. Dark and high carbon dioxide levels can both cause K^+ ions to be pumped out of the guard cells.

Ψ guard cell ↓ ≠ Ψ adjacent cell ↑

A change in turgidity of guard cells causes stomata to open and close. But what is the mechanism that controls the turgidity of the guard cells? Any hypothesis for the mechanism of stomatal opening must account for the more negative water potential of the guard cells during the day, and also why some plants can keep their stomata closed during the day.

One hypothesis involves the movement of potassium ions (K^+) into and out of the guard cells. To open the stomata, adjacent cells pump K^+ ions into the guard cells, as shown in the diagram.

Light and carbon dioxide both affect the opening and closing of stomata. However, they can both be over-ridden by a plant hormone called abscisic acid (ABA). ABA is produced when the plant suffers water stress. This happens when a plant loses much more water through transpiration than it can take in through the roots and is common in plants growing in a desert environment. ABA causes a rapid pumping of K^+ ions from the guard cells into the adjacent cells. This explains how some plants can close their stomata at midday to conserve water. It also explains the complete closing of stomata when the soil is very dry.

APPLICATION: Factors that affect stomatal opening

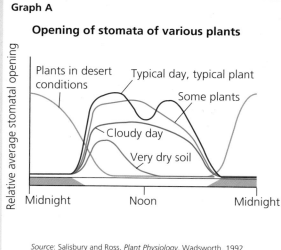

Graph A Opening of stomata of various plants

Source: Salisbury and Ross, *Plant Physiology*, Wadsworth, 1992

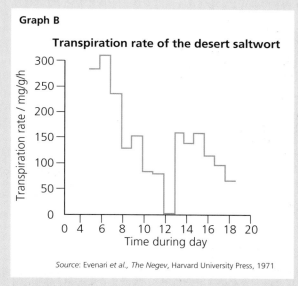

Graph B Transpiration rate of the desert saltwort

Source: Evenari et al., *The Negev*, Harvard University Press, 1971

Graph A shows how external factors affect stomatal opening.

1. Study graph A carefully. Compared with a typical plant on a typical day, what effect does each of the following have on stomatal opening and closing;
   a. very dry soil?
   b. a cloudy day?

   Some plants that are adapted to living in hot environments can close their stomata during the middle of the day, when the temperature is highest, and can open them at night. One such desert plant is the saltwort. The effect of its lunch-time stomatal closing is dramatic: transpiration almost ceases, as Graph B shows.

2. What are the advantages and disadvantages to a desert plant of opening its stomata during the night and closing them around midday?

   Other plants keep their stomata open all day, even at noon (see the light blue curve shown in Graph A). They only close their stomata at night.

3. What type of plant could keep its stomata open all day without suffering the effects of severe water loss?

KEY FACTS

- Leaves must allow the entry of carbon dioxide needed for photosynthesis. Most have **stomata**, pores in the leaf that allow gas exchange to occur.

- Open stomata allow water vapour to diffuse out of the leaf, a process called **transpiration**.

- Water vapour diffuses out through stomata down a water potential gradient between the leaf cells and the air outside.

- This water potential gradient, and so the rate of transpiration, is increased by an increase in air temperature, a decrease in humidity outside the leaf and an increase in air movements around the leaf.

- Stomata open when water enters the guard cells and pushes against their cell walls.

- The stomata of most plants are open during the day and closed during the night. In some desert plants the stomata close during the day to conserve water and open at night to collect carbon dioxide.

13 TRANSPORT SYSTEMS IN PLANTS

13.3 Xerophytes

Plants that live in dry places usually have adaptations to survive long periods of drought. Such plants are called **xerophytes**.

Many desert trees, such as the quiver tree, drop their leaves to stop the whole tree dying from desiccation. This certainly works, as it stops transpiration completely but it is a drastic solution. A tree without leaves can no longer photosynthesise, so it must rely on the store of carbohydrates made during better times.

Cacti also have many adaptations that enable them to survive in deserts:

- The spines on the cactus trap a layer of air that is rich in water vapour next to the plant. This reduces the chances of wind moving the moist air layer away from around the plant. Many cacti have a dense covering of hairs that traps even more water vapour.

Quiver trees in Namibia can shed their leaves and store water in their branches to survive periods of extreme drought. They are called quiver trees because native people cleaned out the soft branches and used them as a quiver (a pouch) for their arrows.

The leaves of the teddybear cactus, found in Arizona, USA, no longer function as photosynthetic organs. Millions of years of adaptation have reduced them to spines.

- The spines and stiff hairs on cacti also help to deter animals that try to eat the cactus to get at the stored water.
- The epidermal cells of cacti have a thick outer layer of wax that reduces water loss by evaporation.
- Many cacti store water in their stems; plants that do this are called **succulents**.

The epidermal cells of cactus stems have chloroplasts. These enable the stems to photosynthesise, since the spine-like leaves have lost this function. The thick waxy layer that prevents evaporation of water from cactus stems also prevents carbon dioxide getting in so the epidermis of a cactus stem has stomata. However, these stomata are closed during the day, so how does the plant obtain the carbon dioxide it needs for photosynthesis? Cacti have a special way of obtaining carbon dioxide for photosynthesis – **crassulacean acid metabolism** (CAM). CAM plants open their stomata at night to absorb carbon dioxide. The epidermal cells combine this gas with an organic compound to form malic acid. This is stored in the vacuole. The large amounts of malic acid that are stored in the vacuoles at night makes a slice of the plant taste sour. During the day the malic acid in the vacuole is broken down to release carbon dioxide, which diffuses to the chloroplasts to be used in photosynthesis.

$$\text{Organic compound} + CO_2 \rightleftharpoons \text{Malic acid}$$

2 Although cacti seem well adapted to life in deserts, their rate of growth is usually very slow. Suggest what factor limits their rate of growth. Explain your answer.

KEY FACTS

- Plants adapted to survive in dry, desert conditions are called xerophytes.
- Some **xerophytes** reduce the area for transpiration by producing smaller leaves during dry periods or by modifying the shape of the leaves.
- Cacti have reduced leaves to spines to conserve water.
- Some desert plants use **crassulacean acid metabolism** (CAM) to collect carbon dioxide. CAM plants open their stomata only at night, store the carbon dioxide absorbed at night as malic acid, then release it during the day for use in photosynthesis.
- Many xerophytes sink their stomata into pits.
- Roots of desert plants often spread out over a wide area.

APPLICATION: Resourceful xerophytes

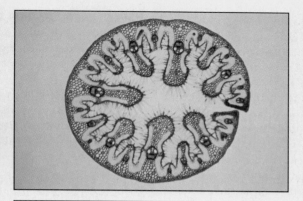

Marram grass usually grows in exposed sandy sites (above). The intricate structure of its hinged leaf is shown in the micrograph (top right) and explanatory diagram (right).

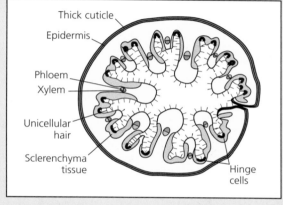

Labels: Thick cuticle, Epidermis, Phloem, Xylem, Unicellular hair, Sclerenchyma tissue, Hinge cells

The diagram below shows the cross-sectional structure of the winter and summer leaf of the grey sagebrush.

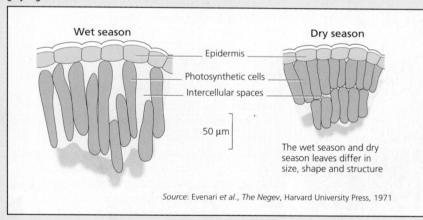

Wet season / Dry season. Labels: Epidermis, Photosynthetic cells, Intercellular spaces. 50 μm.

The wet season and dry season leaves differ in size, shape and structure.

Source: Evenari et al., *The Negev*, Harvard University Press, 1971

Xerophytes are all adapted to dry conditions, but the strategies and adaptations they use differ widely from plant to plant.

The sand dunes on the sea shores in Britain have soil that often contains very little water, but some plants are very well adapted to living there. Perhaps the most successful is marram grass. When the soil is dry hinge cells in the leaf make it roll into a cylinder. The air inside this cylinder quickly becomes saturated with water, resulting in a reduction in transpiration rate. When the soil is moist the leaves uncurl.

A rather strange plant of the *Tortula* species, commonly called the 'all-screwed-up-moss' (right) also lives on sand dunes. It spends much of its life all screwed up but when rain falls its leaves unwind. Some xerophytic plants have different shapes of leaf in wet and dry seasons. Collecting every drop of available water is vitally important to desert plants. Often the roots of these plants spread over a very wide area.

Twisted moss — When rain falls / In drought

1. Suggest two adaptations that reduce the rate of transpiration even when the marram grass leaves are uncurled.

2. Explain how the behaviour of the all-screwed-up-moss helps it to survive dry periods.

3. Explain the advantage to the sagebrush of having smaller leaves during the dry season.

4. Explain why the root systems shown below are xerophytic adaptations.

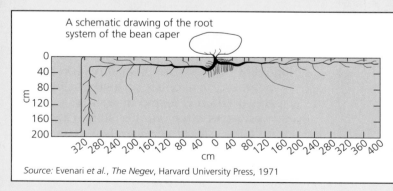

A schematic drawing of the root system of the bean caper

Source: Evenari et al., *The Negev*, Harvard University Press, 1971

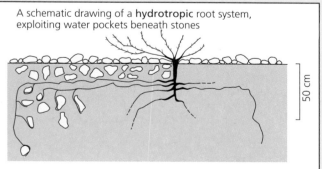

A schematic drawing of a **hydrotropic** root system, exploiting water pockets beneath stones

13 TRANSPORT SYSTEMS IN PLANTS

13.4 Movement of water up the plant

How does water get into the roots and from there up to the leaves? The simple answer is that water enters a plant by osmosis through the root hairs. Whilst this does happen to a limited extent, the bulk of the water that travels through a plant from the soil to the air passes mainly through non-living parts of the plant. There is a continuous flow of water along a water potential gradient that extends from the air outside a leaf right down through the stem to the water in the soil. To see where these non-living parts are we first have to consider the structure of the root

Fig. 3 Root structure

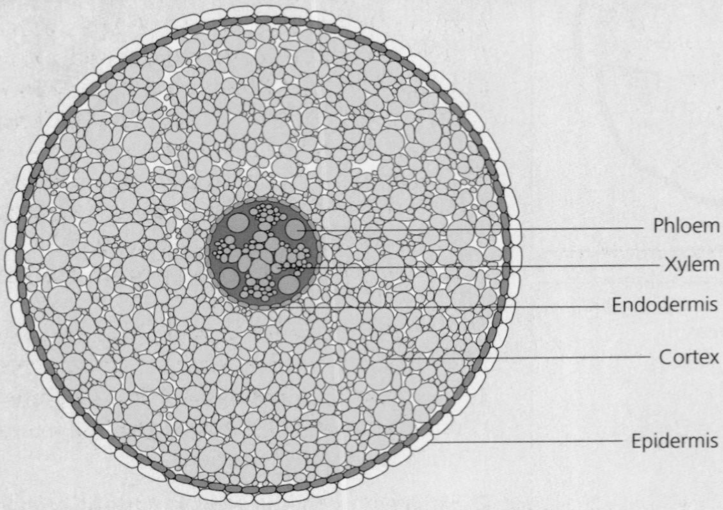

Labels: Phloem, Xylem, Endodermis, Cortex, Epidermis

Diagram (above) and light micrograph (below), showing the cellular structure of a young root. Both are transverse sections through the root of a dicotyledenous plant.

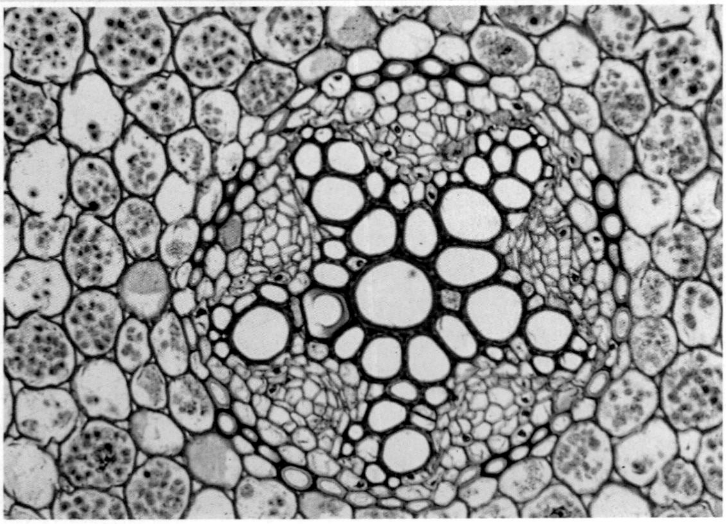

Path of water through the plant root

Fig.3 shows a drawing and photograph of a section through a young root. The main features of a young root are:

- the single outer layer of cells called the **epidermis**; these cells are where water and mineral ions enter the plant; the surface area of many of these cells is increased by projections called root hairs;

- a layer of cells called the **cortex**; these cells are similar to mesophyll cells in a leaf but they do not have chloroplasts;

- the endodermis which is a layer of cells each possessing a **Casparian strip**; this strip is impermeable to both water and mineral ions. It blocks the path to the centre of the root. All substances must pass through the cell surface membrane of the endodermal cells to get deeper into the root. In this way the endodermal cells control which substances pass from the roots to the rest of the plant;

- a central core of vascular tissue consisting of xylem cells which transport water and mineral ions to the rest of the plant and phloem cells which transport organic materials.

Apoplast and symplast

Water passes into and through a root along two main routes, the **symplast** and the **apoplast** (Fig 4). The symplast system involves the living contents of root cells including the membranes, cytoplasm and vacuoles. Minute strands of cytoplasm called **plasmodesmata** pass through the cell walls, connecting the symplasts of adjacent cells. Some water moves by osmosis along a continuous water potential gradient from the soil to the xylem cells in the centre of the root.

The apoplast is also a continuous pathway from the water in the soil to the endodermal cells but, in contrast to the symplast, the apoplast pathway comprises only non-living structures. The apoplast pathway includes the spaces between cells and the cell walls. Spaces between the cellulose fibres in the cell walls allow water to pass easily. When a plant is transpiring a continuous stream of water is literally pulled through the apoplast of the root as far as the endodermal cells, then through the cytoplasm of the endodermal cells into the xylem cells. So, the apoplast and the symplast are separate, continuous systems. Fig. 4 shows the movement of water through the apoplast and symplast systems in a root.

Xylem

Cells that make up the xylem vessels (Fig. 5) have no living contents at maturity. They are therefore entirely apoplast. Water is pulled through the apoplast on its way from the soil water to the air. The driving force is the gradient of water potential between air and soil water. Air has a very negative water potential (approximately -2500 kPa) whereas soil water has a value approaching zero.

Xylem cells join to form continuous tubes that stretch from the roots to the leaves. The tubes formed by the xylem cells function like the water pipes in your home – and, like your plumbing, they work only if there is nothing to block the flow of water. To move water along a stem needs a difference in water potential of 100 kPa for every 10 metres to overcome the forces between the water molecules and the walls of the xylem cells. Add to this a difference of 100 kPa needed to overcome the force of gravity for each 10 metres in height, and a difference of 200 kPa in water potential is required to move water 10 metres up a tree. Even pulling water this short distance up a tree needs a lot of energy. Californian redwood trees can reach heights of more than 110 metres – imagine the energy needed to pull water to that height.

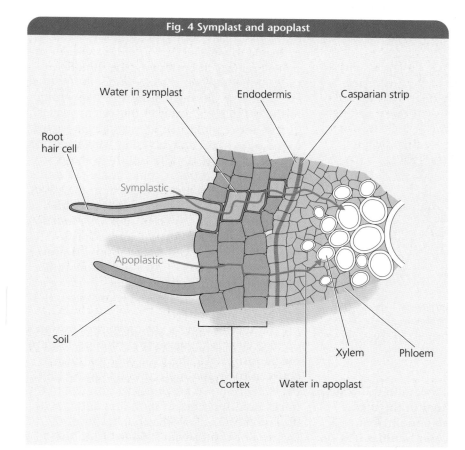

3 If the difference in water potential between the air next to a leaf and the water in the soil is 2500 kPa, what is the theoretical maximum height for a tree?

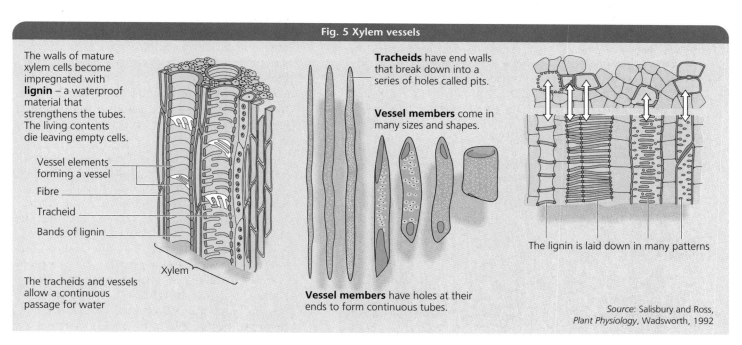

Source: Salisbury and Ross, *Plant Physiology*, Wadsworth, 1992

13 TRANSPORT SYSTEMS IN PLANTS

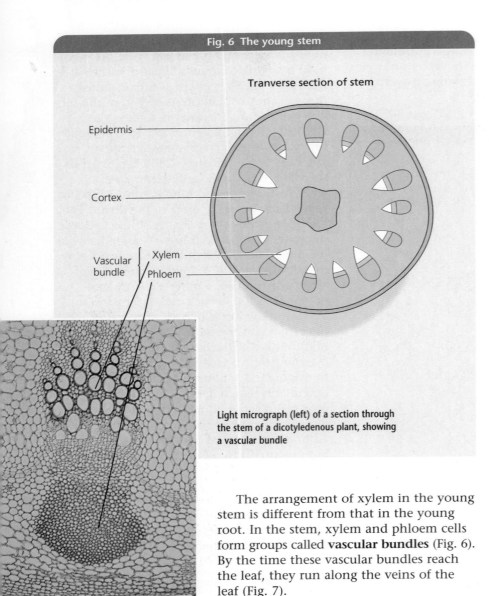

Fig. 6 The young stem

Tranverse section of stem

Light micrograph (left) of a section through the stem of a dicotyledenous plant, showing a vascular bundle

The arrangement of xylem in the young stem is different from that in the young root. In the stem, xylem and phloem cells form groups called **vascular bundles** (Fig. 6). By the time these vascular bundles reach the leaf, they run along the veins of the leaf (Fig. 7).

Cohesion tension

As water is pulled out of a xylem vessel in a leaf, the column of water behind it is pulled upwards (see Fig. 7). As water evaporates from the leaves the water columns in the xylem vessels are pulled upwards because the water molecules stick together. This is because there are forces, called hydrogen bonds, holding one water molecule to another. If you pull a cylinder of modelling clay at both ends, the middle of the cylinder will get thinner and will eventually break. What stops this happening to the columns of water in the xylem vessels?

The water molecules are strongly attracted to the walls of the xylem vessels. This attraction between the xylem walls and the water is so strong that as the water columns are put under tension by evaporation of water from the leaves, they become thinner and actually pull the walls of the xylem vessels inward. One function of the thickening in the xylem walls is to prevent the xylem vessels collapsing. However, the combined effect of the tension on all the xylem vessels in a tree trunk does produce a measurable narrowing of the trunk when the tree is transpiring (Fig. 8).

This mechanism for the movement of water up the xylem is known as the **cohesion-tension mechanism**, since there is a cohesive force between the water molecules themselves and the water columns are under tension due to the forces resulting from evaporation and gravity. As a molecule of water evaporates from the leaf, the next molecule is pulled in to the empty place next to it in the xylem cells, and the water molecule below that is pulled along after the moving water molecule, and so on.

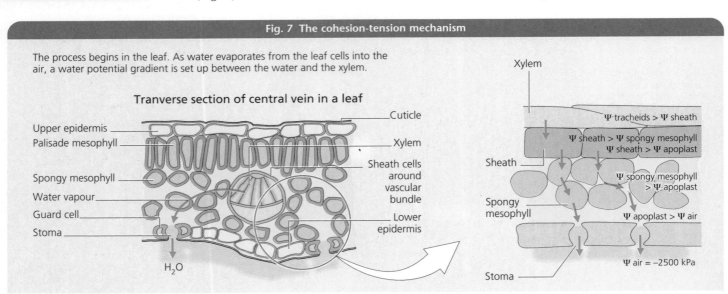

Fig. 7 The cohesion-tension mechanism

The process begins in the leaf. As water evaporates from the leaf cells into the air, a water potential gradient is set up between the water and the xylem.

Ψ tracheids > Ψ sheath
Ψ sheath > Ψ spongy mesophyll
Ψ sheath > Ψ apoplast
Ψ spongy mesophyll > Ψ apoplast
Ψ apoplast > Ψ air
Ψ air = −2500 kPa

4a Which source of energy is used to pull water up a plant?

b The graph in Fig. 8 shows changes in the circumference of three different species of tree during a hot summer day. Which of the trees is best adapted for reducing the rate of transpiration? Explain the reason for your answer.

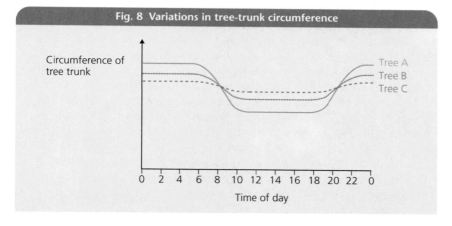

Fig. 8 Variations in tree-trunk circumference

As water leaves the xylem at the top, the xylem cells at the bottom of the column pull water from the cells in the root cortex. The water potential of the root cortex now has a value that is more negative than that of the root epidermis and root hairs. This means that water moves into the cortex cells. The value of water potential for the epidermis is now more negative than that of the soil water and so water moves from the soil into the root.

Water moves across most of the root mainly through the apoplast system. However, the endodermal cells, which form a ring of cells around the xylem and phloem, have a strip of **suberin**, a water-repellent substance, in their walls. This forms the **Casparian strip**, which prevents water passing through. The water and mineral ions have to pass through the symplast of endodermal cells before entering the apoplast of the xylem vessels.

Root pressure

Before leaves develop on a young plant or on a tree in spring, evaporation cannot be the driving force for the movement of water up the xylem. So how does water get to the developing leaves? If the stem of a young plant is cut just above soil level liquid soon begins to exude from the cut end. This liquid has been forced up from the roots by a mechanism called **root pressure**.

When ions reach the endodermis they are moved from the apoplast across the cell membranes by active transport into the symplast. The ions are then moved into the apoplast of the xylem cells, again by active transport. There is now a higher concentration of ions in the xylem cells than in the soil solution. So, the water potential of the contents of the xylem cells is now more negative than that of the soil solution. Water therefore moves by osmosis from the soil solution into the xylem cells to push fluid up the stem. This creates root pressure. Root pressure is only important as a method of moving water into the leaves of young plants at the beginning of the growing season. Usually, it does not provide a large enough force for moving water to the top of tall plants but root pressure can be significant in tropical rainforests where the atmosphere is permanently saturated.

5 Where, in the root, is the partially permeable membrane for the osmotic movement of water that brings about root pressure?

KEY FACTS

- Water moves through a plant, from the soil to the air, along a water potential gradient.
- The air has a much more negative water potential than the water in the soil.
- Water moves from the roots to the leaves mainly through the **xylem vessels** and **tracheids**.
- The **cohesion-tension mechanism** describes how water is pulled up through the xylem to replace the water lost by evaporation from the leaves.
- The energy that drives the movement of water is heat from the Sun which causes water in the **apoplast** of the leaf to evaporate.
- As water is pulled up the xylem, water is drawn across the root **cortex** to replace it.
- Water is drawn out of the soil solution because the root epidermal cells have a more negative value for water potential than the soil solution.
- In young plants, **root pressure** is the mechanism for moving water up to the developing leaves.

13.5 Movement of mineral salts through plants

The roots absorb mineral salts by two main mechanisms. Some mineral ions move passively, dissolved in the water that passes through the apoplast of the root to the endodermis. Some of these ions are then pumped actively into xylem cells. The endodermal cells control which ions enter the xylem, and which do not.

But this method does not account for the enormous difference in concentration of some ions between the root and the external solutions. Table 1 shows that roots absorb different ions at widely variable rates. This suggests that some mechanism other than diffusion and cohesion-tension must be operating.

Essential ions such as phosphate and nitrate are actively and selectively taken up by the epidermal cells. The process of active uptake is particularly efficient in root hair cells. Biologists originally worked this out from experiments with plants and respiration inhibitors. Active uptake needs energy from respiration, so if respiration is inhibited, active uptake should slow down. Fig. 9 confirms that this is the case.

Ringing experiments

Once in the xylem of the root, mineral ions are transported to the rest the plant mainly in the transpiration stream that is carried in the xylem. Biologists originally obtained evidence for this theory from 'ringing' experiments (Fig. 10).

In woody stems the phloem is part of the bark while the xylem forms a central, woody core. Radioactive isotopes can be used to track the movement of substances through plants. In one investigation, two similar plants were used, but one had a ring of phloem removed. The soil in each plant pot was watered with the same volume of a phosphate solution containing phosphate ions labelled with the radioactive isotope ^{32}P. A few hours later the plants were removed from the pots and the amounts of radioactive phosphate in the roots and a leaf were measured with a Geiger counter. The results are shown in Fig. 11. The figures are expressed as a percentage of ^{32}P in the roots.

6 Explain why it is an advantage to the plant that uptake of ions by the root hairs is selective.

Table 1 Ion concentrations in pea plants

Ion	Concentration of ion in external solution /mM	Concentration of ion in root tissue /mM
K^+	1.0	75.0
Na^+	1.0	8.0
Mg^{2+}	0.25	1.5
NO_3^-	2.0	27.0
SO_4^{2-}	0.25	9.5

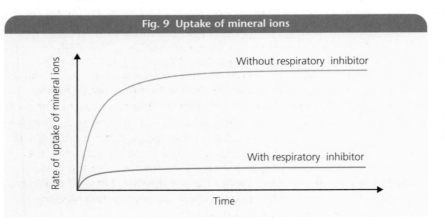

Fig. 9 Uptake of mineral ions

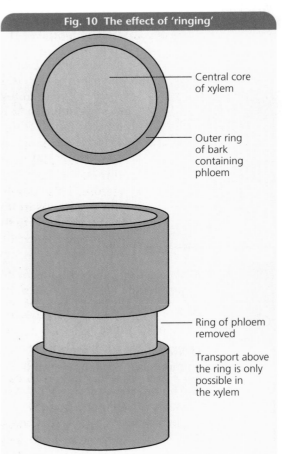

Fig. 10 The effect of 'ringing'

- Central core of xylem
- Outer ring of bark containing phloem
- Ring of phloem removed
- Transport above the ring is only possible in the xylem

Fig. 11 Investigating phosphate transport

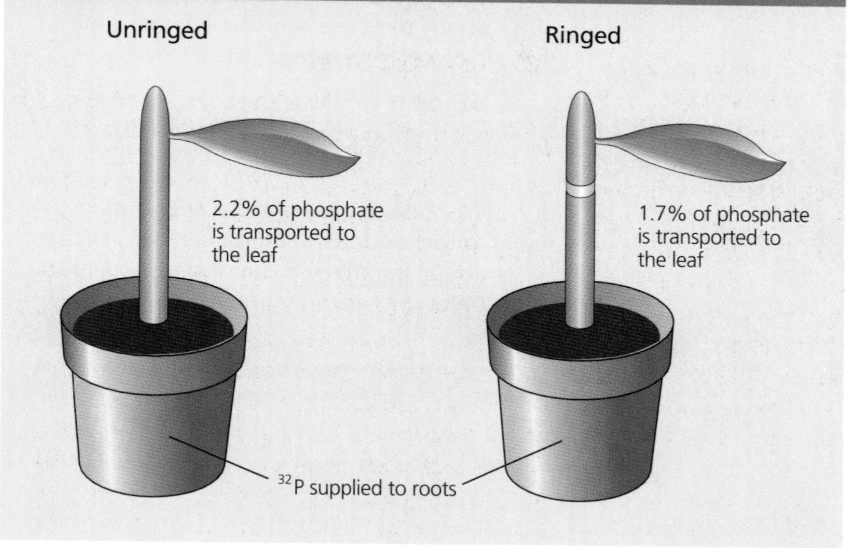

Unringed: 2.2% of phosphate is transported to the leaf

Ringed: 1.7% of phosphate is transported to the leaf

^{32}P supplied to roots

This experiment shows that the upward movement of phosphate is affected only slightly by removing a ring containing the phloem. It follows, therefore, that phosphate travels up the plant in the xylem vessels. Other studies have shown that most mineral ions are transported up the plant in this way.

Experiments where radioactive phosphate was applied to the leaves rather than the roots showed that mineral ions move from the leaves to the roots through the phloem.

KEY FACTS

- Mineral ions enter the root along the apoplast pathway, but selection occurs at the endodermis.
- Active, selective uptake of mineral ions is also carried out by the epidermal cells.
- Active uptake of mineral ions is reduced if respiration in the roots is inhibited.
- Mineral ions are moved up the plant mainly through the xylem, but moved down the plant mainly in the phloem.
- The route taken by mineral ions through the plant has been determined by ringing experiments with radioactive isotopes.

13.6 Movement of carbohydrate

Many plants store large amounts of carbohydrate in underground stems or roots. Potatoes, for example, store starch while sugar beet stores sucrose. This carbohydrate was made by photosynthesis in the leaves of the plant and then transported to the roots for storage. In the potato and the sugar beet, the leaves of the plant are known as the **source** of the carbohydrate and the storage organs are called the **sink**. When needed for new growth, carbohydrate from the storage organs or stems is transported up the stem to provide materials for developing leaves. This time the storage organs are the source and the leaves are the sink. Transport of carbohydrate and other organic compounds in the plant is known as **translocation**.

Phloem vessels

The transport of organic compounds such as carbohydrates occurs in the phloem.

Malpighi first demonstrated the role of the phloem over three hundred years ago. He studied movement of materials from the leaves to the roots. In one experiment he removed a ring of bark from a tree in summer (Fig. 12). After a few weeks a swelling appeared above the ring. He correctly concluded that the swelling above the ring contained substances transported down the stem from the leaves. When he repeated the experiment in winter, no swelling appeared since there were no leaves and no substances could be made.

More recent experiments have shown that the swelling is due to the accumulation of carbohydrates made by the leaves. In one such experiment in 1951, Chen separated the xylem of the geranium from its phloem with waxed paper (Fig. 13). He allowed a leaf above the separated region to photosynthesise using radioactive carbon dioxide. Fifteen hours later he measured the amounts of radioactivity at various positions.

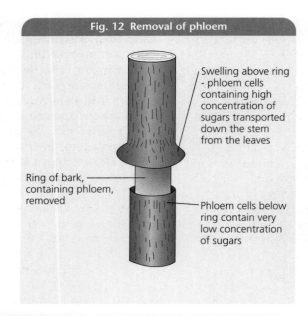

Fig. 12 Removal of phloem
- Swelling above ring - phloem cells containing high concentration of sugars transported down the stem from the leaves
- Ring of bark, containing phloem, removed
- Phloem cells below ring contain very low concentration of sugars

7a What do Chen's results (Table 2) show about the direction of movement of radioactive sugars in the phloem?

b Suggest what was the source and the sink for the sugars in this experiment.

8 What does the presence of abundant mitochondria in companion cells suggest about the mechanism of transport in the phloem? Explain your reasoning.

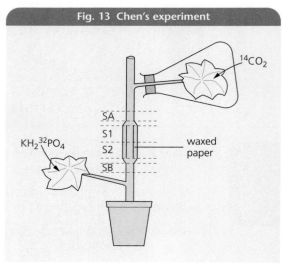

Fig. 13 Chen's experiment

Position of phloem analysed	Radioactivity of carbon and phosphorus in phloem	
	^{14}C/ppm	^{32}P/μg $KH_2{}^{32}PO_4$
SA (above waxed paper)	44 800	186
S1	3480	103
S2	3030	116
SB (below waxed paper)	2380	125

Table 2 Chen's results

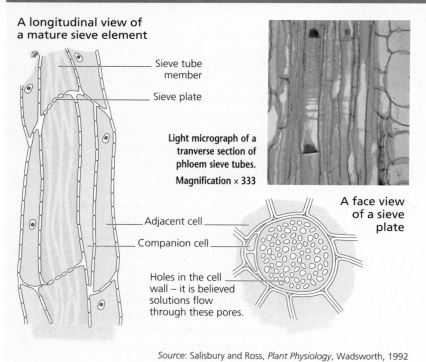

Fig. 14 Phloem structure

A longitudinal view of a mature sieve element
- Sieve tube member
- Sieve plate

Light micrograph of a tranverse section of phloem sieve tubes. Magnification × 333

A face view of a sieve plate

- Adjacent cell
- Companion cell
- Holes in the cell wall – it is believed solutions flow through these pores.

Source: Salisbury and Ross, *Plant Physiology*, Wadsworth, 1992

Transport in the phloem

To understand how phloem transports substances, sometimes in different directions at the same time, you first need to know the essential features of the structure of phloem (Fig. 14). Phloem cells have three important features that distinguish them from xylem cells:

- sieve tubes do not have lignified walls, so their walls are permeable to water and to solutes;

- sieve tube elements usually do not have a nucleus when mature, but they retain most of the organelles of the cytoplasm and so are living;

- each sieve tube element has one or more companion cells associated with it – these companion cells have dense cytoplasm and abundant mitochondria in them.

These adaptations of phloem allow the transport of substances up and down the plant.

Chen's conclusions

Chen's experiment showed that sugars were loaded into the phloem in the leaf, then passed down the phloem towards the roots. How are sugars loaded into the phloem and how do they move down to the roots? The **mass flow hypothesis** attempts to explain this (Fig 15).

Sucrose is actively transported into the sieve tube cells, probably through the symplast of mesophyll and companion cells. This reduces the water potential of the sieve tube elements. Water then diffuses from the surrounding cells into the sieve tube cells.

The increased volume of solution in the sieve tube elements increases the pressure and forces the solution along the sieve tubes. The solution can move very rapidly, particularly in crop plants. Speeds of 660 cm h^{-1} have been recorded in maize plants. Large amounts of solutes can also be transported into sinks. Rates as high as 252 g h^{-1} cm^{-2} of phloem have been recorded in the castor bean plant.

In the sinks, the reverse process takes place. Sucrose is actively pumped out of the phloem into the surrounding cells. The water potential of these cells drops and water leaves the phloem by osmosis. In some xerophytic plants, the sugar is converted immediately into starch in the root cells. Starch has little or no effect on the water potential of the cells.

When the sink is a young leaf, sugars are converted into new cell materials such as cellulose, again effectively lowering the sugar concentration of the sink and keeping the process going. The water is now free to pass from the phloem into the xylem and back to the leaves. Water therefore circulates between the source and the sink (Fig. 16).

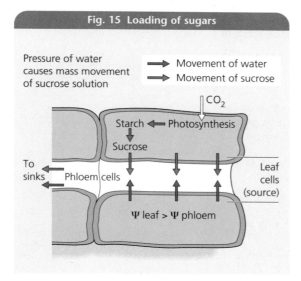

Fig. 15 Loading of sugars

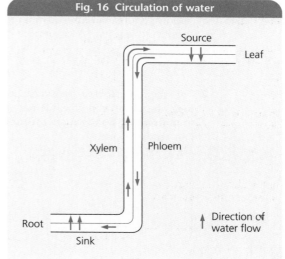

Fig. 16 Circulation of water

KEY FACTS

- Solutes such as sugars and some mineral ions are moved through the plant in the sieve tubes of the phloem.

- These solutes are moved from areas of high concentration, called sources, to areas of low concentration, called sinks. The same area can act as a source and a sink at different times in a plant's life cycle.

- Solutes are moved into sieve tubes by active transport with the help of companion cells.

- The high concentration of sugars in the sieve tubes gives them a very negative water potential, causing water to pass into the sieve tubes by osmosis.

- Water entering the sieve tubes creates a pressure that causes the solution in the phloem to move away from the source towards the sink.

- Mass flow describes the mechanism by which the solutes are transported to the sinks.

- Sugars arriving at a sink are usually converted into storage material or new cell material. This ensures a concentration gradient between source and sink to keep the sugars moving.

13 TRANSPORT SYSTEMS IN PLANTS

EXAMINATION QUESTIONS

1 The 'Two-leaf Hakea' is a plant found in south-west Australia, where the spring is relatively cool and wet but the summer is very hot and dry. The plant produces one type of leaf in spring and a different type in summer. The table shows the average values of a range of measurements taken from the leaves

Characteristic of leaf	Type of leaf	
	A	B
Length / mm	33	55
Maximum width / mm	10	0.8
Surface area / mm^{-2}	292	144
Volume / mm^{-3}	64	63
Cuticle thickness /mm	14	24

a Calculate the surface area to volume ratio for each type of leaf. (2)

b Use the data in the table to explain two ways in which leaf type B is adapted to summer conditions in south-west Australia. (2)

c Suggest and explain the advantages to the plant of producing leaf type A in spring. (2)

BY03 June 1996 Q5

2 The diagram shows a transverse section through a leaf. The plant from which it was taken was growing in normal conditions in an environment where water was readily available.

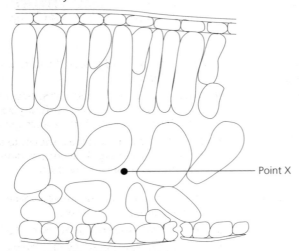

Point X

a Describe and explain how the oxygen concentration at point X would vary over a 24-hour period. (3)

b Explain, in terms of water potential, how:
  i) water is lost from a leaf during transpiration. (4)
  ii) the relative humidity of the atmosphere around the leaf may affect the rate of transpiration. (2)

BY03 February 1997 Q9

3
a The mass-flow hypothesis explains how sugars are moved in the phloem of a plant from a source to a sink. Copy and complete the table with a tick to show which of the plant organs listed act as a source and which act as a sink.

Organ	Can act as a source	Can act as a sink
Terminal bud		
Leaf		
Developing root		

The drawing shows a sugar beet plant. Three of its leaves, are labelled A, B and C. They were treated in different ways as described on the drawing.

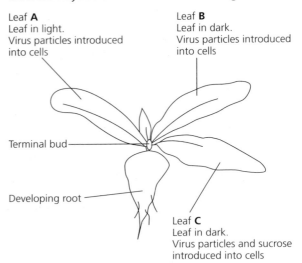

Leaf **A**
Leaf in light.
Virus particles introduced into cells

Leaf **B**
Leaf in dark.
Virus particles introduced into cells

Terminal bud

Developing root

Leaf **C**
Leaf in dark.
Virus particles and sucrose introduced into cells

During mass flow, virus particles may be transported with other substances in the phloem.

b Explain why the virus particles in the sugar beet plant:
  i) will be transported to the root of the plant from leaf A;
  ii) will not be transported to the root of the plant from leaf B. (4)

c Explain whether you would expect the virus particles from leaf C to be transported to the roots. (1)

BY03 June 1997 Q7

KEY SKILLS ASSIGNMENTS

Greening the desert

Marula is a large deciduous tree found in Southern Africa. The tree is highly prized by local people for its fruits. The fruits are plum-sized with a sweet-sour flesh that can be eaten fresh or used to prepare juices and alcoholic drinks. Researchers introduced Marula trees into different sites in the Negev desert in Israel to evaluate its growth under different conditions as shown in the table below. Thirty one-year-old plants were planted at each of four sites.

Site	Temperature extremes	Water supply
Besor	Moderate	Fresh $EC\ 1\ dS\ m^{-1}$
Ramat	Sub-freezing in winter	Brackish $EC\ 3.5\ dS\ m^{-1}$
Qetora	High summer temperatures, warm winters	Brackish $EC\ 3.5 - 4.5\ dS\ m^{-1}$ Ratio of Ca^{++} to Na^+ 1.2
Neot	High summer temperatures, warm winters	Brackish $EC\ 3.5 - 4.5\ dS\ m^{-1}$ Ratio of Ca^{++} to Na^+ 0.8

The plants were drip fertigated every one or two days in summer and every three to five days in winter. In the fourth year the amount of water supplied was determined by the evaporation rate and varied from 17 m^{-3} per tree per year at Besor and Ramat to 25 m^{-3} at Qetora and Neot.

This table shows the mean dimensions of the trees after four years.

Site	Height /cm	Trunk circumference /cm
Besor	533 ± 60	50 ± 2
Ramat	290 ± 25	54 ± 3
Qetora	620 ± 30	58 ± 2
Neot	413 ± 39	40 ± 4

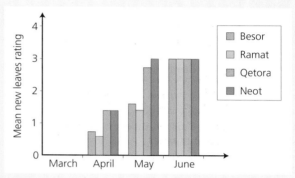

The graph above shows the development of new leaves in the fourth year. The percentage of the tree canopy with new leaves was estimated visually at the middle of each month on the following scale:

0 = no growth   2 = 20 – 60%
1 = less than 20%   3 = 60 – 100 %

3.3 billion hectares of previously useful land are now classified as desert. If we can understand how plants manage their water economy we might be able to reclaim some of the land. The above is an extract from research work on introducing new crops into the desert. (*Adapted from New Crops (pp496–499) Janick and Simon (eds) Wiley, New York*)

1 The salinity (amount of mineral ions) of soil water is measured using electrical conductivity.
   a What is the unit of conductivity?
   b Suggest why electrical conductivity is used to measure salinity rather than chemical methods.
   c Suggest what is meant by 'fertigated'.

2 Suggest why the Marula trees at Qetora and Neot were given much more water than those at Besor and Ramat.

3 a What was the height, after 4 years, of the tallest tree at Qetora?
   b What is the general relationship between the dimensions of the trees, after four years, and the conditions under which they were grown.
   c Use data from the first table shown to suggest why there was such a big difference in dimensions of trees grown at Qetora and Neot

4 The appearance of new leaves showed different patterns at Besor and Neot.
   a Describe the patterns for the two sites.
   b Suggest an explanation for the different patterns.

5 Look at all the data, then prepare a pamphlet to advise farmers in Israel of the best conditions in which to grow Marula trees

6 Use the internet to find mean monthly temperatures and rainfall figures for the Negev desert and to find out the main crops currently grown in the Negev.

Answers to Questions

The answers to text questions and questions within application boxes and extension boxes are given below. Please note that when the question is intended for discussion or extended study, no answer is provided.

Chapter 1

1. The combined magnification of the two lenses in a light microscope is great enough to produce a maximum magnification of only × 1500.

2. 0.0025 nm. In practice the limit due to the wavelength of the electron beam is slightly greater than 0.0025 nm due to technical factors. It is hard, for example, to focus the electron beam perfectly.

3.
   a. Mitochondia, ribosomes, endoplasmic reticulum, golgi body, cell membrane.
   b. Chloroplasts, plasmodesmata.

4. Measured thickness is about 1.5 mm, which is equal to 1 500 000 nm. Magnification is 190 000. Therefore, the actual thickness is 1 500 000 ÷ 190 000. Which equals about 8 nm.

   The length of a phospholipid molecule is half of this, which is 4 nm.

5. The hydrophobic tails of the molecules stick up out of the water and the hydrophilic head rests on the water's surface.

6. The nuclear pores allow large numbers of mRNA molecules that have been transcribed within the nucleus to pass out to the endoplasmic reticulum. Here they are translated by the attached ribosomes to form polypeptides.

7. The folded inner membrane gives a large surface area, allowing more enzymes involved in the final stages of respiration to be attached. This increases the potential rate of respiration and therefore energy release.

8. The electron micrograph is a two-dimensional section taken at one point in the specimen. It cannot reveal all the features of the three-dimensional structure from which the section was taken.

9. Cells that manufacture large amounts of protein need large amounts of rough endoplasmic reticulum because this is where the proteins are translated from the mRNA that is transcribed in the nucleus. Often, the polypeptides produced by the ribosomes need to be processed in some way – by the addition of sugar groups, for example – before the protein becomes fully functional. Such processing takes place in the Golgi body.

10. Protease and lipase enzymes in the Golgi vesicles first digest the membranes of lysosomes and other vesicles. The lysosomes then break down, releasing enzymes that digest other organelles in the cell.

11. Carbohydrates and lipids both contain the elements carbon, hydrogen and oxygen. The proportions of each vary widely between different lipid molecules, but the proportions in a carbohydrate molecule are always 1:2:1.

12. Glycogen is more compact than starch. Reducing bulk is more important to animals than plants because animals usually move about whilst plants are static.

13. Cellulose consists of many repeating units; a phospholipid molecule is made up from 3 units that are all different.

14. **Similarities:** size; shape; origin; both carry our metabolic processes.

    Differences: chloroplasts only present in plant cells; chloroplasts contain chlorophyll and are green, mitochondria don't and are not; mitochondria carry out catabolic reactions; chloroplasts carry out anabolic reactions.

15. Nucleus, mitochondria, chloroplast, ribosome. The nucleus in the bottom fraction, followed by the others in the order above.

16. A prokaryotic cell has no proper nucleus, no mitochondria, no chloroplasts, no endoplasmic reticulum and no Golgi body.

17. Chloroplasts and mitochondria.

Application boxes
Magnification

1a. Image size of diameter of egg = 656 mm. Actual size = 0.1 mm. Magnification therefore equals 656 ÷ 0.1, which gives × 6560.

2. Mean length of single bacteria is about 10 mm. Actual mean length is therefore 10 ÷ 9240, which gives 0.00108 mm. This is 1.08 µm.

3. Bottom left grain is 16 mm diameter. Actual width equals 16 000 ÷ 420, which is 38 µm.

4. Image length is 24 mm. So length of head is about 0.96 mm.

5. Length = 81 ÷ 80 000 = 0.0010125 mm. This is 1012.5 nm.

   Width = 28 ÷ 80 000 = 0.00035 mm. This is 350 nm.

Bacteria have their good points

1. There is no oxygen there to allow aerobic respiration.

2. Mainly cellulose.

3. A long-term study of the incidence of colon cancer in two large groups, one group that takes probiotic bacteria and one group that does not.

Chapter 2

1. Water

2. The greater the concentration of the external solution, the smaller the volume of the red cell.

3. Swollen red cells might not be able to pass through narrow blood capillaries.

4. The dark colour indicates a region of high concentration of solutes from the tea bag; lighter colours indicate regions of low concentrations of these solutes; there is a concentration gradient of these solutes from the dark colour to the lighter colours.

5. Osmosis is a special case of diffusion. It involves the movement of water through a partially permeable membrane.

This membrane allows water molecules to pass through but not solute molecules. The water molecules move along their concentration gradient, but the membrane prevents the solute molecules moving along their concentration gradient.

6 Those on the right hand side since they are less free to move.

7

a $\Psi = (-500 \text{ kPa} + 100 \text{ kPa}) = -400 \text{ kPa}$. Water will move out of the cell since the external $\Psi = -600$ kPa which is more negative than -400 kPa

b Plant cells have cellulose wall that prevents them from bursting.

8

a There is a net movement of water out of the cell at the top left and into the other two cells and a net movement of water out of the cell at the bottom into the cell at the top right.

b Lowest $\Psi = -230$ kPa, Highest $\Psi = -140$ kPa

9 $(-450 \text{ kPa} + 200 \text{ kPa}) = -250 \text{ kPa}$
$(-400 \text{ kPa} + 180 \text{ kPa}) = -220 \text{ kPa}$
Water moves from $\Psi -120$ kPa to $\Psi -250$ kPa, which is more negative

10 X will contain the solution since the cell wall is completely permeable to both solute and solvent molecules.

11 The cell at X will appear swollen since it is turgid and the membrane will be in contact with the cell wall at all points. The cell at Y will not be swollen, but the cell membrane will still be in contact with the cell wall. The cell at Z will be plasmolysed; the cytoplasm and vacuole will have shrunk and there will be some gaps between the cell membrane and the cell wall.

12 The folds provide a larger surface area that increase the rate of diffusion.

13 Molecules move in straight lines but in random directions. Relatively few molecules will be moving along paths that take them between the phospholipid molecules.

14 There are almost an infinite number of gaps between the phospholipid molecules where water molecules can pass through, so the number of spaces does not become a limiting factor to the rate of diffusion. On the other hand there are relatively few carrier protein molecules for glucose in the membrane. Until all of these are in transporting glucose molecules the concentration of glucose molecules is the limiting factor; when all the carrier molecules are transporting glucose molecules, the number of theses carrier molecules becomes the limiting factor for the rate of diffusion.

15 The more mitochondria, the greater the rate of respiration and therefore the more energy available for active transport.

Application boxes
Diffusion and the red blood cell

1 Surface area of spherical cell $= 88 \text{ μm}^2$
Surface area of cylindrical cell $= 121 \text{ μm}^2$

2

a 200 μm per second

b 2 μm per second

3

a Approximately 0.4 milliseconds

b Approximately 2 to 3 milliseconds

4

a Red blood cell 12.1
Spherical cell 8.8

5 The rate of diffusion across the membrane would increase due to an increase in surface area, but the time taken for oxygen to reach the haemoglobin molecules in the centre would increase by a much larger factor.

Ecstasy

1 Sweating involves the loss of both water and salts from the blood plasma. If these are replaced only by water then the Ψ of the plasma will increase. The Ψ will now be higher than the Ψ of the brain cells, so water will diffuse into these cells causing them to swell.

2 Yes. Tell the manager that 'clubbers' need to replace salts lost in sweat just like marathon runners, otherwise there is a risk of illness.

Sports drinks and symports

1

a Because it is isotonic with blood plasma.

b These are oxidised to release energy which is needed for muscle contraction.

2 They both stimulate the uptake of glucose and they contribute to an isotonic solution.

Chapter 3

1

a Flapping its ears will move a current of air over them, increasing the rate of heat loss via conduction and convection.

b The small ears present a reduced surface area, which minimises heat loss.

2

a The shrew is likely to be the more successful since it has the larger surface area in relation to its volume.

b Shrews have a large surface area to volume ratio so they lose energy as heat comparatively rapidly. They therefore need to eat a large amount of food to replenish their energy stores.

3

a The tapeworm exchanges xygen, carbon dioxide and soluble food molecules with its host using diffusion.

b The tapeworm is long and flat. This ensures that substances do not have to diffuse more than 1 mm to enter any of its cells.

4 A fairer unit would be lire of oxygen per minute per kilogram of body mass.

5 $50 \div 130 \times 100 = 38.5\%$.

6 Differences: Humans obtain oxygen from air, fish from water; humans have alveoli, fish have gill filaments; ventilation is tidal in humans, unidirectional in fish.

Similarities: Both have a large are of respiratory surface; both have a good blood supply to the respiratory surface; both ventilate the respiratory surface.

ANSWERS TO QUESTIONS

Application boxes
The walrus and the manatee

1
a Manatee $4 \times \pi \times 0.75 = 9.4$ m^{-2}.
  Walrus $3.5 \times \pi \times 1 = 11.0$ m^{-2}.
  Elephant $3 \times \pi \times 2 = 18.8$ m^{-2}.
b Manatee 1000 kg ÷ 1000 kg m^{-3} = 1.0 m^3.
  Walrus 1700 kg ÷ 1000 kg m^{-3} = 1.7 m^3.
  Elephant 5000 kg ÷ 1000 kg m^{-3} = 5.0 m^3.
c Manatee 9.4 ÷ 1.0 = 9.4.
  Walrus 11.0 ÷ 1.7 = 6.5.
  Elephant 18.8 ÷ 5.0 = 3.8.

2 In water, since water molecules are much closer together than the molecules in air.

3 The walrus has the smaller surface area to volume ratio. This as an advantage because the walrus will lose heat less rapidly than the manatee.

Are women catching up?

1, 2 and **3**: These are practical exercises from which you must draw your own conclusions. If you get into difficulties, please ask your teacher for help.

4 One possible reason might be that men in general have a higher ratio of muscle to body fat. This is true even of athletes.

5 It was assumed, wrongly, that female athletes did not have the stamina for these longer events.

Chapter 4

1
a Yes, because their combined kinetic energy is higher than the minimum activation energy needed for the reaction.
b They react because their combined kinetic energy is still as high as the activation energy required.

2
a A B D E F
b c
c c – d

3
a Carbon, hydrogen, oxygen, nitrogen.
b Nitrogen.
c NH$_2$
d COOH
e COO$^-$, and H$^+$

4
a

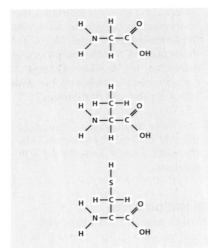

b
NH$_2$(CH$_2$)$_2$COOH + NH$_2$CH$_2$COOH → NH$_2$(CH$_2$)$_2$CONHCH$_2$COOH
  alanine   +   glycine   →   dipeptide
c N – C – C – N – C – C – N – C – C
           peptide bonds

5
a The tangle of collagen rods makes it less likely that the collagen will tear.
a The cornea has to be transparent. The stacking of the collagen rods prevents light being scattered in all directions.

6
a The proteins must be the right size to fit into the membrane; fibrous proteins would be too long. Carrier proteins must be the right shape for the molecules they transport.
b Each is specific to the particular molecule or ion that it carries.

7
a The active site of an enzyme molecule has a particular shape that attracts and accepts only molecules of the correct shape to fit into it. Enzymes are, therefore specific.
b The induced fit hypothesis takes into account the change in shape of the active site as it accepts the substrate molecule(s).

8
a Molecules have low kinetic energy, so they move slowly and are less likely to collide and react.
b 2, 4 and 8.
c The rate of reaction approximately doubles.
d Freezing would not denature the enzyme, so the rate of reaction would be the same.

9
a 57 °C
b Clothes can be washed in warmer water without the enzyme being denatured. there is no need to pre-soak clothes in the washing powder.
c Proteins have large molecules and are not soluble in water. Amino acids are soluble, so once the proteins are broken down the amino acids are washed away.

10
a As the concentration of enzyme increases substrate molecules collide more frequently with active sites and the rate of reaction increases. The maximum rate is reached when all active sites in use all the time. If the enzyme concentration increases so much that there is excess enzyme the rate of reaction cannot increase further. This would normally only occur when substrate concentration is low because enzymes work so fast that only small amounts are needed.
b X Excess of substrate.
  Y Enough substrate to saturate enzyme active sites.
c Increasing the temperature.
d Adding extra substrate because there is already an excess.

11
a The structure of the molecules is very similar, and therefore malonate molecules have a similar shape. The malonate is attracted to the active site but no reaction occurs. This reduces the number of sites available for the succinate to react.
b They do not fit into the active site. Instead they react with a different part of the enzyme molecule.
c The enzyme being inhibited catalyses a reaction that was unique to the pest; the inhibitor has no toxic effect on humans or other organisms; the inhibitor is stable, and does not break down into harmful substances; it can easily be administered to the pests in the right dose; it does not persist or build up to harmful levels in the environment.

Application boxes
Enzymes and extreme pH

1. Your graph should have an optimum around pH 2, and falling to zero at or just before pH 6.

2. The protease is effective over a much wider range of pH, and will work well in alkaline solutions above pH 7.

Using enzymes

1. Fructose is twice as sweet, so only half as much needs to be used for the same sweetness.

2. Cheaper, since cane sugar has to be extracted and processed from cane. Also, maize can be grown in a much wider range of conditions and places.

3. Process takes place more rapidly.

4. Changes the structure of the fructose molecules so that they are rearranged as the isomer, glucose.

5. The enzyme can be reused.

6. The enzyme does not contaminate the syrup and none of it is lost, so the immobilised enzyme in the columns can be used many times.

7. It would be expensive to purify it further, e.g. by treating it several times. The syrup can be used to sweeten slimming foods economically without making their energy value too high.

Chapter 5

1. The molecules and ions are small enough to be absorbed through the cell surface membrane by diffusion or active transport.

2.
a. 3; No. There is no spot opposite its position in the mixture.
b. Q = 0.50, R = 0.30; R is least soluble.

3. Your diagram should be the reverse of Fig. 4 on page 57.

4. Maltose; glucose; glucose and fructose; fatty acids and glycerol; polypeptides/dipeptides/ amino acids; amino acids.

5.
a. B and D have similar concentrations of amylase, at a higher concentration than the standard. C has no amylase.
b. Maltose (or glucose if biological samples also contained maltase).

6. Place samples of each strain of fungus on starch agar plates. Leave under controlled conditions for given time. Test with iodine solution, and measure sizes of clear areas around each fungus.

7. The hydrochloric acid produced by the stomach denatures the enzyme.

8. The endopeptidases create more ends which the exopeptidases can act on. This increases the rate of digestion.

9. Monoglycerides are soluble in the phospholipids that form the main part of the membranes, but monosaccharides and amino acids are not.

10.
a. Mitochondria provide the energy for the active transport of monosaccharides and amino acids from the epithelial cells into the blood capillaries.
b. Since they are soluble in fats they can dissolve easily in the phospholipid membrane and pass through it by diffusion.

11.
a. The tips are growing through the undigested parts of the fruit.
b. Mitochondria provide energy for growth processes and production of enzymes.
c. Oxygen supply needed for respiration in mitochondria.
d. Spores have to be dispersed so that air currents can move them to other food sources.

Application boxes
What is food used for?

1. We now need less energy because of our more sedentary lifestyle, central heating and less outdoor activity.

2.
a. Babies need extra proteins for their high rate of growth of cells and tissues.
b. Marathon runners need large supplies of immediately available energy for muscular activity.
c. In cold climates extra fat is needed for insulation, to reduce heat loss.

3. The man. He would lose heat more rapidly because of his lack of sub-cutaneous fat and therefore needs more energy to maintain body temperature.

4.
a. Adults still need proteins to synthesise materials used in cell replacement and to synthesise short-lived functional proteins such as enzymes and hormones.
b. 60 g of protein fulfils all his needs; excess protein cannot stored and it is broken down in liver and passes out of the body in the urine.
c. The failed kidneys cannot excrete the waste products from protein breakdown. This function needs to be done by a kidney machine. As dialysis is done only 3 times a week, taking in minimal amounts of protein prevents urea building up to dangerous levels.

Using chromatography

1. Glutamic acid, glycine, tyrosine, leucine and proline.

2.
a. 7
b. A and C
c. H, J and K
d. K

Milk intolerance

1. The molecule of lactose has a different shape to that of sucrose. Therefore it does not fit the active site in sucrase.

2. It is particularly common in people of oriental descent, suggesting that the condition is passed from parent to child.

3. The concentration of lactose in the colon may be high. Therefore there the difference in water potential between the contents of the colon and the blood is reduced. Less water is absorbed into the blood by osmosis, so more is contained in the faeces.

Extension boxes
What's the difference between reducing and non-reducing sugars?

1. Maltose is a reducing sugar, but sucrose is non-reducing.

ANSWERS TO QUESTIONS

2 Boiling sucrose breaks it down to glucose and fructose. These are reducing sugars, so Benedict's test now gives a positive result.

3 Prepare glucose solutions of known concentration, e.g. 0.1%, 0.01%. This may be done easily by diluting a stronger solution by fixed amounts. Carry out Benedict's test on each, using the same volumes of both the glucose and Benedict's solutions each time.

Digestion in other animals

1 The microorganisms are retained within the side branch and not swept out by peristaltic movements of the gut.

2
a Digestion of cellulose, producing some free glucose that can be absorbed; recovery of all nutrients when the microorganisms are digested.
b Supply of nutrients; protected environment; rapid rate of growth at body temperature of cow.

3 The position of the appendix at the end of the small intestine means most of the products of digesting the grass would be lost otherwise.

Chapter 6

1
a 1.
b Sperm - 1; fertilised egg - 2; young 16-cell embryo - 32.
c 450.

2
a
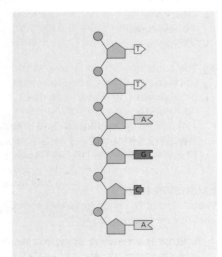

b
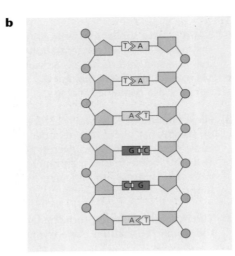

3
a In each organism the percentages of adenine and thymine are the same, and the percentages of cytosine and guanine are the same. This is because they fit together as complementary pairs in DNA, so there must be the same number of adenine and thymine bases, and of cytosine and guanine.
b 24%

4

Strands separate

Old strand

New strand
New strand

Old strand

5
a A two-base code could identify 16 amino acids.
b A three-base code could identify 64 amino acids.

6 Phenylalanine, valine, asparagine, glutamine.

7 Cysteine, asparagine, histidine, valine, histidine.

8 DNA molecules are longer; double-stranded instead of single; have deoxyribose instead of ribose; have thymine instead of uracil.

9 U G C U A A C A C G U G C U C

10
AAC CAG CAC CUC UGC
UUG GUC GUG GAG ACG

11
a 435
b 435
c 14

12
a Substitution, Addition, Substitution, Deletion.
b Val, Val, Ser, Thr, Leu
Val, Ala, Ser, Thr, Leu
Val, Val, Phe, Tyr, Ser
Val, Val, Ser, Thr, Leu
Val, Tyr, Leu, Leu
c One amino acid different
All amino acids changed except first two
No change (CAT and CAG code for same amino acid)
Only first amino acid the same.

13 Only mutations in developing ova in the ovaries could be passed on to children.

14 mRNA molecules transfer genetic information from a gene to a ribosome. Many are produced at each gene. Faulty mRNA molecules will only cause a small number of polypeptide molecules to be faulty in one cell.

Application boxes
Evidence for semi-conservative replication

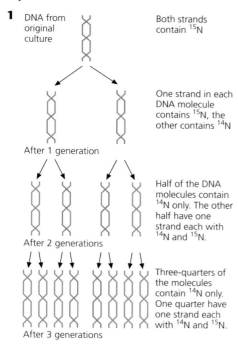

1. DNA from original culture — Both strands contain ^{15}N
   After 1 generation — One strand in each DNA molecule contains ^{15}N, the other contains ^{14}N
   After 2 generations — Half of the DNA molecules contain ^{14}N only. The other half have one strand each with ^{14}N and ^{15}N.
   After 3 generations — Three-quarters of the molecules contain ^{14}N only. One quarter have one strand each with ^{14}N and ^{15}N.

2. Two equal-sized bands, one at the top level (containing ^{14}N only) and one at the lowest level (containing ^{15}N only).

Extension boxes
The genetic code

1.
   a. Lysine, arginine, serine, alanine.
   b. TTC GCG AGA CGT
   c. UUC GCG AGA CGU

2.
   a. The only codon would be UUU, which encodes for phenylalanine.
   b. Lysine.
   c. Serine and leucine. The codons would be alternately UCU and CUC.

Cancer and oncogenes

1. p53 mutates → inactive proto-oncogene → active → uncontrolled cell division.

2.
   a. Substance Y is not made, so no substance Z can be formed from Y.
   b. Enzyme a can still catalyse the production of X from W.
   c. Y would be converted to Z.

Chapter 7

1.
a. There will be twice as much.
b. 92
c. So that the DNA molecules can easily be pulled to each new nucleus in the later stages of mitosis, without getting tangled.
d. So that they attach to the spindle together. The members of each pair are then pulled to opposite poles. This makes sure that each cell has one copy of each chromosome.

2. Metaphase

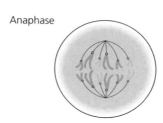

Anaphase

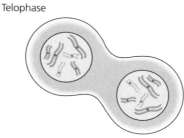

Telophase

3. Asexual reproduction allows the plants to reproduce rapidly and in large numbers. This often enables small numbers to colonise an area very rapidly. There is no need for other plants to be present for pollination. It also means that successful plants retain the features that help them to survive in a particular environment.

4.
a. Mutation.
b. Differences in the environmental conditions.

5. It has no roots to absorb water from the soil and replace lost fluid.

6. The rootstock is a genetically different variety.

7. Many plantlets from one parent; rapid; little space required; environmental conditions can be controlled for healthy and maximum rate of growth; plantlets are disease-free; sterile hybrids can be propagated.

8.
a. The cells in an embryo are produced by mitosis from the fertilised egg and are genetically identical.
b. The fertilised eggs are produced sexually and contain genes from both cow and bull.
c. The donor cow and the donor bull, because the fertilised eggs contain genes from both. The recipient cow acts as a surrogate mother, simply providing the correct uterine environment for the embryo to grow. She makes no genetic contribution.
d. The process would be too complex and expensive. It is more economical to use artificial insemination, which ensures sperm comes from a high quality bull, and to use only cows with desirable characteristics for breeding.

Application boxes
Cell division in an insect embryo

1.
a. 200 minutes
b. 10%, 52.5%, 6.5%, 4%, 27%
c. A very short interphase.

2.
a. Interphase.
b. Telophase.
c. Anaphase.
   Metaphase.
d. Interphase.

3.
a. 16 minutes.
b. 7 minutes.
c. 44 micrometres.
d. X: During metaphase centromeres are attached at the cell equator, half-way between the poles. During anaphase the distance from the poles decreases as the centromeres are pulled towards them.
Y: During metaphase the chromatids are joined at the centromeres. In anaphase the distance between the centromeres of each chromatid

increases as the chromatids are pulled apart. After anaphase, during telophase, the centromeres remain the same distance away from each other at the poles.

Growing Dolly

1
a Sheep A and the cloned sheep.
b To remove all the DNA so that cell X would contain genes from the udder cell only.

2
a 54
b 27
c 54
d 54

Chapter 8

1 If a chromosome was missing, there would be no genetic instructions at all for some features.

2
a 20.
b 14.

3 It has contractile protein filaments that contract and bend its tail, allowing it to move towards an ovum; it has mitochondria situated in the middle piece which can provide energy for movement; its small size makes movement easier and avoids waste of resources; enzymes in the tip of the head enable a sperm to penetrate the coating of an ovum.

4
a Female gametes contain a store of energy that nourishes the developing embryo after fertilisation.
b Less energy and fewer resources are wasted but chances of fertilisation are increased.

5 In birds (and reptiles) the egg is very large since the food store has to last all the way through development. In mammals, the egg is much smaller as the food store needs to sustain the early embryo until it attaches to the wall of its mother's uterus. Further nourishment is then derived through the placenta.

6 The amounts of DNA in the two nuclei are exactly the same.

7 To prevent the growth of bacteria that might cause disease)

8 Protein-digesting enzymes, such as endopeptidases and exopeptidases; lipid digesting enzymes such as lipase.

9
a Stage 2, stage 3.
b Stage 3, stage 2.
c Haploid, because nuclei produced by meiosis contain only one of each chromosome pair.

10 Fertilisation occurs in the oviducts, close to site of egg production; she has a uterus that protects the embryo as it develops; the placenta to provides supplies of nutrients, oxygen, etc.; the pelvis is flexible enough to allow vaginal birth.

11
a Rapid production of large numbers of aphids; no need to find male for fertilisation; rapid spread in areas with plenty of food plants; able to compete successfully with species that reproduce more slowly.
b Introduces variation, so there is more possibility of adaptation to changing conditions.
c Eggs contain food reserves for the early stages of development; sperm do not.

Extension boxes
Ovaries and testes

1 Egg production takes place in the abdomen, sperm production takes place in the testes that are held outside the abdomen. In the human female, all the eggs are present in her ovaries at birth; they simply mature during her reproductive life. Sperm are produced continuously from puberty. Only about 400 eggs are released in a woman's lifetime; millions of sperm are released in each ejaculation. Meiosis in the egg cell is not completed until fertilisation. Sperm cells undergo full meiosis during development in the testes.

2 The testes are held close to the body all the time; therefore, the temperature of the testes may be too high for healthy sperm production.

Chapter 9

1
a Cell membranes are composed of phospholipids. Detergents disrupt lipids.
b Endopeptidase, or exopeptidase, or a combination.

2 Active site on enzyme; only a nucleotide sequence with the right shape will fit into this active site; therefore, the enzyme can only cut the DNA at this position.

3
a A A T T.
b Only complementary bases will join together to produce a molecule with the sugar-phosphate backbones the right distance apart so that the molecule can twist into a regular double helix.

4 To destroy the enzyme; otherwise it would cut the recombinant DNA at the junctions made by the ligase.

5
a Air supplies the oxygen needed by the bacteria for respiration.
b To avoid contamination by any other microbes that might produce other products.
c Respiration in the bacteria releases waste heat energy. The rapid rate of metabolism in the growing culture causes a rapid rise in temperature.
d If not controlled this would kill the bacteria and stop fermentation.

6 The DNA might, for example, be inserted into a section of DNA that forms a gene

7
a The enzyme only cuts the DNA at one particular sequence of bases.
b Each copy of the DNA molecule will be identical. Each will be cut at the same positions, making fragments of the same length.
c The backbone consists of a phosphate–sugar sequence, so every nucleotide has a phosphate group.

ANSWERS TO QUESTIONS

Application boxes

Resctriction endonucleases

1

Bam1	C GGATC	and and	CTAGG C
EcoR11	C GCCTGG	and and	GGACCG C
Hind111	T AAGCT	and and	TCGAA T
Pst1	G CTGCA	and and	ACGTC G

2

TCC and AGGCCTGG

GGACCG and CTGCA

ACGTCGGT and GCCAAGCT

TCGAATC and TAG

3
a 1124
b 7 fragments: 30, 346, 1071, 622, 179, 1366, 748 bases.
c Bal1 and Sna1

Genetic engineering and food

1
a UAGCUGA
b AUCGACU
c Each pair of bases is complementary, so A links to U
d The normal mRNA does not attach to the ribosomes. Therefore the enzyme polygalacturonase is not made. The pectin in the cell walls is not broken down, so the cells do not separate and make the tomatoes go soft.

DNA sequencing

1
a 3. The smallest fragments are carried furthest through the gel.
b 3
c TAACGTCAGCTG
d AUUGCAGUCGAC

Extension boxes
Extracting human genes

1 The bases in the mRNA correspond to specific bases in the DNA strand, i.e. A to T, C to G, G to C and U to A. This will be the order of bases on one of the strands of DNA. The other strand will be the corresponding match with DNA nucleotides.

2 153

Chapter 10

1
a Substitution.
b 394 amino acids, each with a 3 base code gives 394 × 3 = 1182.
c The tertiary structure of the protein has a different shape because the substituted amino acid links up differently during folding. Therefore it does not fit into the active site of the elastase and act as an inhibitor.
d The AAT gene is not affected, so patients continue to produce defective AAT.
e No; someone whose emphysema is caused by smoking has the normal AAT gene already. Giving them treatment with AAT will make no difference.

2
a The high concentration lowers the water potential. Water in the mucus layer tends to be reabsorbed.
b Pancreatic enzymes cannot reach the duodenum. These include lipase, amylase and endopeptidases, so digestion of fats, starch and proteins are all affected. Fat digestion is usually particularly poor because there is no alternative source of lipase.

3 When the CFTR gene is transcribed complementary mRNA is produced. If mRNA is present in a cell it shows that the gene is being expressed.

4 Adenoviruses are more efficient at getting the CFTR gene into the cells and may be more efficient at getting the gene expressed. However there is a possible danger that some unmodified viruses might cause infections, and patients may develop immunity to the viruses.

5 As the cells lining the airways die they are replaced by dividing cells from lower layers which do not contain the genes. Even if the genes were inserted into the cells in lower layers, they would probably not be passed on during cell division because the genes have not been incorporated into the chromosomes. Therefore the epithelial cells have to be resprayed with liposomes or adenoviruses at regular intervals.

6 The DNA might, for example, be inserted into a section of DNA that forms a gene.

7 Gene therapy could introduce anti-sense gene to viral DNA that would block its transrciption and translation.

Application boxes
Genetic testing

All questions are for discussion; there is no definitive answer.

Extension boxes
Gene therapy for SCID

1
a These cells continue to divide so the healthy genes can be passed on to a good supply of cells.
b The white cells will die due to the normal turnover of cells in the body. Most new cells will develop from untreated cells in the patient's bone marrow. Also, the genes are not incorporated into chromosomes and will probably not be effectively passed on during cell division.
c The healthy genes would not be passed on, as only sex cells that develop in the reproductive organs are passed on.

Chapter 11

1
a Oxygen.
b Glucose, amino acids, carbon dioxide, water.
c Oxygen, glucose and amino acids pass into the tissues; carbon dioxide passes into the air in the alveoli.

2 60 ÷ 0.8 = 75

ANSWERS TO QUESTIONS

3 To prevent the valve flaps turning 'inside-out'.

4 In the skin, reduced blood flow through skin capillaries; the skin looks paler and the body conserves heat In the villi, reduced blood flow through capillaries reduces the rate of absorption of soluble food molecules into the blood.

5
a The total cross sectional of capillaries is greater than that of arterioles.
b The total cross sectional area of veins is less than that of venules.
c In the lungs and kidneys because there is a high level of exchange of materials in these organs.
d There is a slow flow rate allowing more time for diffusion; they are thinner, reducing the length of the diffusion path; collectively they provide a very large surface area for diffusion.

6
a $\pi \times (2 \text{ mm})^2 = 12.6 \text{ mm}^2$
$\pi \times (4 \text{ μm})^2 = 50.3 \text{ μm}^2$
b $12.6 \text{ mm}^2 \div 50.3 \text{ μm}^2 = 250\,000$

7
a 95%; 100%; 100%
b 11 kPa
c 100 − 15 = 15%

8 The haemoglobin becomes fully saturated with oxygen even at the low concentrations of oxygen in the surroundings.

9 Most carbon dioxide is carried as hydrogencarbonate ions in the plasma rather than in the red cells.

Application boxes
The active heart

1
a Blood by-passes the lungs, which are non-functional in the fetus.
b Much of the blood does not flow through the lungs and therefore is not oxygenated.

2
a A fall in systolic blood pressure since some blood is forced into the atrium rather than the aorta.
b A fall in diastolic blood pressure since some blood will pass back from the aorta into the left ventricle.

3 The left ventricle pumps blood the whole body (except the lungs) whereas the right ventricle pumps blood only to the lungs.

4
a Multiply your heat rate by 90, then by 24, then by 75 cm^3.
b The body can vary the heart rate according to its activity level.

5
a The contraction of the ventricles, forcing blood against them.
b The contraction of the ventricles, forcing blood against them.
c Ventricular diastole, blood from arteries forcing against them
d The contraction of the atria, forcing blood against them.

Chapter 12

1 To keep the water potential of the blood plasma constant in order to keep red blood cells at a constant size; if the water potential of the blood is less negative than that of the red blood cells, the red blood cells will not shrink.

2 Fats release more energy per unit mass than carbohydrates during respiration. Less water is associated with fat stores than carbohydrate stores, therefore more energy storage with less associated mass.

3
a In the cytoplasm.
s Inside the mitochondria.

4
a There is less energy release from glucose.
b It diffuses into the tissue fluid then enters blood via the capillaries.

5
a about 75%
b about 40%
c Less energy is released from stored energy sources; the build up of lactic acid in the muscle may lead to muscle fatigue.

6 The stretch receptors will no longer be stimulated therefore there will be fewer impulses arriving at the expiratory centre. The inspiratory cells will no longer be inhibited so impulses will be sent to the diaphragm and intercostal muscles and another ventilation cycle will begin.

7 Rate = number of breaths taken per minute.
Depth = volume of each breath taken.

8 Decrease in blood pH → stimulates peripheral receptors → initiates impulses to ventral cells in medulla of brain → initiates mpulses to inspiratory muscles → increases rate and depth of breathing.

9 A greater rate of delivery of oxygen to the muscles and removal of carbon dioxide from the muscles.

10
a A lower rate of heart beat would result in a fall in blood pressure.
b Speeds up heart rate.
c Slows down heart rate.

Application boxes
Putting fat to good use

1 40%

2
a Decreases.
b Increases.

3
a 20 miles.
b 889 miles. In practice it would be much less than this since complete respiration of triglyerides is not possible in the absence of continuous carbohydrate respiration.

Lactic acid

1 About 4 minutes.

2 About 3 minutes.

3 About 3 mm dm^{-3}. It takes time for lactic acid to diffuse out of muscle cells into the tissue fluid and then into the blood capillaries.

4 About 80 minutes.

5 Lactic acid diffuses into the blood capillaries in the muscles, and is then carried by blood to other parts of the body.

ANSWERS TO QUESTIONS

Chemoreceptors and breathing

1

a The aortic bodies are more sensitive at lower concentrations of carbon dioxide but the carotid bodies are more sensitive to increasing concentrations of carbon dioxide.

b The carotid bodies are more sensitive to falling oxygen concentrations.

2 There will be fewer impulses from the chemoreceptors to the inspiratory centre, consequently fewer impulses to the diaphragm and intercostal muscle, resulting in a fall in the rate and depth of breathing.

3 Subject A since low levels of oxygen have resulted in a large increase in ventilation rate.

Chapter 13

1 An increase in temperature increases the energy of water molecules resulting in them moving away from the leaf faster therefore making the water potential of air next to the leaf .more negative. An decrease in humidity decreases the number of water molecules in the air making the water potential of air next to the leaf more negative. An increase in wind moves water molecules away from the leaf faster therefore making the water potential of air next to the leaf more negative.

2 The factor that limits their rate of growth is their reduced surface area for photosynthesis, which produces the carbohydrates needed for growth.

3 2500 kPa ÷ 200 kPa per 10 metres, which equals 12.5 × 10 m = 125 m.

4

a Heat energy from the Sun.

b Tree C since it diameter decreases the least showing that there is the least tension in the xylem vessels resulting from the transpiration pull.

5 The cell-surface membrane of the living cells adjacent to the dead xylem vessel cell.

6 Uptake of ions is against a concentration gradient so energy is required for active transport of the ions. Energy is used to take up only those ions which are essential for the plant's healthy functioning.

7

a Most sugar was transported down the stem.

b The source of sugars is the leaves and the sink the roots.

8 It indicates the an energy-requiring process is involved since abundance of mitochondria is related to energy requirements. The energy is probably used for active transport of sugars into the sieve tubes.

Application boxes
Factors that affect stomatal opening

1

a Very dry soil results in the stomata not opening as wide during the early morning and closing completely before noon.

b A cloudy day results in the stomata not opening as wide as on a sunny day, and beginning to close earlier in the afternoon than on a sunny day.

2 The advantages of opening at night are that carbon dioxide levels in the plant can be replenished without losing much water by transpiration. The disadvantage is that there is no light available for photosynthesis. The advantage of closing the stomata around midday is that less water is lost by transpiration. The disadvantage is that carbon dioxide uptake is reduced at the time when light intensity for photosynthesis is greatest.

3 Plant that lives in very wet conditions, such as a bog plant.

Resourceful xerophytes

1 The stomata being at the bottom of folds in the leaf where there is a high humidity due to slower removal of water vapour molecules by air currents. The hairs reduce the flow of air, increasing the humidity at the leaf surface. Both of these make the water potential of the air less negative, reducing the rate of diffusion of water molecules. Both of these make the water potential of the air less negative, reducing the rate of diffusion of water molecules.

2 Screwing-up reduces the surface area from which transpiration could occur. Many stomata will be on the inside of the curves, resulting in a higher internal humidity and protection from air currents.

3 Smaller leaves give a smaller surface area and fewer stomata, both of which will reduce the rate of transpiration, but some photosynthesis will take place.

4 Some roots are nearer the surface of the soil, so any rainfall will reach them quickly. Some roots go very deep, in case there is any water deep in the soil. Other roots grow towards moisture trapped under stones.

Glossary

Acidosis
A decrease in the pH of blood caused by increased number of hydrogen ions.

Activation energy
The minimum kinetic energy required to overcome the repulsion between molecules to enable a reaction to take place between them.

Active site
The position on an enzyme molecule to which substrate molecules attach temporarily during a chemical reaction.

Active transport
Movement of substances in or out of cells that requires energy.

Adaptation
Special feature of an organism that makes it well suited to its environment. Also a special structure within an organism that suits it to its function.

Addition
A gene mutation that results from the insertion of an extra nucleotide.

Adenine
One of the four bases in the nucleotides of DNA and RNA.

Adenoviruses
A group of viruses in which DNA is the nucleic acid. Several types infect the organs of the human breathing system.

Aerobic respiration
Respiration in the presence of oxygen.

Alkalosis
An increase in blood pH caused by a decrease in the concentration of hydrogen ions.

Alleles
Different forms of a gene that occupies a specific gene locus. Different alleles have slightly different sequences of nucleotide bases.

Alveolus
An air sac in the lung where gas exchange takes place.

Amino acids
Monomers which join together to form polypeptides and proteins; each amino acid contains an amino group (NH_2) and an acid group (COOH).

Anaerobic respiration
Respiration in the absence of oxygen.

Angina
A pain caused by oxygen starvation of the heart muscle.

Anther
The male reproductive organ in a flowering plant; it produces pollen.

Anticodon
The sequence of three bases on a tRNA molecule that binds to a codon on an mRNA molecule during translation.

Aorta
The main artery of the body that carries oxygenated blood from the left ventricle.

Apoplast
The pathway through plants that consists of cell walls and intercellular spaces.

Artery
A blood vessel the carries blood away from the heart.

Asexual reproduction
Form of reproduction in which a single parent organism produces offspring that are genetically identical to itself.

Assay
Method of determining the quantity or concentration of a substance in biological material by comparing its activity with that of a standard sample.

Atheroma
A fatty deposit in the wall of a blood vessel.

ATP – adenosine triphosphate
A molecule used as a temporary energy store. Energy is stored in a bond that links adenosine diphosphate to inorganic phosphate.

Atrioventricular valve
A valve that prevents backflow of blood from a ventricle to an atrium.

Atrium
An upper chambes of the heart. The mammalian heart has two atria. They receive blood into the heart and pump it to the ventricles.

Balloon angioplasty
A technique to widen blood vessels that have been narrowed by an atheroma.

Basic
A substance that reacts chemically with an acid, neutralising it.

Bile
An alkaline fluid produced by the liver that passes into the duodenum.

Blood plasma
The liquid part of blood in which blood cells are suspended.

Bohr effect
The shift to the right of the haemoglobin dissociation curve caused by the high concentrations of carbon dioxide that occur in the muscles.

Breathing rate
The number of breaths taken per minute.

Bronchiole
A narrow air-tube in the lung; a subdivision of a bronchus.

Bronchus
A wide tube carrying air to the lung; a subdivision of the trachea.

Buffer
A chemical system that maintains a relatively constant pH.

Capillary
A blood vessel with a wall one cell thick. Its function is to exchange materials with the tissues.

Carbonic acid
The weak acid formed when carbon dioxide reacts with water.

Carbonic anhydrase
The enzyme that catalyses the reaction between carbon dioxide and water that forms carbonic acid.

Carboxyl group
The acid group, –COOH.

Cardioacceleratory centre
A group of cells in the medulla of the brain that initiate impulses that increase heart rate.

Cardioinhibitory centre
A group of cells in the medulla of the brain that initiate impulses that decrease heart rate.

Carrier protein
An intrinsic protein molecule in a cell membrane that carries molecules such as glucose across the membrane.

Cartilage
A firm, flexible tissue used both in support and in movement.

GLOSSARY

Casparian strip
A strip of suberin in the cell wall of an endodermal cell in a plant root. It diverts materials passing through the root from the apoplast pathway so that they continue in the symplast of the endodermal cells.

Catalyst
A substance that increases the rate of a reaction. It does this by lowering the activation energy required for the reaction to get going. The catalyst itself remains unchanged at the end of the reaction.

Cell division
The division of a cell to form two daughter cells.

Cell surface membrane
The double phospholipid membrane that forms the outer boundary of the cytoplasm of a cell.

Cellulose
A carbohydrate that is a major component of plant cell walls. It is a polysaccharide formed from β-glucose.

Central receptors
Receptors found in the brain itself, rather than in the rest of the body.

Centrifuge
An instrument that spins samples at high speed. It is used to separate organelles.

Centromere
The point at which two chromatids are joined during the early stages of cell division, and where the spindle fibres attach.

Channel protein
An intrinsic protein molecule in a cell membrane that transports ions across the membrane.

Chemical digestion
The breaking down food compounds by enzyme action and hydrolysis.

Chemoreceptors
Receptors sensitive to concentration of chemicals in, for example, blood.

Chloride shift
The movement of chloride ions from plasma into red cells to balance the movement of hydrogencarbonate ions from red cells into plasma.

Chlorophyll
A green pigment that occurs in the chloroplasts of plant cells. It absorbs light energy and then transfers it to other compounds during the process of photosynthesis.

Chloroplast
An organelle in plant cells that is the site of photosynthesis.

Chromatid
One of the two chromosome strands visible during prophase of mitosis. Chromatids remain attached at the centromere.

Chromatin
The DNA molecules in a nucleus that is not undergoing cell division. It stains as a dark mass of material without visible chromosomes.

Chromosome
One of the thread-like structures in nucleus, consisting a single very long DNA molecule that carries genes, together with associated proteins. Chromosomes contract and become more visible during cell division.

Chylomicrons
Tiny protein-coated droplets formed as fats are digested. They contain triglycerides and passfrom the intestine into lymph vessels.

Circulatory system
The system that transports substances around the body in liquid that flows through tubes of varying diameters.

Clone
Genetically identical cells or individuals produced by asexual reproduction.

Codon
A sequence of three nucleotide bases in the sense strand of a DNA molecule. Codons act as codes for particular amino acids during the assembly of a polypeptide.

Cohesion-tension mechanism
The mechanism that pulls water up xylem vessels when water evaporates from leaves.

Competitive inhibitor
Substances that block the action of an enzyme because they have a very similar molecular shape to that of the enzyme's substrate.

Concentration gradient
The difference in concentration of a substance that occurs between two areas.

Cornea
The clear tissue in front of the eyeball.

Countercurrent flow
Two liquids flowing past each other.

Countercurrent multiplier
A system that uses countercurrent flow to maximise the rate of exchange of a substance between two liquids.

Crassulacean acid metabolism
A series of reactions in some species of desert plant in which carbon dioxide absorbed during the night is converted to malic acid. Carbon dioxide released is used for photosynthesis during the day.

Cristae
Folds of the inner membrane in mitochondria on which the later stages of aerobic respiration occur.

Cyclin
Protein produced in cells during interphase. It stimulates DNA replication and hence cell division.

Cystic fibrosis
A genetic disorder caused by a mutant allele in which a channel protein that normally transports chloride ions is faulty. Excess mucus accumulates in the lungs and digestive system.

Cytoplasm
The part of a cell that contains organelles other than the nucleus, vacuole and cell surface membrane.

Cytosine
One of the four nucleotides present in DNA and RNA.

Deletion
A gene mutation caused by the loss of a nucleotide.

Denaturation
A change in the tertiary structure of a protein molecule and so its three-dimensional shape. When an enzyme is denatured, eg by high temperature, it is no longer able to function.

Deoxyribonucleic acid (DNA)
The genetic material that codes for genes. It contains nucleotide bases attached to a sugars and phosphate groups and assembles to form a double helix.

GLOSSARY

Deoxyribose
The sugar in the nucleic acid, DNA.

Diaphragm
A muscular sheet that separates the thorax from the abdomen and that is used to ventilate the lungs.

Diastole
Relaxation of heart muscle.

Differential centrifugation
Use of a centrifuge to separate cell organelles of different mass by disrupting cells and collecting the sediments produced at different spin speeds.

Differentiation
Development of a young, unspecialised cell into a mature specialised cell.

Diffusion
The spread of particles from areas of high concentration to areas of lower concentration. Diffusion is a passive process that results from the random movement of molecules.

Digestion
Enzyme-controlled breakdown of the large molecules in food into smaller molecules that can be absorbed.

Dipeptide
Substance formed when two amino acid molecules join together.

Diploid
A diploid cell has two sets of chromosomes in homologous pairs.

Dissociation
The splitting up of a molecule into two or more parts.

DNA polymerase
The enzyme that joins nucleotides together during DNA replication.

Double helix
The three-dimensional structure of DNA; two polynucleotide strands coil round each other in a double helix.

Duct
Tube through which substances such as bile and pancreatic enzymes pass from their organ of origin to where they are needed.

Duplication
A gene mutation in which a sequence of nucleotides is repeated.

Electrochemical gradient
A difference in concentration of charged particles (ions) between two areas.

Emulsification
The break down of large fat droplets into much smaller ones.

Endocytosis
The process that a cell uses to take in substances by folding its cell surface membrane around them.

Endopeptidase
A protein-digesting enzyme that splits proteins into shorter polypeptide chains.

Endoplasmic reticulum
A series of phospholipid membranes that extend through most of a cell. Often abbreviated to ER.

Endothelial cell
A cell found on an internal body surface.

Enzyme
A protein that catalyses reactions in living organisms.

Enzyme-substrate complex
The temporary compound formed when substrate molecules link with the active site of an enzyme during a chemical reaction.

Epithelial cells
Cells that cover the surface of an organ.

Eukaryotic cell
A cell that contains membrane-bound organelles and a true nucleus.

Exopeptidase
An enzyme that breaks the peptide bond at the end of polypeptide chains and releases the terminal amino acids.

Expiration
Breathing out. The opposite of inspriation.

Expiratory cells
A group of cells in the medulla of the brain that initiate impulses which inhibit the inspiratory centre, causing expiration.

External intercostal muscle
Muscles between the ribs that contract to bring about inspiration.

Facilitated diffusion
Diffusion of molecules such as glucose across the cell surface membrane. Substances that are transported in this way are taken from one side of the membrane to the other by specific protein carrier molecules that span the membrane.

Fenestrations
Gaps between adjacent cells that allow rapid diffusion of materials from one cell to another.

Fertilisation
The fusion of the two gamete nuclei during sexual reproduction.

Fibrinogen
A blood protein involved in clotting

Fibrous proteins
Structural proteins with long molecules with a regular secondary structure, found, for example, in hair, cartilage and bone.

Fick's law
Rate of diffusion is equal to;
$$\frac{\text{Surface area} \times \text{Concentration difference}}{\text{Thickness of membrane}}$$

Flagellum
A long 'whip-like' structure used for locomotion by some single-celled organisms.

Fluid-mosaic structure
The continually changing pattern of proteins in a phospholipid membrane.

Gated channel
An intrinsic protein molecule in a membrane that can 'open' to allow the passage of specific ions, or 'close' to prevent their transport.

Gene
A specific section of a DNA molecule that codes for a particular polypeptide.

Gene therapy
Treatment of a genetic disorder by inserting copies of a healthy gene into the cells of a person with a faulty gene.

Genetic code
The triplet sequence of nucleotides that encodes genetic information in nucleic acids.

Genetic fingerprinting
Technique in which the polymerase chain reaction is used to amplify DNA from a small sample and to match the genetic sequence with a sample from a particular individual.

Genetic probe
Short section of nucleic acid, often radioactively labelled, used to locate the position of a particular gene in an organism's chromosomes.

Genome
All the genes in a single individual.

Germline
Cells in an organism which give rise to the sex cells.

Gamete
A sex cell, such as an ovum and a sperm, that fuses with another sex cell during sexual reproduction.

Globular proteins
Proteins such as enzymes in which the molecules have an almost spherical structure and whose function depends on this specific three-dimensional shape.

Glucagon
A hormone produced by the pancreas in response to low blood sugar levels. It stimulates the enzymes in the liver that convert glycogen to glucose.

Glycogen
A storage polysaccharide; a polymer of glucose that, in mammals, is stored mainly in the liver and muscles.

Glycogenolysis
The breakdown of glycogen to glucose.

Glycolysis
The first reactions in respiration in which glucose is partly broken down, releasing a small amount of energy.

Glycoprotein
A protein molecule with an attached sugar group.

Golgi body
An organelle that processess proteins produced by ribosomes attached to the endoplasmic reticulum.

Grafting
Method of artificial propagation in which a cutting from a desired variety is fused to the rootstock of a variety that grows more easily or vigorously.

Grana
Double phospholipid membranes in chloroplasts that contain chlorophyll.

Guanine
One of the four bases in the nucleotides of DNA and RNA.

Guard cells
The two cells that surround and therefore control the width of stomata.

Haemoglobin
The red pigment in red blood cells that transports oxygen.

Haemoglobinic acid
The acid formed when hydrogen ions combine with haemoglobin

Haploid
A haploid cell has only a single set of chromosomes, one from each homologous pair.

Hemicellulose
A complex carbohydrate material that occurs in plant cell walls.

Homogenate
The suspension of organelles produced when cells are disrupted in an homogeniser.

Homogeniser
An instrument used to break up cells. This is done to bring organelles into suspension.

Homologous chromosomes
A pair of chromosomes with the same sequence of genes and therefore the same overall size and shape. They occur in a diploid nucleus.

Hydrolysis
The reverse of a condensation reaction, in which polymers are broken apart by the addition of water.

Hydrophilic
Water-loving; a molecule that can mix with water.

Hydrophobic
Water-hating; a molecule that cannot mix with water.

Hydrostatic pressure
Pressure that results from the force of a liquid on a surface.

Hypertonic
A solution that has a lower concentration of water molecules than another solution.

Hyphae
Thin threads that make up moulds and many other fungi.

Hypotonic
A solution that has a higher concentration of water molecules than another solution.

Hypoxia
A condition in which the concentration of oxygen in the blood is too low.

Induced fit hypothesis
An explanation of how enzymes work. This theory takes account the change in shape of the active site that occurs as a substrate molecule fits into it to make an enzyme-substrate complex.

Inspiration
Breathing in.

Inspiratory cells
A group of cells in the medulla of the brain that initiate impulses which bring about inspiration.

Insulin
A hormone produced by the pancreas in response to high blood sugar concentrations. It stimulates enzymes in the liver that convert glucose to glycogen.

Internal intercostal muscles
Muscles between the ribs that contract to bring about expiration.

Interphase
The stage in the cell cycle when a nucleus shows no visible sign of dividing. Replication of the cell's DNA occurs towards the end of interphase.

Intrinsic protein
A protein moleculethat spans a phospholipid membrane.

Inversion
A gene mutation in which a sequence of nucleotides is turned round.

Isomers
Molecules with the same molecular formula, but with different structural formulae.

Isotonic
A solution that has the same concentration of water molecules as another solution.

Krebs cycle
A series of chemical reactions that take place in the mitochonrdial matrix during aerobic respiration.

Lacteal
A branch of alymph vessel in the centre of a villus in the small intestine.

Ligase
An enzyme that joins sections of DNA together. In nuclei it helps to repair damage to the DNA, and it is widely used by genetic engineers, eg to insert a gene into a bacterial plasmid.

Ligation
A reaction catalysed by ligase in which two sections of DNA are joined together.

Lipid
A molecule made up from glycerol and fatty acids.

Liposomes
Tiny lipid droplets.

Lock and key hypothesis
An explanation of how enzymes work based on the idea that a substrate molecule of the correct shape (the 'key') fits into an active site on an enzyme of a corresponding shape (the 'lock'). It has been superceded by the induced fit hypothesis.

Locus (gene locus)
Specific position on the DNA molecule of a chromosome that is occupied by a particular gene.

Lumen
The space inside an enclosed area, such as the cavity inside the gut.

Lymph
A fluid, derived from tissue fluid, which drains from tissues into the lymphatic system.

Lymph node
A part of the lymphatic system that contains large numbers of white blood cells.

Lymphatic
One type of vessel found in the lymphatic system.

Lysosome
Vesicle produced by the Golgi body that contains digestive enzymes.

Marker gene
Gene deliberately put into a cell along with a donor gene to make it possible to identify cells that have taken up the donor gene.

Mass transport
The bulk movement of fluids in the transport system of an organism.

Matrix
The fluid-filled space inside mitochondria and chloroplasts.

Maximum breathing capacity
The maximum amount of the air that can be moved into and out of the lungs in a given time.

Mechanical digestion
Breaking down large lumps of food into smaller ones by physical action such as the grinding motion of teeth and muscular churning of the stomach.

Medulla oblongata
The part of the brain that controls unconscious activities such as the rate of breathing.

Membrane
The double-layered phospholipid structure that surrounds a cell and that forms part of many cell organelles.

Mesosome
An in-folding of the inner surface membrane of a prokaryotic cell.

Messenger ribonucleic acid (mRNA)
Complementary copy of the sense strand of DNA in a gene that passes from the nucleus to the ribosomes, where it is used to assemble the amino acids in the correct sequence to form a protein.

Metabolic pathway
A series of chemical reactions, controlled by enzymes, in which one substance is converted to another.

Metabolism
All the chemical reactions that take place in the cells of an organism.

Micrometre
One thousandth of a millimetre; 10^{-6} metre.

Micropropagation
Growing plantlets from small samples of plant material using tissue culture techniques.

Microvilli
Very thin finger-like projections on the surface of epithelial cells in the intestine.

Mitochondrion
An organelle in eukaryotic cells that is the site of aerobic respiration.

Mitosis
Cell division in which all the chromosomes are copied and each daughter cell has the same number of chromosomes as the parent cell.

Molecular pump
An intrinsic protein molecule in a membrane that can move molecules through the membrane against a concentration gradient.

Monomer
A small molecules which can join up in large numbers to form a long molecule called a polymer.

Monosaccharide
A type of carbohydrate that contains a single sugar molecule. Glucose and maltose and examples of monosaccharides.

Motile
Able to move by its own means of propulsion, as a sperm can.

Mucosa
The inner layer of the gut wall.

Mutagen
A substance or agent, such as radiation, which increases the frequency of gene mutation.

Mutation (gene)
A change in the sequence of the nucleotide bases in the DNA of a gene, which causes a change in its code.

Mycelium
The mass of hyphae that form the body of a mould or other fungus.

Myocardial infarction
A heart attack caused by a blockage in a vessel supplying heart muscle

Nanometre
One thousandth of a micrometre; 10^{-9} metre.

Non-competitive inhibitors
Substances with molecules that attach to and alter the shape of an enzyme molecule so that the active site can no longer bind with the substrate.

Non-reducing sugars
Sugars, such as sucrose, that are not readily oxidised and therefore give a negative result with Benedict's solution.

Nucleic acid
DNA and RNA, the substances that carry genetic information in long chains of nucleotides.

Nucleoid
The collection of nucleic acid strands found in a prokaryotic cell.

Nucleotide
One of the monomers that make up the nucleic acids DNA or RNA. Nucleotides contain a sugar (deoxyribose or ribose), a phosphate group and a base.

Nucleus
The organelle in eukaryotic cells which contains many of the nucleic acid molecules.

Operculum
The flap covering the gills in some types of fish.

Optimum temperature
The temperature at which a reaction proceeds at its maximum rate.

GLOSSARY

Organelle
An area in a cell that carries out a particular function.

Osmosis
The net movement of water molecules through a differentially permeable membrane from a region of higher (less negative) water potential to a region of lower (more negative) water potential.

Ovary
Female reproductive organ that produces female gametes.

Ovulation
The release of a mature egg cell from an ovary.

Oxidising agent
A substance that readily releases oxygen or accepts electrons.

Oxygen deficit
The amount of oxygen required after exercise to convert the remaining lactate produced during exercise to glycogen.

Oxygen dissociation curve
A graph showing how the percentage saturation of haemoglobin varies with the external oxygen concentration.

Oxyhaemoglobin
A haemoglobin molecule that has combined with oxygen.

Partially permeable membrane
A membrane that allows some substances to pass through but not others.

Passive
A process that does not require an external energy source is said to be a passive process.

Pectin
Soluble polysaccharides found in plant cell walls.

Peptide bond
The link between two amino acid molecules, formed by a condensation reaction.

Peripheral receptors
Receptors in parts of the body other than the brain and spinal cord.

Peristalsis
The regular succession of muscular contractions that moves food through the gut, and other substances through other tubes in the body.

Phosphatase
An enzyme produced in the liver that catalyses the hydrolysis of glycogen to glucose.

Phospholipid
A molecule which has two fatty acid molecules and one phosphoric acid molecule attached to a molecule of glycerol.

Photosynthesis
A process in which light energy is used to produce carbohydrates.

Plantlets
Tiny plants grown in tissue culture from single cells or very small pieces of tissue.

Plasmid
A small ring-shaped DNA molecule found in the cytoplasm of bacteria.

Plasmodesmata
Minute strands of cytoplasm that penetrate the cell walls of adjacent cells, connecting the cytoplasm of these cells.

Plasmolysis
When so much water diffuses out of a plant cell that gaps appear between the cell wall and the cell membrane.

Polymer
Large molecule formed when many small molecules called monomers join together.

Polymerase chain reaction (PCR)
Technique for producing large numbers of copies of a particular DNA sequence from a sample containing only one or a few molecules of the DNA.

Polynucleotide
Long chain of nucleotide monomers joined together, as in DNA and RNA.

Polypeptide
A substance with molecules made by joining large numbers amino acids into a chain.

Polysaccharide
A polymer molecule made by joining together many monomers called monosaccharides.

Polyunsaturated
A fatty acid containing many double bonds between carbon atoms.

Primary structure (of a protein)
The specific sequence of amino acids in a protein.

Prokaryotic cell
Organisms whose nucleic acid molecules are not contained in a nucleus.

Promoter genes
Genes that initiate the process of transcription, thus, for example, ensuring that a gene transferred into a bacterium produces its product.

Protein
A polymer molecule made by joining together many monomers called amino acids.

Proto-oncogene
Genes that control normal cell growth and division.

Pulmonary artery
The blood vessel carrying deoxygenated blood from the right ventricle of the heart to the lungs

Quaternary structure (of a protein)
A combination of polypeptides and non-protein groups which makes a complex molecular structure, as in haemoglobin.

Recombinant DNA
DNA produced by genetic engineering in which one DNA sequence is combined with another, e.g. a gene is inserted into a plasmid.

Recovery oxygen
The amount of oxygen needed after exercise to replenish the stores of ATP and glycogen in the liver and muscles.

Redox reaction
A reaction in which one substance is oxidised and loses electrons while another is reduced and gains electrons.

Reducing agent
A substance that readily removes oxygen from or adds electrons to another.

Reducing sugars
Sugars (monosaccharides and several disaccharides) that are readily oxidised and therefore give a positive result with Benedict's solution.

Replica plating
Production of several cultures from a master culture of bacteria.

Replication (DNA)
The process of producing identical copies of a DNA molecule.

Glossary

Resolving power
The power of a microscope to distinguish between two adjacent objects.

Restenosis
The formation of another a second atheroma in a blood vessel that has been treated by balloon angioplasty.

Restriction endonucleases
Enzymes used in genetic engineering that cut DNA molecules at specific points.

Reverse transcriptase
Enzyme that synthesises a complementary strand of DNA on a section of mRNA.

Rf value
In chromatography, distance moved by a spot of substance.

Ribosome
An organelle concerned with protein synthesis.

RNA polymerase
An enzyme that joins RNA nucleotides together and forms mRNA during transcription.

Root pressure
The movement of fluid into the xylem of a root resulting from active transport of ions into the xylem from living cells adjacent to it.

Rough endoplasmic reticulum
Endoplasmic reticulum that has ribosomes attached to it.

Rumen
A large bag-like structure in the stomach of cattle that contains microorganisms that help to digest the cellulose in grass.

Saprophytes
Organisms, such as moulds, that feed on dead organic matter.

Saturated
A fatty acid which has no double bonds between carbon atoms.

Secondary structure (of a protein)
The coiling, or pleating, of the primary amino acid chain in a protein.

Semi-conservative replication
The process in which the strands of a DNA molecule are separated, and new strands are formed alongside each from free nucleotides. The new DNA molecules consist of an original strand and a new one built onto it.

Semilunar valve
A valve that prevents backflow of blood from an artery into the ventricle of the heart.

Sense strand
The strand of a DNA molecule along which mRNA is formed during transcription, and which in effect carries the code used in the synthesis of a polypeptide.

Serum
Blood plasma that has had the 'clotting' protein fibrinogen removed.

Sink
The part of the plant that removes substances from phloem cells during translocation.

Smooth endoplasmic reticulum
Endoplasmic reticulum which has no ribosomes attached to it.

Sodium dodecyl sulphate
Detergent used during gene transfer to break down cell membranes.

Solvent
A substance in which other substances dissolve, e.g. water.

Source
The part of the plant that loads substances into phloem cells during translocation.

Spindle
The network of protein fibres in a cell that chromosomes attach to during cell division.

Stenting
Using a wire grid to keep open an artery that had been previously blocked by an atheroma.

Sticky ends
Uneven ends of a DNA molecule produced when a restriction endonuclease cuts the two strands at slightly different positions, leaving one strand a few nucleotides longer than the other.

Stoma (plural stomata)
Minute holes, surrounded by guard cells that facilitate exchange of gases between plants and the atmosphere.

Stretch receptors
Nerve cells that are sensitive to changes in tension.

Stroma
The fluid filled cavity of a chloroplast.

Submucosa
The layer of the gut wall below the mucosa that separates it from the muscular layers.

Substitution
A gene mutation in which a nucleotide is replaced by one with a different base.

Substrate
A substance which is acted on by an enzyme and takes part in a reaction.

Substrate specificity
Only molecules of one, or a small number of, substrates can fit into the active site of an enzyme.

Suberin
A waxy material used as a waterproofing agent in some plant cells.

Succulents
A plant with a swollen appearance due to the storage of large amounts water in specialised tissues.

Supercoils
The tightly packed coils of DNA formed when chromosomes contract during the early stages of cell division.

Supernatant
The overlying liquid produced during centrifugation.

Superovulation
Hormonal stimulation of a mammal so that the ovaries release larger numbers of mature eggs than normal.

Surfactant
A substance that can reduce the surface tension of a liquid.

Symplast
A pathway in plants involving the cytoplasm and vacuole of their cells

Systole
Contraction of heart muscle.

Tertiary structure (of a protein)
The folding of the secondary structure of a protein to form a specific three-dimensional shape, held together by weak chemical bonds, eg hydrogen bonds and sulphur bonds.

Testis
Male reproductive organ that produces male gametes.

Thorax
The upper part of the body commonly called the chest.

Thylakoid
The phospholipid membranes in chloroplasts that contain the chlorophyll molecules.

Thymine
One of the four bases that occur in the nucleotides of DNA, but not in those of RNA.

Tidal
The flow of air in and out of the lungs is described as tidal.

Tissue fluid
A fluid derived from blood plasma that acts as an intermediary between blood and tissue cells.

Trachea
The air-tube that carries air from the throat to the bronchi in the lungs.

Tracheids
A type of water conducting cell in the xylem of plants.

Transcription
The process in which the genetic code in DNA is copied to produce mRNA.

Transfer RNA (tRNA)
Small molecules of ribonucleic acid that bring the correct amino acid to the ribosome during protein synthesis.

Transformation
Process in which recombinant DNA is taken up by a cell.

Transgenic organism
Organism that has had a gene from another species inserted in order that it produces a useful product, such a sheep that secretes human alpha-1-antitrypsin.

Translation
The process in which amino acids are joined together to make a polypeptide, using the code on mRNA. It takes place on the ribosomes.

Translocation
The movement of substances in a plant.

Transpiration
The loss of water vapour from the aerial parts of a plant.

Turgid
Swollen due to the entry of water.

Turgor pressure
Pressure resulting from the entry of water into a cell.

Turnover number
The number of substrate molecules that can be acted upon by a single molecule of an enzyme in one minute.

Ultrastructure
The detailed structure of cells that reveals the organelles.

Unsaturated
A fatty acid with some double bonds between carbon atoms.

Uracil
The base that replaces the thymine of DNA in the nucleotides of RNA.

Vegetative propagation
Method of reproducing crop plants by growing them from vegetative parts, eg growing potato plants from tubers.

Vein
A blood vessel that returns blood to the heart.

Vena cava
The main vein that returns deoxygenated blood to the right atrium of the heart.

Ventilation
The movement of air or water over a respiratory surface.

Ventral group
A group of cells in the base of the medulla of the brain.

Ventricle
One of two lower chambers of the heart, pumps blood out of the heart.

Villi (singular – villus)
Extensions of the wall of the small intestine that project into the lumen and much increase the surface area for absorption.

Vital capacity
The maximum amount of air that can be breathe out of the lungs in one breath.

Water potential
The ability of water to move. By convention, pure water is given a water potential of 0 and solutions are given a negative value. Symbol Ψ. Unit kPa.

Xerophytes
Plants adapted to living in dry conditions.

Zygote
The cell formed when two gametes fuse during fertilisation; a fertilised egg.

Index

α-glucose 19
α-glycosidic links 18
AAT (alpha-1-antitrypsin) 144-5
ABA (abscisic acid) 186
abdominal muscles 46
abscisic acid (ABA) 186
absorption
    active transport for 36
    of digestion products 82
        amino acids 80
        from lipids 81
        glucose 34, 80
    water 82, 87
acclimatisation 53
acetylcholine 179, 180
acid groups, amino acid 57
acidosis 163
acids
    stomach 80, 84
    see also acid groups; amino acids; lactic acid; nucleic acids
acrosome 123
activation energy 55-6
active sites 60
active transport 36-7
    and absorption 80, 82
    and ATP 15
    and sugars, in plants 197
active uptake of minerals 194
ADA (adenosine deaminase) 150
adaptations
    chemical adaptation 84
    and digestion 83-5
    female adaptations 126-7
    of human gut 76
    to dry conditions 188, 189
addition in gene mutation 100
adenine 91, 92, 96
adenosine deaminase (ADA) 150
adenosine diphosphate (ADP) 171
adenosine triphosphate see ATP
adenovirus replication 148
adenoviruses 148, 151
adipose tissue 169
ADP (adenosine diphosphate) 171
aerobic respiration 171, 173
    in mitochondria 15
agar plates 78
air pressure, altitude sickness and 53
alkaline environment, intestinal 81, 84
alleles 90
alpha-1-antitrypsin (AAT) 144-5
altitude sickness 53
altitude training for sports 53
alveolar membranes 10
alveoli 43, 45
    in emphysema 144-5
amino acid chromatography 73
amino acids
    absorption 36, 82
    in cystic fibrosis 146
    in digestion 80
    in DNA 95
    as monomers 57
    in protein synthesis 97-8, 99
    in proteins 57-8, 59
amino groups 57

amylases 64
    in digestion 77, 78, 80
        measuring activity 78
amyloglucosidase 64
anaerobic respiration 25, 171, 173
anaphase 107
angina 160
animal cells 9
animals
    cells 9
    cloning of 106, 113-14, 117
    digestion in 79
    genetically modified 144-5
    glycogen storage in 19
    temperature control 40-2, 69, 87
    see also earthworms; fish; insects; mammals
antagonistic actions 180
anthers 120
anticodons 97
aorta 156, 158
    chemoreceptors in 174
    pressure receptors in 180
aphids 127
apoplast 190-1, 193
apples, grafting of 111
arteries 155, 158, 160
    aorta 156, 158, 174, 180
    carotid artery 174, 180
    pulmonary artery 155-6
    receptors in 174, 180
arterioles 158
artificial blood 167
asexual reproduction 106
    in aphids 127
    disadvantages of 125
    vegetative propagation 110-11
    see also cloning
assay 78
atheroma 160
ATP (adenosine triphosphate)
    and exercise 168, 173
    in mitochondria 15
    in polypeptide formation 97
    in respiration of glucose 171
atria 155, 177
atrial systole 177
atrioventricular node (AVN) 177, 178, 179, 180
atrioventricular valves 155, 156
AVN (atrioventricular node) 177, 178, 179, 180

β-glucose 19
babies
    heart defects of 157
    lungs of 45
bacteria
    enzymes from 54, 64
    in gene transfer 132, 134, 135, 136
    as prokaryotes 22, 23
    see also micro-organisms
balloon angioplasty 160
base pairing 91, 92, 93
bases, nucleotide 91, 92, 97
    and gene mutation 100
    investigating sequence 140

    in RNA 96
    triplet sequence of 95
basic, amino groups are 57
Benedict's solution 70, 71
bile 77, 81
biochemical tests for food 70, 71
biuret solution 70
blood 164
    artificial blood 167
    cells see blood cells
    composition 164
        changes in exercise 175
    pH of 162, 163
        and chemoreceptors 175, 176
    and serum 165
    see also blood system
blood cells
    radiation damages 101
    red see red blood cells
    white 150, 165
blood glucose 130, 169, 170
blood plasma 26-7, 163
blood pressure and flow 156, 158-9
blood system 155-60
    and exchanging materials 161-5
    see also heart
blood vessels 158-60
    see also arteries; capillaries; veins
blue-green algae 25
body cells, chromosomes in 89-90
body membranes 10
body temperature 40-2, 69, 87
Bohr effect 162
bonds
    and action of catalysts 55-6
    glycosidic bonds 18, 19, 71
    peptide bonds 57, 80, 97
brain
    chemoreceptors in 174, 175
    and heartbeat control 178, 179
    oedema in 33
breathing 45-7
    control of 174-6
    ventilation in fish 48-9
breathing rate 46-7
bronchi 46
bronchioles 46
buffers 163
bundle of HIS 177

cacti teeth 188
CAM (crassulacean acid metabolism) 188
cancer
    cervical cancer 6
    and mutations 101
    and oncogenes 102
    possible gene therapy for 151
    skin cancer 101, 102
capillaries 155, 158-9, 164
capsule 23
carbohydrates 18-19
    in diet 69
    digestion of 77, 78, 80
    movement in plants 195-7
    see also disaccharides; monosaccharides; polysaccharides; sugars

carbon dioxide
  in blood 162, 179
  and cell evolution 25
  and chemoreceptors 174, 175
    and heartbeat control 180
  and gas exchange 43, 45
    in fish 49
    in plants 50, 185, 188
  and scuba diving 179
carbonic acid 162
carbonic anhydrase 162
carboxyl groups 57
cardiac cycle 156
cardioaccelleratory centre 178, 179, 180
cardioinhibitory centre 178, 179, 180
carotid artery 174, 180
carrier proteins 35, 36, 82
  see also intrinsic proteins
cartilage 46
Casparian strip 190, 193
catalysts 55-6
  enzymes as 54
cell cycle 107-9
cell division 6
  controlling 109
  in insect embryos 109
  meiosis 118, 119
  mitosis 107-8
    DNA copied in 89, 93
  and mutations 101, 102
  and proto-oncogenes 102
cell membranes 8, 10, 12
  and absorption 34-5
  in cystic fibrosis 146
  of ova 123
cell size 7-8
cell walls, plant 18, 20
cells 6-9
  animal 9
  blood see blood cells
  body cells, chromosomes in 89-90
  companion cells 197
  differentiation 6
  and enzymes 54, 55
  epithelial see epithelial cells
  eukaryotic 22, 25
  evolution of 25
  exchanging materials 28-37
  expiratory cells 174
  guard cells 50, 185, 186
  muscle cells, respiration in 171-3
  organelles see organelles
  plant cells 9, 18-21, 32-3, 131
  prokaryotic cells 22-3, 25
  sex see sex cells
  stretch receptor cells 174
cellulase 79
cellulose 18, 19
  in animal digestion 79
  in diet 78
  in plant cell walls 18, 20
central receptors 175
centrifuges 22
centromeres 108
cervical cancer 6
CFTR (cystic fibrosis transmembrane regulator) 146, 147

CFTR genes 146, 147, 148, 149
channel proteins 35
  in cystic fibrosis 146
chemical adaptation 84
chemical digestion 74
chemical reactions, enzymes and 55-6
chemoreceptors 174-5, 176, 180
chewing 68, 74
chloride ions
  in blood 26-7, 162
  in cystic fibrosis 146, 147
  in sports drinks 27, 37
chloride shift 162
chlorophyll 20
chloroplasts 8, 20, 21
  in cacti 188
  in cell evolution 25
cholesterol 81
chromatids 108
chromatin 13, 89, 107
chromatograms 72
chromatography 70, 72-3
chromatography tanks 70
chromosomes 13, 89, 107-8
  diploid and haploid numbers 118-20
  homologous 90
  human 88, 89, 119
  in sex tests 118
chylomicrons 82
circulatory system 155
cloning 106
  of animals 106, 113-14, 117
  future of 117
  of humans 106, 117
  of plants 106
  see also asexual reproduction
clotting, blood 165
codons 95
cohesion-tension mechanism 192
collagen 59
colon 87
companion cells 197
competitive inhibition 63
concentration
  and enzyme activity 63
  and osmosis 30
  and water potential 31
concentration gradient
  active transport works against 36
  and diffusion 28-9
condensation reactions
  of amino acids 57
  of fatty acids 11
  of glucose 18, 71, 169
continuous peritoneal dialysis (CPD) 39
corneal membranes 10
cortex, root 190
countercurrent flow 48
countercurrent multiplier 48-9
cows, digestion in 79
CPD (continuous peritoneal dialysis) 39
crassulacean acid metabolism (CAM) 188
Crick, Francis 91
cristae 15
crop plants 110-12
  and genetic engineering 130, 137, 142

culturing gene product 131, 135
cuttings 111
Cyanobacteria (blue-green algae) 25
cyclin 109
cystic fibrosis 144
  and gene therapy 146-9
cystic fibrosis transmembrane regulator (CFTR) 146, 147
cytoplasm 6
cytosine 91, 92
cytosol 16

degenerate codes 99
deletion in gene mutation 100
denaturation 61, 62
deoxyribonucleic acid see DNA
deoxyribose 91
desert conditions 184, 199
  stomatal closing in 186, 187
  xerophytes in 188-9
dextrins 64
diabetes mellitus 130
dialysis 39
diaphragm 46, 47
diastole 156
diet 69, 87, 183
  see also food
differential centrifugation 22
differentially permeable membranes 30
  in kidney dialysis 39
differentiation, cell 6
diffusion 28-9
  and absorption 82
  facilitated see facilitated diffusion
  in gills 48
  in lungs 43, 45
  in small organisms 42-3
digestion 68, 74-87
  absorbing products see absorption
  in animals 79, 83-5
  bacteria for 23
  extracellular digestion 85
  saprophytic digestion 85
digestive enzymes 76-81
digestive glands 76, 77
  see also liver; pancreas
digestive organs 74-82, 84
dipeptidases 80
dipeptides 57
  absorption 82
  in digestion 77, 80
  hydrolysis of 74
diploid number 118-19, 123
disaccharides 18, 83
  in digestion 78
  see also maltose; sucrose
dissociation of oxygen 161
DNA (deoxyribonucleic acid) 13, 89-90
  investigating 138-40
  and mutagenic agents 101
  in prokaryotes 22-3
  replication see replication
  structure of 91-4
  see also genes
DNA polymerase 93
DNA sequencing 140

INDEX

Dolly the sheep (clone) 106, 114
donation, blood 167
double helix of DNA 92
drugs, ecstasy 33
dry areas
    xerophytes in 188-9
    see also desert conditions
ducts 76
duodenum 76, 78, 80, 81
    for absorption 82
duplication in gene mutation 100

earthworms, gut of 84
ECG (electrocardiograms) 178
EcoR1 132
ecstasy 33
eggs (ova) 126
    in asexual reproduction of aphids 127
    in cloned organism production 114
    human 119, 120, 121
        in fertilisation 123
        in vitro fertilisation 117, 124
    in transgenic organism production 145
elastase 144
electrical impulses in heartbeat control 177, 178, 179, 180
electrocardiograms (ECG) 178
electrochemical gradient 35
electron microscopes 8
electrophoresis 139
elephants, body temperature in 40-1
embryo transplantation 113
embryos
    cell division in 109
    and in vitro fertilisation 117, 124
emphysema 144
emulsification 81
emulsion test 70
endocytosis 17
endopeptidase 77, 80, 84
endoplasmic reticulum (ER) 8, 16-17
endosymbiotic theory 25
endothelial cells 158, 159
energy
    activation energy 55-6
    and active transport 36
    and active uptake of minerals 194
    and catalysis 54, 55-6, 61
    conversion in plants 20, 21
    and diffusion 28
        facilitated 35
    for exercise 168-70
    and respiration of glucose 171
energy sources 168-70
    for sperm mobility 124
environment, genetically modified foods and 142-3
enzyme-substrate complexes 60
enzymes 54-67
    bacterial 54, 64
    and concentration 63
    in digestion 76-81
    in fertilisation 123
    in food manufacture 64
    measuring activity 78
    and pH 62
    and temperature 51
epidermis, root 190

epithelial cells 6
    in intestines 34, 78, 79, 80, 82
    lining airways 146, 147, 148, 151
ER (endoplasmic reticulum) 8, 16-17
erythropoietin 53
eukaryotes 25
    and isolating DNA 131
eukaryotic cells 22, 25
exchanges
    and the blood system 161-5
    gas exchange 43-9
    in plants 50, 185, 188
    heat exchange 40-2
    see also transport
excretion by kidneys 36, 39
exercise
    blood composition changes in 175
    and breathing 46-7, 174-5
    and diet 183
    energy for 168-70
    and oxygen uptake 173
    and training 47, 172
    at altitude 53
    see also sports
exhibitionism, male 126
exopeptidases 77, 80
expiration 46
expiratory cells 174
extinct animals, re-creating 105
extracellular digestion 85
eye colour 90

facilitated diffusion 34-5
    and absorption 80, 82
    in capillaries 163
Factor VIII, genetic engineering for 130, 145
farm animals
    cloning of 106, 113-14
    digestion in 79
    transgenic 144
fats 11
    digestion of 77
    storage 169
fatty acids
    absorption 82
    condensation reactions of 11
    in digestion 77, 81
    in exercise 169, 170
female gametes see eggs
females
    adaptations for reproduction 126-7
        see also sexual reproduction
    women in sport 44, 118
fenestrations 159
fermentation in gene transfer 135
fertilisation 118, 119, 123-4
fertility 129
    and in vitro fertilisation 117, 124
fetuses
    and genetic testing 149
    heart of 157
fibre, dietary 78
fibrinogen 165
fibrous proteins 59
Fick's law 29
fish, gas exchange in 48-9
flagella 23
flowering plants 110
    ovaries and anthers in 120
    pollen 121

fluid mosaic structure 10
follicle stimulating hormone (FSH) 113
food 68-74
    and energy for exercise 168
    genetically modified see genetically modified foods
    investigating 70-3
    see also crop plants; diet; digestion
food manufacture, enzymes in 64
forensic science 138
fructose 18, 64, 71
    absorption 82
    sperms use 124
FSH (follicle stimulating hormone) 113
fuel, pollution-free 54
fungi, feeding in 84-5

galactose 82, 83
gall bladder 81
gametes 118-22
    differences between male and female 121, 126
    in fertilisation 123-4
    in gene therapy 151
    human 119, 120, 121
    and mutations 100
    see also eggs
gamma rays, mutations and 101
gas exchange 43-7
    in fish 48-9
    in plants 50, 185, 188
gated channel proteins 35
gender
    in sport 44, 118
    see also females; males
gene locus 89
gene mutation 100-2
gene product, obtaining 131, 135
gene therapy 144, 151, 153
    and cystic fibrosis 146-9
    for SCID 150
genes 88, 89
    and base sequence 92
    in cell division 93
    CFTR genes 146, 147, 148, 149
    coding for proteins 95-8
    marker genes 134
    promoter genes 134
    see also genetic engineering
genetic code 88-104
genetic engineering 130-43
    cloning 106, 113-14, 117
    manipulating genes 130-7
    and medicine 144-53
    transferring genes 131-7
    see also genes
genetic fingerprinting 138, 139
genetic probes 136
genetic testing 149
genetically modified animals 144-5
genetically modified foods 130-1, 137, 142-3
genetics 91
    see also genes; genetic engineering
genome 88
germline gene therapy 151
gills 48-9
glands, digestive 76, 77
globular proteins 59
glucagon 169
gluconic acid 54, 71

INDEX

glucose 18
   blood 130, 169, 170
   condensation reactions of 18, 71
   diffusion through membranes 35
   in digestion 77, 78, 80, 83
      absorption 34, 82
   in exercise 168, 169
   in food manufacture 64
   as a monomer 19
   and pollution-free fuel 54
   respiration of 171
glucose dehydrogenase 54
glucose isomerase 64
glucose-sodium symport protein 37
glycerol 11
   in digestion 77, 81, 82
glycogen
   in diet 69, 78
   for exercise 169, 171, 173
   as polysaccharide 18, 19
glycogenolysis 171
glycolysis 171
glycoproteins 144
glycosidic links (bonds) 18, 19, 71
Golgi body 17
grafting 111
grana 20, 21
guanine 91, 92
guard cells 50, 185, 186
gut
   earthworms 84
   human 74-7, 78-82, 84
      epithelial cells in 34, 78, 79
      and water absorption 87
gut wall 75

haem 161
haemoglobin 58, 161-2
haemoglobinic acid 163
haemophilia 130, 145
Hall, Dr Jerry 117
haploid number 119, 123
health, genetically modified foods and 142
heart 155-6
   controlling 177-80
   diseases of 154, 160, 178
   use of lactate 77
heart block 177
heart, secondary 159
heartbeat, controlling 177-80
heat exchange 40-2
hemicelluloses 20
Hill, Robin 22
HIV (human immunodeficiency virus) 136
homogenate 22
homogenisers 22
homologous chromosomes 90
Human Genome Project 88
human growth hormone 130, 136
human immunodeficiency virus (HIV) 136
human reproduction 118-29
   fertilisation 119, 123-4
      in vitro 117, 124
   fertility 117, 124, 129
   gametes 119, 120, 121, 126
      in fertilisation 123-4
   gender roles 126-7
   meiosis 119
   ovaries 122
   ovulation 122
   testes 122

humans
   blood *see* blood
   breathing 45-7, 174-6
   chromosomes 88, 89, 119
   cloning 106, 117
   digestion 68, 74-87
   exercise *see* exercise
   food 68-74, 168
   genes
      extracting 136
      transfer to sheep 144, 145
   gut 74-7, 78-82, 84
      epithelial cells in 34, 78, 79
      and water absorption 87
   lungs 43-7
      and cystic fibrosis 144, 146-9
      and emphysema 144
   reproduction *see* human reproduction
   respiration 171-3
humidity, transpiration and 185
hydrocarbon chains 11
hydrogen, pollution-free fuel and 54
hydrogen bonds 58
   in cell division 93
   in cohesion-tension mechanism 192
   in DNA 91, 92
hydrogen peroxide 56
hydrogenase 54
hydrolysis 74
   in digestion 76, 77, 80, 81
   of dipeptides 74
   of glycogen 169
   of triglycerides 169, 170
hydrophilic 12
hydrophobic 12
hydrostatic pressure 164
hypercapnia 179
hypertonic solutions 30
hyphae, mould 85
hypotonic solutions 30
hypoxia 176

ileum 76, 82
impulses *see* electrical impulses
in vitro fertilisation
   in genetic engineering 145, 151
   human 117, 124
induced fit hypothesis 60
industrial uses of enzymes 54
infertility 129
   and in vitro fertilisation 117, 124
inheritance 91
inhibition 63, 64
inner membrane, mitochondrion 15
insects 40, 109
   aphids 127
inspiration 46
inspiratory cells 175
insulin 95, 98, 136
   and blood glucose 130, 169
intercostal muscles 46, 47
intermittent peritoneal dialysis (IPD) 39
internal adaptations 84
internal intercostal muscles 46, 47
internal pacemaker 177
interphase 107
intestines 76-82, 84
   epithelial cells in 34, 78, 79
   and water absorption 87

intrinsic proteins 34-5, 36, 80
   symports 37
   see also carrier proteins
inversion in gene mutation 100
iodine solution 70
ion concentration *see* concentration
ionised particles, mutations and 101
ions *see* mineral ions
IPD (intermittent peritoneal dialysis) 39
isolation (gene transfer) 131
isomers 18
isotonic solutions 27
   and osmosis 30
isotonic (sports) drinks 26-7, 34
IVF *see* in vitro fertilisation

keratin 59
kidneys 36, 39
Krebs, Hans 22
Krebs cycle 22

lactase 83
lactate ions 171-2, 173
lacteal 82
lactic acid 47, 171-2
lactose, intolerance to 83
leaves 185, 192
   gas exchange in 50
   as source 195
life cycle of mammals 119
ligase 101, 132
ligation (gene transfer) 131, 132-3
light
   and photosynthesis 20, 21
   and stomatal opening 186
light microscopes 6, 7-8
lignin 20, 191
lipase 77, 81
lipids 1
   in cell membranes 10
   in cystic fibrosis 147, 151
   in diet 69
   digestion of 77, 81
   tests for 70
   and water 12
   see also phospholipids
liposomes 147, 151
liver 77
   in energy storage 169, 170
   recycles lactate 172, 173
lock and key hypothesis 60
locus, gene 89
lumen 74
lungs 43-7, 144, 146-9
lymphatic system 82, 164, 165
lysosomes 17

magnification 7-8, 14
maize, genetic engineering and 130
male gametes 119, 120, 121, 126
   in fertilisation 123-4
males
   adaptations for reproduction 126-7
      see also sexual reproduction
   exhibitionism 126
Malpighi, Marcello 195
maltase 77, 78, 80
maltose 18, 71
   in digestion 77, 78

INDEX

mammals
  body temperature 40-2, 69, 87
  cloning of 106, 113-14, 117
  digestion in 79
  gametes of 121
  heat exchange in 40-2
  life cycle of 119
  reproduction 126-7
    see also human reproduction
  transgenic 144-5
  see also humans
manatees 42
marker genes 134
marram grass 189
mass transport systems 154
  in humans 154-67
  in plants 184-99
mating rituals 126
matrix, mitochondrion 15
maximum breathing capacity 47
measurement, units of 7
mechanical digestion 74
medicine
  and *diabetes mellitus* 130
  and fertility 117, 124, 129
  and genes 144-53
medulla oblongata
  chemoreceptors in 174, 175
  and heartbeat control 178, 179
meiosis 118, 119
membranes 8, 10
  cell see cell membrane
  differentially permeable membranes 30, 39
  diffusion through 28-9
  nuclear 13
Meselson, Matthew 94
mesophyll 50
mesosomes 23
messenger RNA see mRNA (messenger ribonucleic acid)
metabolic pathways 55
metabolic water 87
metabolism 55
metaphase 107
micro-organisms
  in cellulose digestion 79
  see also bacteria; fungi
microfibrils, cellulose 19, 20
micrometers 7
micropropagation 112
microscopes
  electron microscopes 8
  light microscopes 6, 7-8
microvilli 78, 80
  see also villi
milk
  from transgenic sheep 144, 145
  and lactose intolerance 83
Miller, S.L. 25
mineral ions
  active transport of 36
  exchange in capillaries 163
  facilitated diffusion of 35
  movement through plants 194-5
  in plasma 26-7
  in sports drinks 26-7, 37
  see also chloride ions; phosphates; potassium ions; sodium ions

mitochondria 8, 15-16
  in cell evolution 25
  in sperm 124
mitosis 107-8
molecular pumps 36
monoglycerides 81, 82
monomers 19
  amino acids as 57
  nucleotides as 91
monosaccharides 18
  in digestion 78
  see also fructose; glucose
motility 121
moulds 84-5, 125
mRNA (messenger ribonucleic acid) 17, 96, 97-9
  in genetic engineering 136, 137
mucosa 75, 76
muscle cells, respiration in 171-3
muscle fatigue 47
muscle layer of the gut 75
  adaptations of 76
muscles
  and breathing 46-7
  and energy storage 169, 170
mutagenic agents 101
mutations 100-2, 125
mycelium 85
myocardial infarction 160
myoglobin 59

nanometres 7
non-competitive inhibition 63, 64
non-reducing sugars 70, 71
noradrenaline 179, 180
nuclear membranes 13
nuclei 6, 13
  in cell division 108
nucleic acids 13, 89
  see also DNA; RNA
nucleoli 13
nucleotide bases see bases
nucleotides 91
  in DNA replication 93
  in translation 97

oedema 33
oesophagus 74, 76, 84
oils 11
oncogenes 102
opercula 49
optimum temperature, enzyme 61
organelles 10-17
  differential centrifugation to investigate 22
  in microscopes 6, 8
  of plant cells 18-21
organisms
  adaptations and digestion 83-5
  carbohydrate uses in 19
  see also animals; eukaryotes; fungi; plants; prokaryotes
organs 8
  digestive organs 74-82, 84
  reproductive 120, 122, 123
  see also brain; heart; kidneys; lungs
osmosis 30-3
  in capillaries 163
  into roots 190, 193
  in plant cells 32

outer membrane, mitochondrion 15
ova see eggs
ovaries 120, 122
oviducts 123
ovulation 122
oxidising agents 71
oxygen
  in blood 161-2
    at altitude 53, 176
  and breathing rate 46-7
  and cell evolution 25
  and chemoreceptors 175
    and heartbeat control 179, 180
  and gas exchange 43, 45
    in fish 48-9
    in plants 50
oxygen deficit 173
oxygen dissociation curve 161
oxyhaemoglobin 161-2, 163

pacemakers 177
palisade mesophyll 50, 192
pancreas 76, 77, 80, 81, 84
  and blood sugar control 169
partially permeable membranes 30, 39
particles, ionized, mutations and 101
passive process, diffusion as 28
PCR (polymerase chain reaction) 138
PD (peritoneal dialysis) 39
pectins 20, 137
pegs and sockets mechanism 172
pepsin 80
peptide bonds 57, 80, 97
peripheral receptors 175
peristalsis 74
peritoneal dialysis (PD) 39
pH
  of blood 162, 163, 175, 176
  and buffers 163
  and enzyme activity 62
Pharmaceutical Proteins 144
phloem 194, 195-7
phosphatase 169
phosphates
  in DNA 91, 92, 93
  movement in plants 194-5
phospholipids
  from digestion 81
  in membranes 10, 12, 13
    and absorption 34-5
photosynthesis 20, 21
  obtaining carbon dioxide for 185, 188
plant cells 9, 18-21
  and isolating DNA 131
  osmosis in 32
  water potential in 32-3
plantlets 112
plants
  cells 9, 18-21, 32-3, 131
  cloning of 106
  crop see crop plants
  flowering 110, 120, 121
  gas exchange in 50, 185
  leaves 50, 185, 192, 195
  micropropagation 112
  reproduction
    asexual 106, 110-12
    sexual 110, 120, 121
  starch storage in 19

222

INDEX

transport systems 184-99
    carbohydrates 195-7
    mineral salts 194-5
    water 190-3
    xerophytes 188-9
plasma, blood 26-7, 163
plasma membrane *see* cell membrane
plasmids 132, 133, 135
plasmodesmata 190
plasmolysis 32
polar bears, body temperature in 40-1
pollen 121
polygalacturonase 137
polymerase chain reaction (PCR) 138
polymers 19
    polynucleotides as 91
    polysaccharides as 19
    proteins as 57
polynucleotides
    in DNA replication 93
    as polymers 91
polypeptides 58, 59
    digestion of 77, 80
    producing 89, 97-8
polysaccharides 18
polyunsaturated fats 11
potassium ions, stomatal opening and 186
potatoes 110
pregnancy, genetic testing in 149
pressure
    air, and altitude sickness 53
    blood 156, 158-9
    and water potential 31
pressure potential 32
pressure receptors 180
primary structure of proteins 57
prokaryotes 22-3, 25
    and isolating DNA 131
prokaryotic cells 22-3, 25
promoter genes 134
prophase 107
proteases 62
proteins
    in the body 57
    carrier 35, 36, 82
    in cell membranes 10
    channel proteins 35
    in diet 69
    digestion of 77, 80, 84
    gated channel proteins 35
    glucose-sodium symport protein 37
    intrinsic 34-5, 36, 80
        symports 37
    as polymers 91
    structure of 57-9
    synthesis 89, 95-9
    tests for 70
proto-oncogenes 102
pulmonary artery 155-6
Purkinje fibres 177, 178
pyruvate 171, 172, 173

quaternary structure of proteins 58

R_f value 72
rabbits, cellulose digestion and 79
radiation 101
radioactive tracers
    in DNA investigations 139, 140
    in plant transport investigations 194-6
    in replication investigations 94

receptors *see* chemoreceptors; pressure receptors
recombinant DNA 133
recombinant gene technology 136
recovery oxygen 173
red blood cells 161-2
    and altitude 53
    and diffusion 29
    and mineral ion concentration 27
    and osmosis 30
redox reactions 71
reducing agents 71
reducing sugars 70, 71
replica plating 134
replication 93-4, 108
    adenovirus replication 148
    mistakes in 100-1
reproduction
    asexual *see* asexual reproduction
    sexual 118-29
        human *see* human reproduction
    of plants 110, 120, 121
resolving power 7-8
respiration
    and active uptake of minerals 194
    aerobic 15, 171, 173
    anaerobic 25, 171, 173
    and energy sources 169
    and metabolic pathways 55
    in muscle cells 171-3
respiration inhibitors 194
restenosis 160
restriction endonucleases 131-2, 133, 138
restriction (gene transfer) 131-2
reverse transcriptase 136
ribonucleic acid *see* RNA (ribonucleic acid)
ribose 96
ribosomes 13, 16, 17, 97, 98
ringing experiments 194-5
RNA polymerase 97
RNA (ribonucleic acid) 13, 95-6
    mRNA 17, 96, 97-9
    in genetic engineering 136, 137
    tRNA 97-8
root pressure 193
roots 190, 193
    adaptations to dry conditions 189
    as sink 195
Roslin Institute 114
rough endoplasmic reticulum (rough ER) 16-17
rumen 79

saliva 74
salivary glands 76, 77
SAN (sino-atrial node) 177, 178, 179, 180
sand dunes, plants in 189
Sanger, Frederick 95
saprophytes 84
saprophytic digestion 85
saturated fats 11
SCID (Severe Combined Immunodeficiency) 150
scrotum 122
scuba diving 179
SDS (sodium dodecyl sulphate) 131
secondary heart 159
secondary structure of proteins 58
secondary thickening, plant cell wall 20
selection (gene transfer) 131, 134

semi-conservative replication 93-4
semilunar valves 155
sense strands 97
serum 165
Severe Combined Immunodeficiency (SCID) 150
sex cells 118-22
    differences between male and female 121, 126
    in fertilisation 123-4
    in gene therapy 151
    human 119, 120, 121
    and mutations 100
    see also eggs
sex tests 118
sexual reproduction 118-29
    advantages of 125-7
    of plants 110, 120, 121
shape, organism 40-2
    see also surface area
sheep
    cloned (Dolly) 106, 114
    transgenic (Tracey) 144
sieve tubes, phloem 196, 197
single sugars *see* monosaccharides
sink (plants) 195, 197
sino-atrial node (SAN) 177, 178, 179, 180
size, organism 40-2
skin cancer 101, 102
small intestines 76-82, 84
    epithelial cells in 34, 78, 79
    and water absorption 87
smooth endoplasmic reticulum (smooth ER) 16-17
sodium dodecyl sulphate (SDS) 131
sodium ions
    absorption of 82
    in blood plasma 26-7
    in sports drinks 26-7, 37
solute potential 32
solutions
    hypertonic 30
    hypotonic 30
    isotonic 27, 30
    water potential *see* water potential
solvents
    in chromatography 72
    water as 87
source (plants) 195, 197
soya beans, genetic engineering and 130
specific heat capacity 87
speed, enzyme 56
sperm 119, 120, 121, 126
    in fertilisation 123-4
sperm counts 129
spongy mesophyll 50, 192
sports
    sex tests in 118
    and training 47, 172
    at altitude 53
    women's performance 44
    see also exercise; training
sports drinks 26-7, 34, 37
Stahl, Franklin 94
starch 18, 19
    conversion to fructose 64
    digestion of 77, 78, 80
    tests for 70
starch agar plates 78
stems, plant 192

INDEX

stenting 160
sticky ends 132, 133
stomach 76, 77, 78
    adaptations 84
    and protein digestion 80
stomata 50, 185-7
stretch receptor cells 174
stroma 20, 21
suberin 20, 193
submucosa 75, 76
substitution in gene mutation 100
substrate specificity 63
substrates, enzymes and 60, 63
succulents 188
sucrose 18, 71
    in plants 195, 197
sugars 18
    in blood 130, 169, 170
    catalytic breakdown 55-6, 64
    in digestion 77, 78, 82
        lactose intolerance 83
    in DNA, deoxyribose 91
    exchange of, in capillaries 163
    glucose *see* glucose
    maltose 18, 71, 77, 78
    non-reducing and reducing 70, 71
    in plants 195, 197
    in semen 124
supercoils 108
supernatant 22
superovulation 113
superweeds, genetic modification and 142
surface area
    and diffusion 29
    and gas exchange in leaves 50
    and heat transfer in mammals 40-1
    and volume ratio 41
surfactants 45
sweating 87
symplast 190-1, 193
symports 37
systole 156
    atrial systole 177
    ventricular systole 177

T-lymohocytes 150
tapeworms, exchanging materials and 42-3
telophase 107
temperature
    body 40-2, 69, 87
    and enzyme activity 51, 54
    and proteins 54
    and transpiration 185
tertiary structure of proteins 58
testes 120, 122
thorax 46
thrombosis 160, 178
thylakoids 20, 21
thymine 91, 92, 97
tidal ventilation 46
tissue fluid 163-4, 165
tissues 8
tomatoes, genetic engineering and 130, 137
trachea 46
tracheids 191
Tracy the sheep (transgenic) 144, 145

training 47, 172
    at altitude 53
    see also exercise; sports
transcription 96-7
transfer ribonucleic acid (tRNA) 97-8
transformation (gene transfer) 131, 134
transgenic organisms 144-5
translation 97-8
translocation 195-7
transmembrane regulator, cystic fibrosis 146, 147
transpiration 185-7
    and the apoplast system 190-1
    and cohesion-tension mechanism 192-3
    and mineral transport 194-5
    and xerophytes 188-9
transport
    cellular mechanisms 28-37
        see also active transport; diffusion; osmosis
    mass transport 154
        in humans 154-67
        in plants 184-99
        see also exchanges
trees, desert 188, 199
triglycerides 11
    in diet 69
    in digestion 81, 82
    hydrolysis of 169, 170
    storage in the body 169
triplet sequence of bases 95
tRNA (transfer ribonucleic acid) 97-8
tumours *see* cancer
turgidity in cells 30, 32, 186
turgor pressure 32
turnover number, enzyme 56
twins 106, 113
two-way chromatography 73

ultrastructure 8
ultraviolet light, mutations and 101
units of measurement 7
unsaturated fats 11
uracil 96, 97
Urey, H.C. 25

vagina 123
valves
    in hearts 155, 156
    in veins 159
variation 90, 100, 110, 125
vascular bundles 192
vascular tissue, root 190
vasomotor centre 180
vectors 132
    adenovirus vectors 148
vegetative propagation 110-11
veins 155, 158, 159
vena cava 159
ventilation 45-7
    control of 174-6
    in fish 48-9
ventilation cycle 174
ventral group 175
ventricles 155, 177
ventricular systole 177

venules 158
vesicles 17
vessel membranes 191
villi 82
    s*see also* microvilli
viruses
    adenoviruses 148, 151
    in gene therapy 148, 150
    genetic material of 136
vital capacity, lung 47
vitamins, absorption 82
volume, ratio to surface area 41

walruses 42
water
    absorption 82, 87
    in diet and digestion 87
    exchange in capillaries 163
    heat exchange in 41-2
    importance of 87
    metabolic water 87
    in plants 185, 190-3, 197
        and osmosis 32-3
        see also transpiration
water potential 31
    and osmosis 31-3
    and tissue fluid 164
    and transport in plants 185, 190, 191-2, 197
water potential gradient 185
water stress 186
Watson, James 91
white blood cells 150, 165
wilting 32
wind speed, transpiration and 185
women
    and sport 44, 118
    see also females

X-rays, mutations and 101
xerophytes 188-9, 197
xylem 190-4
    cells 20, 190, 191
    vessels 191, 192

zygotes 119, 123